Principles of Sustainabl

T0230797

This book provides an introduction to sustainable aquaculture practices, focusing on how we develop social, economic and environmental resilience.

Aquaculture has seen phenomenal worldwide growth in the past 50 years, and many people view it as the best solution for the provision of high-quality protein to feed the world's growing population. This new edition has been fully revised and updated to reflect new developments in the field and includes new case studies. Focusing on developing more sustainable aquaculture practices and aquatic food systems, the book provides a toolbox of approaches to support widespread adoption and appropriate adaptation of regenerating aquaculture strategies, ensuring that it has practical relevance for both students and professionals. Drawing on a range of case studies from around the world, the book shows where progress, in terms of developing ecologically sound and socially responsible forms of aquaculture, has been made. The book is based on extensive evidence and knowledge of best practices, with guidance on appropriate adaptation and uptake in a variety of environmental, geographic, socio-economic and political settings. Concentrating on low-impact aquaculture systems and approaches, which have minimal adverse effects on the environment, the book also emphasizes socially responsible and equitable aquaculture development to enhance the natural resource base and livelihoods.

Principles of Sustainable Aquaculture is essential reading for students and scholars of aquaculture, fisheries, marine and water resource governance, and sustainable agriculture and sustainable food systems more broadly. It will also be of interest to professionals working in the aquaculture and fisheries industries.

Stuart W. Bunting, PhD, is an independent researcher, consultant and author who has over 25 years of experience in the fields of aquaculture, food systems, agriculture, biodiversity conservation and fisheries management. He has worked on projects across the globe and undertaken assignments sponsored by international organisations, such as WorldFish, the FAO, the Global Innovation Fund, Asian Development Bank, European Commission, The Crown Estate and the UK Government. He sits on the editorial boards of *Frontiers in Aquaculture*, *Frontiers in Sustainable Food Systems* and the *International Journal of Agricultural Sustainability*.

Praise for this edition

"Expanded sustainable aquaculture production is essential to provide healthy nutrition for the growing global population, and integration of scientific, environmental, and socio-economic considerations is key to its acceptance and success. Bunting provides a detailed, thoughtful, and readable assessment that sets the stage for responsible and sustainable growth and resilience. It should be required reading for researchers, students, managers, and policy makers, and anyone else with an interest in sustainable aquaculture."

Sandra E. Shumway, *Editor-in-Chief,* Reviews in Fisheries Science and Aquaculture *and* Journal of Shellfish Research

Praise for the first edition

"A synthesis dealing with the social, economic and environmental resilience aspects of this most important food production sector is long overdue. Stuart Bunting has the experience and insights needed to ground the work in field evidence and distil the principles on which future sustainable development of the sector can be based."

Dr Malcolm Beveridge, *Director of Aquaculture and Genetics at the WorldFish Center, Zambia*

"*Principles of Sustainable Aquaculture* is a timely and monumental review of what is required for aquaculture to fulfil its potential to contribute to socially and environmentally sustainable development. It covers the diversity of systems of both developed and developing countries, the theory of and practice for planning and management of the sector, and is replete with real world case studies."

Dr Peter Edwards, *Emeritus Professor, Asian Institute of Technology, Thailand, and Advisor, Sustainable Farming Systems Program, Network of Aquaculture Centres in Asia-Pacific*

"*Principles of Sustainable Aquaculture* presents a valuable landscape of trends in the World's fastest growing food production sector. Giving a systems perspective, it is a timely balance to the host of recent books on the technological challenges in this dynamic sector, and will support practitioners and policy makers to better understand both the basic underlying concepts and emerging best practices."

Professor David Little, *Institute of Aquaculture, University of Stirling, UK*

"Enhanced with Notes, References, and an Index, *Principles of Sustainable Aquaculture: Promoting Social, Economic and Environmental Resilience* is a seminal work, very highly recommended for professional, governmental, corporate, and academic Environmental Studies reference collections in general, and Agricultural Studies supplemental reading lists in particular."

Library Bookwatch

Earthscan Food and Agriculture

Principles of Sustainable Aquaculture

Promoting Social, Economic and Environmental Resilience

Second Edition

Stuart W. Bunting

LONDON AND NEW YORK

from Routledge

Designed cover image: © Getty

Second edition published 2024
by Routledge
4 Park Square, Milton Park, Abingdon, Oxon, OX14 4RN

and by Routledge
605 Third Avenue, New York, NY 10158

Routledge is an imprint of the Taylor & Francis Group, an informa business

© 2024 Stuart W. Bunting

The right of Stuart W. Bunting to be identified as author of this work has been asserted in accordance with sections 77 and 78 of the Copyright, Designs and Patents Act 1988.

Trademark notice: Product or corporate names may be trademarks or registered trademarks, and are used only for identification and explanation without intent to infringe.

First edition published by Routledge 2012

British Library Cataloguing-in-Publication Data
A catalogue record for this book is available from the British Library

ISBN: 978-1-032-37970-8 (hbk)
ISBN: 978-1-032-37967-8 (pbk)
ISBN: 978-1-003-34282-3 (ebk)

DOI: 10.4324/9781003342823

Typeset in Times New Roman
by MPS Limited, Dehradun

Contents

Preface

Principles of Sustainable Aquaculture aims to provide students, practitioners and the diverse and varied stakeholders committed to developing sustainable aquaculture practices with evidence and knowledge of better practices and guidance on appropriate adaptation and uptake promotion in other environmental, geographic, socio-economic and political settings. This book has a focus on emerging themes of low-impact and integrated aquaculture systems, socially responsible and equitable aquaculture development, and aquaculture to enhance the natural resource base and foster sustainable livelihoods. Drawing on pertinent case studies from around the world, the objective is to show where progress in terms of developing ecologically sound and socially responsible forms of aquaculture has been made and where promising approaches are being lost owing to unchecked driving forces and pressures and policies and attitudes that promote unsustainable practices. A toolbox of approaches to support widespread adoption and appropriate adaptation of sustainable aquaculture strategies is presented and reviewed and opportunities to promote an enabling environment and appropriate policies to facilitate wider adoption and uptake are highlighted.

In compiling this work I have drawn extensively on my experiences of working in South and Southeast Asia and hope that the ingenuity and diversity of production systems that have been developed in this region are apparent. Irrespective of where aquaculture development is proposed or where practices are already established, I hope that the evidence base presented here and associated methods could assist in promoting a new paradigm of ecologically sound, socially responsible, economically viable and resilient aquaculture development.

Acknowledgements

I would like to thank Professor Patrick Smith and Mr Robin Wardle of Aquaculture Vaccines Ltd., Saffron Walden for giving me a first insight to the world of commercial aquaculture and exciting avenues it had to offer. Having elected to study at the Institute of Aquaculture, University of Stirling, Scotland, I was fortunate to be supervised during my PhD studies by Professor James Muir and Dr Malcolm Beveridge, and I owe them both a great debt. I would like to thank all my colleagues from previous research posts and projects for their encouragement and support. I'm especially grateful to the numerous stakeholders who have contributed their knowledge and insights to a greater understanding of the pressures facing aquatic resources globally and the poor and marginal communities that depend upon them. I would like to acknowledge those who have given their permission to reproduce materials here and to especially thank Dr Nesar Ahmed, Dr Roel Bosma and Dr Varunthat Dulyapurk. In particular, I would like to acknowledge the Resource Centres on Urban Agriculture and Food Security Foundation (RUAF) for permission to reproduce materials here and to note that this book contains public sector information from the United Kingdom's Department for International Development licensed under the Open Government Licence v1.0. Finally, I would like to thank my friends and family for their support and patience over the years, in particular Caroline, Emily, Charlotte and Thomas for their love and encouragement.

1 Introduction

Key points

The aims of this introductory chapter are to:

- Define the status and characteristics of the current aquaculture sector.
- Review outcomes of the Blue Food Assessment of aquatic food systems and highlight key findings to guide sustainable development globally.
- Describe how aquaculture can appropriate environmental goods and services.
- Review the main physical, chemical and biological interactions, as well as the social and economic outcomes, of aquaculture.
- Discuss the social, economic and environmental implications of badly managed and inefficient production practices and unregulated and poorly planned aquaculture growth.
- Highlight international calls for improvements in the management and planning of aquaculture and concerted action on behalf of all stakeholders.
- Assess the potential contribution of sustainable aquaculture to the United Nations' Sustainable Development Goals by 2030.
- Provide an overview of the origins of sustainable aquaculture and recent initiatives to identify and promote more ethical and sustainable practices.
- Describe the sustainable livelihoods approach to assessing the situation of producer groups and communities and discuss the concepts of vulnerability and resilience.
- Introduce the food system framework and highlight the critical role of food environments in influencing decision-making amongst consumers and buyers of aquatic foods (see de Bruyn et al., 2021) and consider the influence of intra-household choices.
- Introduce the One Health framework and describe its potential in achieving holistic assessments and improvements in animal, environmental and human health and consider whether a One Nutrition framework could help achieve greater progress when it comes to animal and human nutrition outcomes and help avoid losing nutrients to the biosphere.
- Review prospects for transition away from conventional production modes to sustainable aquaculture development pathways.

DOI: 10.4324/9781003342823-1

Aquaculture defined

Defining aquaculture is important for several reasons. In principle, it permits a distinction to be made between farming and fishing and the relative contribution and trends of each in meeting rising demand for fish from a growing number of consumers. Aquaculture was defined by the Food and Agriculture Organization of the United Nations (FAO) as:

> [F]arming of aquatic organisms, including fish, molluscs, crustaceans and aquatic plants. Farming implies some form of intervention in the rearing process to enhance production, such as the regular stocking, feeding, protection from predators, etc. Farming also implies individual or corporate ownership of the stock being cultivated. For statistical purposes, aquatic organisms that are harvested by an individual or corporate body that has owned them throughout their rearing period contribute to aquaculture, whilst aquatic organisms that are exploitable by the public as common property resources, with or without appropriate licenses, are the harvest of fisheries.
>
> (FAO, 1995a: iii)

There are some difficulties and ambiguities in differentiating between fish coming from culture and capture activities, challenging the notion that there can be a clear distinction between aquaculture and fisheries. The advent of culture-based fisheries, fisheries enhancement programmes supported by restocking activities, deployment of artificial reefs to enhance aquatic habitats and species biomass, corralling of species such as tuna in marine cages and ongrowing of small, indigenous freshwater species for food and juvenile marine fish for the ornamental trade by small-scale fisheries challenge the notion that there can be a clear distinction between aquaculture and fisheries. Problems of classification and reporting are compounded further as in many situations the origins and means of production of fish may not necessarily be recorded or surveillance systems may not be comprehensive or particularly trustworthy. Substantial amounts of production from small-scale aquaculture and fisheries may be consumed within households or distributed amongst extended families or local communities and consequently would not be recorded as part of monitoring programmes centred on wholesale or retail markets. Calls for capacity-building and strengthening of surveillance systems were reiterated and further articulated through the FAO Sub-Committee on Aquaculture (FAO, 2010a).

Despite potential limitations and concerns, evidence regarding the scale and distribution of production originating from aquaculture is important for policymakers to inform planning and resource allocation, direct environmental monitoring and extension services and guide education and skills training activities. National, regional and global aggregate aquaculture data can elucidate trends with possible resource allocation, transboundary, public health, trade or food security implications. Growth stagnation or production declines might point to constraints from seed and feed supplies, limited site availability, disease problems, other specific research and development needs or market demand being satisfied and growing competition from other producers.

Rapid production increases for selected species might herald opportunities for input providers and producers elsewhere with a competitive advantage or signal to authorities the need for targeted monitoring and policy development to protect the environment, safeguard benefits of production and protect the interests of other resource users and stakeholders. Information on the diversity and significance of aquaculture to global food and, increasingly, on feedstock for biofuel production, pharmaceutical and chemical industries and vertically integrated aquaculture production is critical to the sector itself, helping promote recognition of responsible aquaculture as a legitimate and important user of aquatic resources. Greater awareness amongst decision- and policymakers could help ensure appropriate policy formulation and aquaculture sector planning, resource allocation in support of research and innovation, capacity-building and education provision.

Criticism was levelled at the earlier FAO definition which appeared to exclude, or fail to acknowledge at least, groups other than individuals and corporate bodies engaged in aquaculture. Contrasting with the earlier FAO definition, Beveridge and Little (2002) suggested that if there is intervention to increase yields and either ownership of stock, or controls on access to and benefits accruing from interventions, this should be classed as a form of culture. Households, nuclear and extended families, community-based organisations, religious groups, charities, non-governmental organisations (NGOs), coopera- tives and governments, amongst others, may assert ownership over cultured stocks. Collective management of common property resources is emerging as an important means to share access and distribute benefits derived from aquatic ecosystems including seasonal water bodies, lakes, rivers, highland and coastal wetlands and floodplain fisheries. Constituted stakeholder groups, often designated as resource user or self-help groups, are increasingly tasked with planning culture activities, regulating access and distributing benefits. Producer clusters are increasingly regarded as essential to promoting adoption and effective implementation of better management practices and standards for sustainably or responsibly cultured products (see chapter 8).

Appropriate collective management arrangements are of particular impor- tance where sustainable aquaculture practices such as fisheries enhancement and culture-based fisheries programmes and small-scale, low-impact, inte- grated and multi-trophic aquaculture development are proposed as a suitable mechanisms for empowering local communities, permitting them to capitalise on productive ecosystem services, but where equitable management is likely to depend on community-based organisations that retain ownership or at least the right to exploit cultured aquatic plants and animals. Implications of local communities having access to components of integrated aquaculture systems are discussed further in chapter 4 and elsewhere (see Bunting et al., 2013).

Elucidating the rationale for the earlier aquaculture definition promulgated by FAO (1995a: iii), it was noted that it was 'made as far as possible on the basis of common usage and closely parallels the practical distinction made between hunting and gathering on the one hand, and agriculture on the other'. Moreover, the authors acknowledged possible limitations, and it was anticipated that 'the definition between fishing and aquaculture will be sharpened as knowledge of aquaculture practise is improved' (FAO, 1995: iii). Differentiation between fisheries and aquaculture products is increasingly important for buyers and consumers who wish to exercise choice based on knowledge of prevailing culture and fishing practices. Explicit information on the means and location of production is now mandatory for seafood in Europe.

Proliferation of fisheries enhancement, ranching and culture-based fisheries programmes, however, has resulted in further blurring of the divide between fisheries and aquaculture. Fisheries enhancement is the practice of releasing captive-reared juveniles to bolster wild stocks or recreational and capture fisheries landings, but where culture and fisheries components are considered separate enterprises, with limited benefit sharing or cost recovery. Culture-based fisheries and ranching, where juveniles are introduced to large water bodies, whether natural (lakes, floodplains and rivers) or manmade (reservoirs, rice-fields, canals and irrigation systems) or seeded and released in coastal or marine areas (lagoons, bays, reefs and open-sea) depend on ownership of stock being retained, access to stocks being controlled or establishment of benefit sharing arrangements. Stocked environments may be under varying degrees of management, with objectives to establish new stocks or replace or supplement wild stock (Thorpe, 1980; Muir, 2005) safeguarding livelihoods and providing income and employment opportunities (Liu et al., 2019). Large-scale interventions such as marine protected areas with associated habitat enhancement and artificial reef deployment to conserve and enhance wild stocks constitute interventions to manage aquatic systems that might be considered within the broad realms of aquaculture development, as might proposals for geo-engineering, such as applying fertiliser to coastal and open-sea areas and harvesting the resulting biomass.

Global aquaculture development

Following a brief introduction to traditional, low-impact aquaculture practices, notable developments in the aquaculture sector over the second half of the last century are summarised below. An assessment concerning the global status of contemporary aquaculture relative to capture fisheries landings follows, and major producing nations and countries with rapidly growing aquaculture sectors are reviewed. Aquaculture in China dominates production statistics and is consequently considered separately to better assess emerging trends with potential to influence global aquaculture development.

On the origins of aquaculture

Proto-aquaculture practices emerged in Egypt 4000 years BP, in coastal lagoons in Europe in 2400 BP and in China in 2300 BP (Beveridge and Little, 2002). Raising fish in ponds and culturing shellfish on seashores spread throughout Europe with the Romans and later religious orders. In China, carp farming in ponds evolved into an intrinsic component of intricate and finely balanced integrated farming systems, notably dike–pond farming systems. Celebrated practices whereby mulberry (*Moras* spp.) was cultivated using nutrient-rich pond sediments, the leaves fed to silkworms and waste from silk production and processing was returned to fishponds stocked with a poly-culture of Chinese carp, thus making the most efficient use of available feeding niches, were widely adopted in the Pearl River Delta. Aquaculture in freshwater ponds remains a locally important source of fish throughout central Europe, and pond-based production, often integrated into wider farming and societal systems, continues to dominate inland aquaculture production in Asia. Traditional aquaculture practices have made a significant contribution to poverty alleviation, through improved food security, enhanced protein, healthy fat and micro-nutrient availability, income generation and employment.

Locally appropriate and important traditional aquaculture practices evolved elsewhere, notably: rice–fish culture in mountainous Southwest China, Japan and Java; small-scale cage culture in Indonesia; *ahupua'a* integrated aquaculture ecosystems in Hawaii; extensive fish culture in coastal lagoons in Asia and Europe; and wastewater-fed aquaculture in several countries in Asia and in Eastern Europe. Trout eggs and farming practices originating in Europe were introduced to South America, the Indian sub-continent and other territories by colonialists to establish trout populations, principally for angling and food production in temperate highland areas. Consequently, modest trout production and coldwater aquaculture practices became established in several remote upland areas throughout Asia and Latin America, and the burgeoning middle classes across these continents may provide renewed stimulus to high-value coldwater fish culture. Movements of people and processes of information exchange resulted in knowledge of emerging commercial aquaculture techniques such as artificial propagation of carp and Chinese hatchery technology in Southeast Asia and scallop spat collection on mesh bags in Japan during the 1960s, cage-based salmon farming in Scotland and Norway during the 1970s, intensive shrimp culture throughout the tropics in the 1980s, cage-based seabream and seabass production in the Mediterranean in the 1990s, and super-intensive pangasius culture originating in Vietnam in the 2000s being rapidly transferred across national boundaries and disseminated regionally.

Aquaculture development depends on three fundamental principles: control of reproduction, control of growth and elimination of natural mortality agents (Beveridge, 2004). Reviewing domestication of aquatic animals, Bilio (2007) noted that of 202 aquatic species recorded as being cultured in 2004, around a fifth (42 species) could be classified as domesticated. Far more species have been domesticated for aquaculture than for commercial livestock of poultry farming, and calls to focus resources and efforts on enhancing the yield and quality of already domesticated aquatic species are growing. Recent decades have seen rapid advances in key fields such as nutrition, genetics, engineering, physiology and biochemistry, resulting in significantly improved yields and quality across a range of commercially farmed species. Whether or not these practices and products are sustainable, acceptable to stakeholders and fulfil emerging environmental, social and ethical requirements of consumers constitutes a fundamental area for enquiry.

Contemporary status

Aquaculture production systems and associated value chains and food systems sustain millions of diverse and vibrant livelihoods, including among poor and marginal communities (Hernandez et al., 2018; Short et al., 2021; Bunting et al., 2023). The number of people directly employed in the aquaculture sector in 2020 was 20.7 million, and production (including aquatic plants) was worth US$105.8 billion (FAO, 2022). Globally, aquaculture is increasingly important in sustaining food supplies (FAO, 2022) and has great potential in safeguarding and enhancing food and nutrition security (Thilsted et al., 2016; Gephart et al., 2021). Landings from marine and freshwater capture fisheries are relatively static at around 80 and 10 million tonnes per year, respectively (Table 1.1). Global assessments indicate that in 2019 only 64.6 per cent of fishery stocks were

Table 1.1 Aquaculture and fisheries development and use trends globally in million tonnes (mt)[*]

Category	Average per year		Per year		
	2000s	2010s	2018	2019	2020
Aquaculture production (mt)					
Inland	25.6	44.7	51.6	53.3	54.4
Marine	17.9	26.8	30.9	31.9	33.1
Total	43.4	71.5	82.5	85.2	87.5
Capture fisheries (mt)					
Inland	9.3	11.3	12.0	12.1	11.5
Marine	81.6	79.8	84.5	80.1	78.8
Total	90.9	91.0	96.5	92.2	90.3
Use and consumption					
Non-food use (mt)	25.0	19.3	22.2	19.3	20.4
Human consumption (mt)	109.3	143.2	156.8	158.1	157.4
Population (billions)	6.5	7.3	7.6	7.7	7.8
Average per capita availability (kg)	16.8	19.5	20.5	20.5	20.2

Source: Developed from FAO (2022).

Notes
[*] Excludes production of aquatic higher plants and seaweeds.

exploited within biologically sustainable levels (FAO, 2022). The proportion of stocks that were being exploited at biologically unstainable levels had risen from 10 per cent in 1974 to 35.4 per cent in 2019. It was noted, however, that 82.5 per cent of capture fisheries landing by volume globally were derived from stocks assessed as biologically sustainable (FAO, 2022).

Despite rapid global population growth, reaching 7.8 billion in 2020, aquaculture production per capita increased from 0.7 kg in 1970 to 7.6 kg on average in the 2010s (Table 1.1; FAO, 2022). In concert with capture fisheries landings, per capita fish supplies have increased to around 20 kg (Table 1.1). Such averages, however, mask distinct national variations, regional dietary variations within countries and differences between poor and rich consumers in similar areas (Naylor et al., 2021; Bunting et al., 2023). Per capita consumption in China was 4.2 kg y^{-1} in 1961 and increased to 40.1 kg y^{-1} in 2019, whilst low-income countries suffered a decrease in per capita consumption of 0.2 per cent over the same period (FAO, 2022).

Aquaculture production (including aquatic higher plants and seaweeds), driven by technical innovation, market opportunities, high prices and human resources development, enabling institutional environments and investment incentives, increased at an average annual rate of 6.7 per cent between 1990 and 2020, reaching 122.6 million tonnes annually (FAO, 2022). During this period production of aquatic animals increased at an average rate of 6.5 per cent per year, reaching 87.5 million tonnes in 2020, and accounted for 49.2 per cent of total fish production globally (Table 1.1). Algae cultivation (dominated by seaweed) reached 35.1 million tonnes in 2020, and between 1990 and 2020, algae production increased at an average rate of 7.3 per cent per year (FAO, 2022).

Twenty countries with the highest annual aquaculture production in 2020 (excluding aquatic plants) are listed in Table 1.2 together with average annual growth rates with a 2010 baseline. A new conglomeration of the top-twenty aquaculture producing countries globally, described as the F20 countries, was proposed in the plenary session of the 9th Asian Fisheries and Aquaculture Forum to highlight the growing significance and future potential of the aquaculture sector (Williams, 2011).

Promoting dialogue and knowledge exchange between major aquaculture producing countries could be encouraged to address common problems and to promote shared objectives in pursuit of global development goals. It might be argued, however, that countries in the F20 are at different stages in the development of their aquaculture sectors, with distinct research and development needs related to the array of species being cultured and prevailing environmental, social and economic conditions.

Countries in Africa with sizable (in the top-ten nations on the continent by production volume in 2020) and in several cases rapidly growing (e.g. over 10 per cent growth per year) aquaculture sectors – for example, Ghana, Tanzania, Tunisia, Zambia and Zimbabwe (Table 1.3) – may be susceptible to environmental degradation or emerging animal and public health risks. This may warrant particular attention in terms of monitoring and establishing appropriate

Table 1.2 F20 countries with highest aquaculture production[*] in '000 tonnes in 2020

Country	Production ('000 t y^{-1})		Annual growth (%) 2010–2020
	2010	2020	
China	35,513	49,620	3.4
India	3786	8636	8.6
Indonesia	2305	5227	8.5
Viet Nam	2683	4601	5.5
Bangladesh	1309	2584	7.0
Egypt	920	1592	5.6
Norway	1020	1490	3.9
Chile	701	1486	7.8
Myanmar	851	1145	3.0
Thailand	1286	962	−2.9
Philippines	745	854	1.4
Ecuador	273	775	11.0
Brazil	411	629	4.4
Japan	718	599	−1.8
Korea, Republic of	476	566	1.8
Iran (Islamic Republic of)	220	481	8.1
United States of America	497	448	−1.0
Türkiye	168	421	9.7
Cambodia	60	399	20.9
Mexico	126	279	8.2

Source: Developed from analysis of the FishStatJ database (FAO, 2023) for this assessment.
Notes
[*]   Production of fish, crustaceans and molluscs is included.

Table 1.3 Countries in Africa with highest aquaculture production[*] in 2020 and average annual growth rates from 2010 to 2020

Country	Production (tonnes y^{-1})		Average annual growth (%) 2010–2020
	2010	2020	
Egypt	919,585	1,591,896	5.6
Nigeria	200,535	261,711	2.7
Uganda	95,000	123,897	2.7
Ghana	10,200	64,010	20.2
Zambia	10,290	45,670	16.1
Tunisia	5447	23,396	15.7
Kenya	12,154	19,981	5.1
Tanzania, United Republic of	454	17,475	44.1
Zimbabwe	2782	15,425	18.7
Sudan	0	9850	3.5
Malawi	2631	9393	13.6
Mali	2083	7686	13.9

(*Continued*)

Table 1.3 (Continued)

Country	Production (tonnes y^{-1})		Average annual growth (%)
	2010	*2020*	*2010–2020*
Rwanda	100	7055	53
South Africa	3133	6038	6.8
Madagascar	6886	5466	−2.3
Algeria	1759	5436	11.9
Côte d'Ivoire	1700	4620	10.5
Congo, Democratic Republic of the	3025	3590	1.7
Cameroon	570	3556	20.1
Mauritius	568	3298	19.2

Source: Developed from analysis of the FishStatJ database (FAO, 2023) for this assessment.

Notes

* Production of fish, crustaceans and molluscs is included.

planning and regulatory frameworks. Other attributes of national aquaculture production that might be considered when proposing groupings to promote joint working and knowledge sharing include prevailing culture conditions and common species being cultured, rates of growth, similar levels of intensification and numbers of people employed per production unit or comparable proportions or amounts of production entering international trade.

China accounts for the largest proportion of global aquaculture output (Table 1.2). In 1999, fish production within the country from aquaculture (22.8 million tonnes) surpassed that from fishing (17.8 million tonnes) for the first time. Growth in the aquaculture sector has been sustained, increasing on average by 3.4 per cent between 2010 and 2020 (Table 1.2). Aquaculture output in the country in 2008 was more than double that from fisheries, whilst a higher proportion of fish production goes directly for human consumption, 86 per cent in 2008 (Table 1.4), as compared with 78 per cent in the rest of the world (FAO, 2010b). Consequently average per capita fish consumption between 2005 and 2008 was maintained around 30 kg per year.

Aquaculture in earthen ponds still dominates production globally (FAO, 2022), and in China, pond-based culture in 2008 accounted for 70.4 per cent of freshwater aquaculture, with average production of 6.8 t ha^{-1} annually (FAO, 2010b). A significant area under rice production in China, 1.47 million ha in 2008, was producing fish at an average rate of 0.79 t ha^{-1} y^{-1}, potentially contributing to food security and nutrition locally. In contrast, aquaculture of small carp, estimated at around 1 million tonnes in 2008, was undertaken to provide feed for culture of high-value mandarin fish (*Siniperca chuatsi*; FAO, 2010b). Intensification of previously integrated aquaculture production, notably the dike–pond systems of the Pearl River Delta, occurred as producers focused on culturing more valuable aquatic products in response to growing demand from burgeoning numbers of affluent consumers. Recent

Table 1.4 Aquaculture and fisheries development and use trends for China in million tonnes

Category	Nature	2005	2006	2007	2008	2009[*]
Aquaculture production						
	Inland	17.3	18.5	19.7	20.7	22.1
	Marine	10.8	11.3	11.7	12.1	12
	Total	28.1	29.9	31.4	32.7	34.1
Capture fisheries						
	Inland	2.2	2.2	2.3	2.2	2.2
	Marine	12.4	12.5	12.4	12.5	12.7
	Total	14.6	14.6	14.7	14.8	14.9
Use and consumption						
	Non-food use	5.9	6.1	6.9	6.7	6.8
	Human consumption	36.9	38.3	39.2	40.8	42.3
	Population (billions)	1.3	1.3	1.3	1.3	1.3
	Average per capita (kg)	28.4	29.5	30.2	31.4	32.5

Source: Developed from FAO (2010b).

Notes
[*] Provisional figures.

assessments in China, however, have indicated that authorities and producers are tending towards less intensive production modes to better protect the environment and optimise resource use efficiency (Newton et al., 2021).

Global assessment insights to guide sustainable aquaculture development

Concerns have been voiced that aquaculture development globally may be tending towards unsustainable development pathways. This was exemplified by the situation in China where the government has taken action to prohibit excessive aquaculture development in waterbodies susceptible to nutrient enrichment or when they are prioritised for other uses (Zou and Huang, 2015; SeafoodSource, 2016). Production has instead transitioned to less intensive management strategies (Newton et al., 2021). Pressing issues have been noted, too, with species-specific production systems that depend on practices that are harmful to animal welfare and high-profile cases in the seafood sector globally that human rights abuses have been witnesses in supply chains and that international trade in feed ingredients may be exacerbating food and nutrition insecurity in poor and marginal communities (Thiao and Bunting, 2022). Given the importance of aquaculture to global food supplies and the rapid growth of the sector in selected geographies and jurisdictions the Blue Food Assessment (BFA) was commissioned to examine the most pressing challenges and to highlight promising areas for future research and development assistance (Blue Food Assessment, 2023). Core elements of the BFA are summarised in Table 1.5, and key findings and research and

Table 1.5 Focal points of the BFA and key findings to guide future research and development

Focal point	Key findings	References
Potential contributions of aquatic foods to nations	Aquatic foods are central to the culture, livelihoods, food and nutrition security and economies of many nations; safeguarding and promoting access to aquatic foods could help alleviate omega-3 and vitamin B_{12} deficiencies in vulnerable communities in countries across Africa and South America; cardiovascular disease rates and greenhouse gas emissions associated with ruminant meat consumption in countries in the Global North could be reduced with moderate consumption of aquatic foods with lower environmental impacts.	Crona et al. (2023)
Environmental performance of aquatic foods	Bivalve mollusc and seaweed production show the best environmental performance; carp culture generated the lowest greenhouse gas and nutrient emissions; more efficient food conversion ratios enhanced the environmental performance of fed species (across all life cycle assessment impact categories); culture of salmon and trout used land and water resources most efficiently; achieving higher fish yields lowers land and water use.	Gephart et al. (2021)
Nutritional aspects of aquatic foods and potential contributions to human health	Socio-cultural and sub-national factors strongly influence demand for aquatic foods; significantly higher aquaculture production by 2030 could make aquatic foods more affordable and help individuals avoid micronutrient deficiency (166 million cases).	Golden et al. (2021)
Rights and representation for justice in aquatic food systems	Injustices were noted across typical food systems; production and consumption are constrained by lack of accountability and voice, formal education and wealth; gender inequality makes aquatic foods less affordable; policies often fail to address gender-based and political barriers to consumption and production.	Hicks et al. (2022)
Demand and supply for aquatic foods	Over half (55 per cent) of fish consumption globally occurs in just ten countries; due to rising incomes and population growth, demand will increase to over 100 million tonnes by 2050.	Naylor et al. (2021)
Diversity of stakeholders in aquatic food value chains	Individuals (especially small-scale actors) across value chains face multiple threats (i.e. climate change impacts, environmental hazards, political instability and adverse social and economic pressures); constraints to effective adaptation were noted.	Short et al. (2021)
Climate change risks and resilience	Nationally, climate change presents a high degree of risk to aquatic food supplies; resilience can be bolstered through enhanced governance and action to achieve poverty alleviation and gender equality.	Tigchelaar et al. (2022)

development priorities to guide sustainable aquaculture development globally are highlighted.

Appropriate aquaculture development, it is widely felt, is the only means to ensure current levels of per capita fish consumption globally and can constitute an efficient means to increase protein supplies for a growing population. But cultured fish is not necessarily a ready or cost-effective substitute for aquatic foods from wild capture fisheries or low-cost and relatively environmentally benign production of chicken and eggs. Furthermore, the environmental conditions and ecosystem services required to undertake aquaculture may not be suitable or accessible, and the capacity and skills of prospective producers may be lacking. Given the cost, complexities and practical limitations to offshore marine finfish culture, it has been suggested that increased production from freshwater systems is most likely to contribute to food and nutrition security globally (Belton et al., 2020).

Aquaculture appropriates a range of ecosystem services and, where demand exceeds the environmental carrying capacity, adverse consequences result. Environmental impacts attributed to poorly planned and managed aquaculture development include physical disruption to the environment, habitat loss, fragmentations and land-use change, alteration to flow regimes and chemical parameters in receiving waterways, eutrophication, shifting trophic status and interactions, problems with escapees and impacts from pathogens and disease agents. Consequences and salient features of environmental pressures that may be caused are summarised in Table 1.6 and described and discussed further in chapter 2.

Assessments regarding the impact of specific aquaculture units or farm sites, often an emotive issue when environmental, animal welfare and social justice concerns are at the fore, must be undertaken rationally based on the best available data (Boyd, 1999), invoking the precautionary principle and appropriate impact and risk assessments and informed through interactive stakeholder participation; otherwise, the debate risks being distorted by stakeholders with different agendas, values, beliefs and priorities.

Prompting a paradigm shift

International agencies and institutions have recommended aquaculture development to increase the production and availability of aquatic foods and generate employment and livelihoods opportunities (World Bank, 2007; FAO, 2010b; Belton et al., 2016). Nutrition-sensitive aquaculture can contribute to household and community food and nutrition security and facilitate more efficient water and nutrient management (Thilsted et al., 2016; Bunting and Edwards, 2018). Integrated production practices can sustain diversified incomes and livelihood strategies, thus reducing vulnerability (Belton and Azad, 2012; Bunting et al., 2023). Disadvantaged groups such as ethnic minorities and women from poor communities can benefit from aquaculture development and associated livelihood opportunities in processing and service provision (Brugere et al., 2001;

Table 1.6 Negative impacts associated with poorly planned and managed aquaculture development

Impact	Consequence	Key features	Reference
Physical	• modified hydrology	• reduced flow rates, modified channel morphology and flow regimes and salinisation of surface waters in coastal areas	Beveridge and Phillips (1993), Phillips et al. (1993), Tran et al. (1999)
	• sedimentation	• reduced interstitial water flow, increased embeddedness and sediment anoxia	
Chemical	• pollutants	• excreted ammonia toxic to invertebrates; waste nutrients lead to hyper-nutrification; therapeutants and their residues affect non-target organisms, leading to antibiotic-resistant strains	NCC (1990), Phillips et al. (1993), Weston (1996), Davies et al. (1997)
	• reduced oxygen concentrations	• on-farm respiration reduces oxygen levels, leading to exclusion of sensitive species; biological and chemical oxygen demand further depletes oxygen levels	Kelly and Karpinski (1994), Gillibrand et al. (1996)
Nutrient enrichment	• eutrophication	• increased phytoplankton and periphyton production near cages, including possible stimulation of toxic algal blooms; increased epiphyte growth downstream of land-based farms; elevated respiration during decomposition	NCC (1990), Bonsdorff et al. (1997), Selong and Helfrich (1998)
Shifting trophic status	• modified species assemblages	• elimination of pollution-sensitive invertebrates and fish; increased abundance and biomass of tolerant species and eutrophication leading to trophic cascades	Oberdorff and Porcher (1994), Loch et al. (1996), Selong and Helfrich (1998)
Escapees	• predation, competition and ecological impacts	• ecosystem disruption through foraging and consumption of native flora and fauna; escapees breed with resident populations, leading to genetic degradation	Welcomme (1988), Arthington and Bluhdorn (1996), Bardach (1997)
Disease and parasites	• loss of native species	• viruses, bacteria and parasites infest native populations; exotic parasites may also devastate non-resistant indigenous populations	Arthington and Bluhdorn (1996), McAllister and Bebak (1997)
Self-pollution	• reduced production and product quality	• upwelling of anoxic water causing fish-kills in cages; reduced water quality leading to disease outbreaks and stimulation of toxic algal blooms	Lumb (1989), Black et al. (1994), Corea et al. (1998)

(Continued)

Table 1.6 (Continued)

Impact	Consequence	Key features	Reference
Restricted amenity	• decline in capture fisheries and water quality	• competition and disease can damage capture fisheries; sedimentation and plant growth restrict water flow in navigation and irrigation canals; reduced water quality affects access of livestock and humans to water, causing social unrest	NCC (1990), Phillips et al. (1993), Primavera (1997), Tran et al. (1999)
Reduced function-ality	• loss of ecological functions	• discharged wastewater can degrade ecosystems, leading to habitat loss, decreased diversity, restricted storage capacity for nutrients and water and disruption to flows of environmental goods and services	Burbridge (1994), Robertson and Phillips (1995)
Impacts on option and non-use values	• reduced perception of aquatic resources	• degraded aquatic environments and stakeholder conflicts lead to a negative perception of aquaculture, with reduced values attributed to the ecosystem	Turner (1991), Folke et al. (1994), Muir et al. (1999)

Source: Adapted from the review by Bunting (2001).

Kibria, 2004; Pant et al., 2014; Short et al., 2021). Parties to the resolution *Transforming Our World: The 2030 Agenda for Sustainable Development* adopted by the United Nations have committed to achieving seventeen Sustainable Development Goals by 2030 (United Nations, 2015). Comprehensive assessments have demonstrated that sustainable aquaculture development could contribute to achieving the Sustainable Development Goals (ASC, 2022; Bunting et al., 2022).

Despite considerable social and economic benefits associated with responsible aquaculture, poorly planned and managed development can have various negative environmental, economic and social impacts (Table 1.6). Specific aquaculture developments can cause resentment and sometimes precipitate conflict, especially where traditional or *de facto* rights of way and access to fishing grounds or water resources are disrupted. Poorly planned and managed aquaculture development can have serious negative impacts on native species and associated fisheries. Culturing non-native species, which can escape and outcompete or predate native species, is a particular problem; emphasis should therefore be given to using native species and, ideally, strains that evolved under local conditions. Aquaculture development can facilitate the spread of pathogens and disease, affecting both cultured and wild stocks. Loss of native species is counterproductive in the long-term development of aquaculture as native stocks constitute an important genetic source for current and future aquaculture. Therefore, stakeholders ranging from policymakers and aquaculture development agencies through to farmers and communities must have an interest in conserving wild stocks.

To achieve significant growth globally, the 'high-road' scenario presented by the FAO (2022: 220) would require production from marine and inland aquaculture to increase by 11.9 and 8.5 million tonnes per year, respectively. A sustained increase of 10.1 million tonnes per year from marine capture fisheries would also be needed. Under the high-road scenario, apparent per capita supply of aquatic foods might be expected to increase to an average of 25.5 kg per year by 2050 (Table 1.7).

Unsustainable exploitation of aquatic resources, whether marine, brackish or freshwater, is evident in many countries and often stems from a lack of understanding of the true ecological and economic values of such resources. Consequently, several intergovernmental agencies and international organisations have called for improvements in planning and managing to achieve sustainable aquaculture development (FAO, 1995b; World Bank, 2007). The origins of sustainable development thinking and practice relating to aquaculture are reviewed below. Enhanced stakeholder participation is at the core of many strategies, and agendas formulated to promote sustainable aquaculture development and constraints and opportunities associated with interactive stakeholder participation are discussed in chapter 3 and chapter 7.

Table 1.7 Plausible scenarios for future aquaculture production and capture fishery landings (million tonnes, mt) globally

Sector	Predicted yield $(mt\ y^{-1})$ for scenarios indicated below		
	Low road	*Business as usual*	*High road*
Aquaculture			
Inland	75.6	89.9	98.4
Marine	45.3	50.1	62
Total aquaculture	120.8	140	160.3
Capture fisheries			
Inland	10.1	13	13.5
Marine	65.8	85.4	95.5
Total capture fisheries	75.8	98.3	109
Total aquaculture and capture	196.7	238.3	269.3
Aquatic foods for direct human consumption	180.5	217.4	248.2
Global aquatic food supply per capita $(kg\ y^{-1})$	18.5	22.3	25.5

Source: Developed from FAO (2022).

Sustainability: definition and practice

Sustainable development has been enshrined in the lexicon of international governance, policymaking, environmental management and natural resources conservation following the publication of *Our Common Future*, the Report of the World Commission on Environment and Development (United Nations, 1987). Chaired by Gro Harlem Brundtland, the commission findings are often referred to as the Brundtland Commission Report. The Commission was tasked with preparing 'a global agenda for change' proposing 'long-term environmental strategies for achieving sustainable development by the year 2000 and beyond', recommending 'ways concern for the environment may be translated into greater co-operation', considering 'ways and means by which the international community can deal more effectively with environment concerns' and defining 'shared perceptions of long-term environmental issues and the appropriate efforts needed to deal successfully with the problems of protecting and enhancing the environment, a long-term agenda for action during the coming decades, and aspirational goals for the world community' (United Nations, 1987: 11).

Within the context of the report, concern is evident that attempts to protect the environment in 'isolation from human concerns have given the very word "environment" a connotation of naivety in some political circles' and that

the word 'development' has also been narrowed by some into a very limited focus, along the lines of 'what poor nations should do to become richer', and thus again is automatically dismissed by many in the international arena as being a concern of specialists.

(United Nations, 1987: 13)

By way of rebuttal, the Commission simply noted, 'But the "environment" is where we all live; and "development" is what we all do in an attempt to improve our lot within that abode. The two are inseparable' (United Nations, 1987: 14). Summarising their deliberations, it was noted that 'sustainable development is development that meets the needs of the present without compromising the ability of future generations to meet their own needs' (United Nations, 1987: 54). Concerning needs, it was noted that essential needs of poor people globally should receive priority, whilst the state of technology and social organisation impose limitations on the ability of the environment to meet current and future needs.

Although hugely influential and widely cited, the Brundtland Commission Report's conception of sustainable development was criticised in certain quarters as unworkable and leading to confusion. A decade and a half later, the *Report of the World Summit on Sustainable Development* reiterated that 'fundamental changes in the way societies produce and consume are indispensable for achieving global sustainable development' (United Nations, 2002: 13). Paragraph 10 of the plan of implementation called on states to 'improve policy and decision-making at all levels through, inter alia, improved collaboration between natural and social scientists, and between scientists and policy makers' and 'make greater use of integrated scientific assessments, risk assessments and interdisciplinary and intersectoral approaches' (United Nations, 2002: 56). Furthermore, the plan calls for efforts to 'promote the integration of the three components of sustainable development – economic development, social development and environmental protection – as interdependent and mutually reinforcing pillars' and reiterates that 'poverty eradication, changing unsustainable patterns of production and consumption and protecting and managing the natural resource base of economic and social development are overarching objectives of, and essential requirements for, sustainable development' (United Nations, 2002: 1).

Recognition that sustainable development is founded on three interdependent components, economic development, social development and environmental protection (Figure 1.1), is widespread amongst policymakers and natural resources managers.

Practices and procedures to achieve sustainable development are less well defined, with their application neither uniform nor consistent. Sustainable development necessarily demands adoption of the precautionary approach to avoid negative environmental impacts and protect biodiversity. Recognition that economic development is an interrelated component of sustainable development and that the two are not incompatible is essential. Understanding that the environment underpins economic activity and sustains livelihoods is also central to comprehending sustainable development pathways.

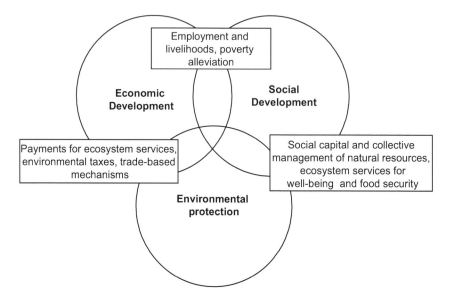

Figure 1.1 Interrelated pillars of sustainability theory and practice.

Sustainable aquaculture: origins and incarnations

Calls for responsible, ethical, low-impact and sustainable aquaculture have come from numerous sources. Despite sustainable development being high on the political agenda during the 1980s, progress in implementing practical strategies to achieve sustainable development in agriculture and aquaculture was initially slow. National- and state-level environmental protection and regulatory mechanisms were generally inadequate when faced with novel and evolving demands associated with rapidly growing aquaculture sectors. Attention and resources were drawn to dealing with local environmental impacts and exerting greater control over antibiotic and chemical use as knowledge of resistance and residues emerged. Consideration of broader ecological, biodiversity, social, economic and trade implications of aquaculture development were, however, beyond the scope and remit of many planning and regulatory agencies.

Principles of responsible aquaculture development were enshrined and elucidated in Article 9 of the *Code of Conduct for Responsible Fisheries* adopted by the FAO (1995b). Guidance covered planning, legislation, livelihoods impacts, environmental assessments, transboundary considerations, conservation of genetic resources and maintaining the integrity of aquatic communities and ecosystems, promotion of responsible aquaculture practices and active stakeholder participation, animal and farm health management practices, waste management and food safety.

For European consumers, products certified as organic or ecological may well be regarded as synonymous with sustainable farming practices and offer a practical means of avoiding negative ecological impacts owing to irresponsible practices, whilst supporting progressive producers with shared environmental ideals. Organic or ecological aquaculture in Europe is governed by Council Regulation (EC) No 834/2007 (European Union, 2007), which establishes common objectives and principles for organic production that are implemented by national certification bodies. Tailored standards have been developed, often guided by expert groups and steering committees, for selected species and production system. Not all organic certification bodies cover aquaculture, and whilst the Soil Association, UK and Naturland, Germany have developed schemes, these have proved to be controversial. Concern has been expressed that farming salmon in cages is an anathema to the principles of organic agriculture, with notable areas of contention including release of untreated process water, high stocking densities and animal welfare concerns, risks from escapees predating and competing with native populations, parasite and disease transfer to wild stocks and curbing the migratory instinct and behaviour of the fish.

Management and husbandry practices covered by Naturland standards include site selection and interaction with surrounding ecosystems; species and origins of stock; breeding, hatchery management; design of holding systems, water quality, stocking density; health and hygiene; oxygen supply; organic fertilising; feeding; transport, slaughtering and processing; smoking; and social aspects. Supplementary regulations address issues relevant to specific farming systems and species, including culture of trout, salmon and other salmonids in ponds or net cages; rope-based mussel culture; and pond culture of shrimp.

Organic production and standards are focused on environmental and ecological aspects, with less attention paid to working conditions. Producers in countries with separate legislation on these aspects must abide by that, but in several countries representing major centres of aquaculture production, little attention is given to health and safety at work and employment terms and conditions. Fair trade schemes have emerged as a practical means to address social and economic imperatives for small-scale producers; although environmental concerns are recognised, conditions imposed on producers are perhaps not as stringent as for organic schemes.

Animal welfare is another area where certification schemes are being developed to meet growing demands from consumers and retailers for ethical alternatives to conventionally produced fish and livestock products. Central to most schemes to safeguard animal welfare are five freedoms stipulated by the Farm Animal Welfare Committee (FAWC, 2011) and adapted previously by the Royal Society for Prevention of Cruelty to Animals (RSPCA, 2010) to form the basis of the 'freedom foods' certification scheme for aquaculture products (Table 1.8). Particular issues covered by the aquaculture guidelines include freshwater life-stages and management, management and stock handling, husbandry practices, equipment and environmental quality, feeding, health, transport, slaughter and wider environmental impacts.

Table 1.8 Five freedoms and generic principles for ethical animal production and adapted principles for aquaculture

Freedom	Generic principles	Aquaculture principles
Freedom from hunger and thirst	• by ready access to fresh water and a diet to maintain full health and vigour	• by permitting access to an environment in which fluid and electrolyte balance can be maintained and an appropriate high-quality diet
Freedom from discomfort	• by providing an appropriate environment including shelter and a comfortable resting area	• by maintaining water at an appropriate temperature and chemical composition and providing well-designed enclosures or tanks, with shading if necessary
Freedom from pain, injury or disease	• by prevention or rapid diagnosis and treatment	• by avoiding situations likely to cause pain, injury or disease; by rapid diagnosis and treatment of disease and humane killing
Freedom to express normal behaviour	• by providing sufficient space, proper facilities and company of the animal's own kind	• by providing appropriate space and environment for the species
Freedom from fear and distress	• by ensuring conditions and treatment which avoid mental suffering	• by minimising stressful situations such as handling or predator attack as far as possible, making gradual changes in husbandry and water quality and humane slaughter

Source: Developed from FAWC (2011) and RSPCA (2010).

Products meeting organic, fair trade and animal welfare standards are intuitively regarded as being sustainable, but this raises questions concerning how to assess the absolute and relative sustainability of different production practices. Organic and animal welfare–conscious production may not necessarily include assessments of greenhouse gas emissions and global warming potential or social considerations such as poverty alleviation, food security, gender equality, employment generation, equitable resource allocation or social capital enhancement. Broader assessments of environmental, social and economic sustainability indicators for aquaculture have been conducted; for example, the European Community–funded CONSENSUS project developed a suite of over seventy indicators through a process of stakeholder workshops and consultations.

Dialogues initiated by the World Wildlife Fund have similarly developed a series of standards through a prolonged process of stakeholder consultation with the intention of passing the final standards to a newly established Aquaculture

Stewardship Council to administer and regulate implementation. This scheme parallels that of the Forest Stewardship Council for wood products and Marine Stewardship Council for fisheries landings. Other organisations have developed schemes to certify aquaculture products and market chains as responsible, and this has resulted in competition amongst schemes and their respective advocates. Producers face the dilemma of which schemes to adopt. Furthermore, transaction costs and administrative requirements may be beyond poor and small-scale producers, although cluster and group formation may help in this situation. These issues are discussed further in chapter 8.

Given the changing constellation of certification schemes, evolving scientific knowledge-base and transient consumer perceptions and demand, it might be surmised that producers, buyers and policymakers are increasingly uncertain about which scheme to adopt. This situation may further challenge consumers when it comes to buying decisions. Several large retailers are now committing to selling only sustainably sourced fish and seafood products, and this may ensure consumer confidence. Fish and seafood consumers are not a homogenous group, however, and some may favour locally produced food to reduce food miles or products bearing a commitment to reducing carbon footprints or purchase directly from producers and smaller suppliers, thus guaranteeing freshness and with the prospect of detailed knowledge of the product's origins. Reputation associated with longstanding producers and suppliers and provenance of selected products may be taken as indicators of sustainability. Price, taste and perceived quality may be critical to other consumers.

Sustainable livelihoods approaches

According to Carney (1998: 4), 'A livelihood is sustainable when it can cope with and recover from stresses and shocks and maintain or enhance its capabilities and assets both now and in the future, while not undermining the natural resource base'. Furthermore, this author noted that 'a livelihood comprises the capabilities, assets (including both material and social resources) and activities required for a means of living' (Carney, 1998: 4). To contextualise the notion of development and research focused on achieving sustainable livelihoods, several agencies including CARE, OXFAM and VSO developed conceptual models depicting the main constituents of a livelihood and external influences that might represent threats or work to support beneficial outcomes (Ashley and Carney, 1999). These authors provided a comprehensive comparative assessment of the origins and early application of sustainable livelihoods approaches by the international NGOs mentioned above as well as the United Nations Development Programme and the UK Government's Department for International Development (DFID). The sustainable livelihoods framework (Figure 1.2) formulated by DFID has found widespread application, and further information on how to navigate and implement this framework is provided in online guidance sheets (DFID, 2001).

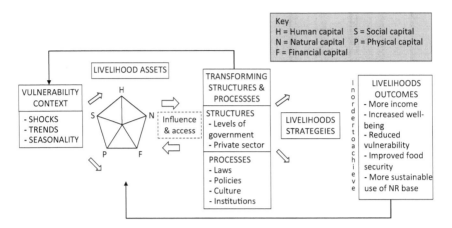

Figure 1.2 DFID sustainable livelihoods framework.

Source: DFID (2001).

Although seeming to refer to an individual's occupation and situation, assessments within sustainable livelihoods frameworks routinely consider access to assets, prevailing policies and legislation, institutional arrangements and threats from the perspective of households, groups or communities. Various additions and refinements to this framework have been proposed since its inception, but in essence it provides a valuable starting point for structuring joint assessment and decision-making. Central to the majority of sustainable livelihoods frameworks is a constellation of assets or capitals. Commonly, five core asset categories or capitals are identified upon which people build their livelihoods:

- financial (e.g. savings, earned income, state benefits and remittances);
- human (e.g. skills, knowledge, ability to labour and health);
- natural (e.g. land, water, erosion protection and biodiversity);
- physical (e.g. infrastructure, tools and equipment);
- social (e.g. networks, connectedness, memberships and relationships of trust, reciprocity and exchange).

Other categories have been proposed, notably cultural and political capital, in certain contexts, to highlight the critical role power plays in governing political and economic livelihood outcomes (Baumann and Subir, 2001).

Many frameworks culminate in livelihood outcomes as these encompass both positive and negative consequences of strategies adopted. Development initiatives and research projects often focus on promotion or analysis of positive livelihood outcomes or achievements, more income, increased well-being, reduced vulnerability, improved food security, more sustainable natural resources use, enhanced social relations and status, dignity and (self-) respect.

However, there are trade-offs inherent in transitions to alternative livelihood strategies, and these must not be ignored or underestimated.

A central premise of sustainable natural resources management is that conservation and enhancement of social capital are critical to ensuring continued social–ecological resilience, enabling collective management of resources and action to mitigate undesirable trends and shocks and to be able to cope with, and recover from, setbacks and disasters (Pretty, 2003). This raises questions such as: Which other types of capital accumulation might have to be foregone or sacrificed to invest in social capital maintenance and accumulation? What constitutes a desirable or acceptable level of social capital? Might excess endeavours to build social capital have negative connotations, creating resentment or fatigue? Is investment in social capital equally worthwhile in all cultures and communities? Social capital comprises a diverse array or relationships and associations, at various locations and scales; consequently, greater understanding concerning the types and portfolios of social capital that contribute to resilience is required. Inter- and intra-household and group differences in social capital are also likely to be manifest and affect the resilience of both individual households and communities. Identification of practices and policies that tend to conserve social capital and encourage accumulation will be critical.

The preceding discussion highlights a major limitation to most sustainable livelihoods frameworks, notably, that decision-making processes, resource allocation decisions, strategising and goal setting by individuals or within households and communities and motivations, incentives and disincentives are not included as distinct and important governing factors. Approaches to budgeting and planning being employed by primary stakeholders are often overlooked. Higher level legal frameworks, organisational arrangements and planning strategies are often assessed within the realms of policies, institutions and processes by means of institutional assessments and policy and legislative reviews. Assessment within the confines of a sustainable livelihoods framework often fails to give prominence to the diversity of livelihood activities or from a gendered perspective, including investment of time and resources often by females, children and older adults of households in reproductive tasks, domestic work and community-organising activities (Moser, 1989). Remittances, loans made, credit received and social norms, obligations, taboos and corruption are not easily accounted for but undoubtedly play a major, yet largely overlooked or ignored, role in people's livelihoods.

Livelihoods diversification is often considered a means to achieving more secure livelihoods, although entailing transaction costs, resulting in exposure to new hazards and unspecified risks and not necessarily being straightforward (Ellis, 1998). Diversification in the context of rural livelihoods was defined as 'the process by which rural households construct an increasingly diverse portfolio of activities and assets in order to survive and to improve their standard of living' (Ellis, 2000: 15). Diversification can enable financial or environmental risks, for example, to be spread across activities; it can help improve income distribution, enable women to engage in novel activities that are the traditional reserve of men and help to reduce pressure on the natural resources base. Within agricultural households especially, income and expenditure are often highly seasonal and generally do not coincide; hence, diversification in majority (developing) and minority (developed) world farming communities is often promoted as one means to improve cash flows, spread risks and capitalise on assets that may be under-exploited or idle for large parts of the year.

Livelihoods diversification can lead to greater inequality and divert investment away from sectors such as agriculture, potentially impacting poor and marginal communities disproportionately. Transition to activities and jobs not directly dependent on the natural resource base may lead to less concern and vigilance over the state of the environment and reduce interest and investment in maintaining and enhancing agroecosystems that provide employment, produce food and generate other benefits for landowners and managers, local communities and visitors. Care is needed, however, as landscapes managed for agriculture, forestry, fisheries and aquaculture may sustain modified and reduced portfolios of ecosystem services as compared with natural systems (or ecosystems where the influence of humans is minimal or expunged). Externalities associated with intensive agricultural practices, for example, may outweigh benefits of employment, food production and agroecosystem services (Pretty et al., 2005).

Accumulated traditional ecological knowledge (TEK) or ecoliteracy can be eroded and lost as a consequence of reduced natural resource dependence, wealth or financial capital accumulation and entering formal schooling (Pilgrim et al., 2007). Ecoliteracy loss could compromise the assessment of alternative livelihood strategies or appropriateness of novel cropping strategies and farming techniques and technology, potentially leading to uptake of inappropriate, externalising technologies that do not fit with the asset base of farmers, damage the environment and result in less secure, more vulnerable livelihoods. Transition to more intensive agricultural production mediated by generic technological packages developed by scientists and agro-industry can result in lock-ins (technology, credit and marketing), stifle innovation locally and prevent farmers from reverting to traditional coping strategies when faced with uncertain or unfavourable environmental, social or economic conditions (Pretty, 2008).

Development agencies and governments in many countries are working to promote and encourage rural communities and farming households to adopt rainwater harvesting to counter water shortages and safeguard against increased variability in rainfall patterns predicted as a consequence of global climate change. For communities in coastal areas, such as those living in the Sundarbans mangrove forest area, West Bengal, India, rainwater harvesting is being promoted in concert with livelihoods diversification (horticulture, floriculture and aquaculture) to capitalise on enhanced water availability and newly constructed water storage facilities. The strategy is supported by government departments, and resources are being channelled through the National Rural Employment Guarantee Act, 2005, to enable communities to implement works to develop rainwater harvesting and storage structures; indeed, within the NREGA, 2005, eligible activities are listed in priority order, and '(i) water conservation and water harvesting' is mentioned first (GoI, 2005: 13).

Knowledge of recent inundation of communities in the Sundarbans with seawater during Cyclone Aila in May 2009, however, raises concerns that investment in rainwater harvesting and storage and increased dependence on appropriated freshwater resources may increase vulnerability. Livelihoods diversification for shrimp seed collectors, one of the poorest groups in the Sundarbans, is being promoted with activities such as handicrafts, apiculture and small businesses being recommended by government and non-governmental agencies. Given that these people are poor, often illiterate and lack money to invest in new ventures, the challenges of facilitating livelihoods diversification with disparate groups and communities across a wide and sometimes inaccessible area are immense. Resulting livelihood strategies and household and community activity portfolios will be complex and governed by assets, markets, programme assistance and threats and opportunities people perceive.

Vulnerability

Sustainable livelihoods frameworks account to some extent for threats, commonly referring to the vulnerability context that encompasses shocks (natural disasters, famine, economic crashes, war) and trends (changing resource levels, input and commodity prices, climate change) that affect people's livelihoods negatively. Within the DFID sustainable livelihoods framework guidance notes, it is stated that 'vulnerability context refers to the seasonality, trends, and shocks that affect people's livelihoods' (DFID, 2001, section 4.8). Potential impacts of hazards (seasonality, trends and shocks) are generally regarded to be a factor of the extent of exposure and sensitivity of livelihoods. However, this is moderated by the ability or capacity to adjust or change to cope with anticipated or actual stress and pressures, often referred to as 'adaptive capacity'. Vulnerability has subsequently been defined as the outcome of this combination of potential impact and adaptive capacity.

Hazards encompass extreme climatic events including hurricanes, tornadoes and cyclones and natural disasters such as tsunami, earthquakes, volcanic eruptions, wildfires, floods and landslides. Disease, parasites and pandemics and famines and droughts also constitute major hazards. Anthropogenic factors are often responsible for exacerbating natural disasters. Working Group I of the IPCC (2007c: 10) reported that 'most of the observed increase in global average temperatures since the mid-20th century is very likely due to the observed increase in anthropogenic greenhouse gas concentrations', whilst

Stern (2007: 56) noted that 'climate change threatens the basic elements of life for people around the world – access to water, food, health, and use of land and the environment'. The villainous role of the state in curtailing freedoms and failed governance in promoting and actually being a pre-requisite for famines was highlighted by Sen (2000). Human-induced hazards can include chemical pollution, inadequate sanitation and communicable diseases, notably HIV/ AIDS, which affects millions of people globally. Armed conflict, terrorism and war affect large numbers of people globally and have adverse effects on all social groups, including non-combatants, and often result in environmental degradation, food insecurity and human rights abuses.

Although often used interchangeably for hazards, risks are more properly defined as the potential of a hazard to cause harm. Consequently, a risk assessment might identify a range of hazards, but the level of risk from each will vary, perhaps with location, season or anticipated frequency of occurrence. Within hazard analysis circles, hazard analysis critical control points assessments are standard practice, notably in the pharmaceutical and food production and processing sectors. Once hazards are identified and quantified, potential critical control points are identified and associated management procedures, protective barriers and safety measures are implemented. Thus, the level of exposure to hazards can be reduced, improving environmental, animal and public health. Hazard analysis critical control points assessments might be best suited to businesses and organisations, but people, households and communities are constantly engaged in less formal, ad hoc and innate hazard analysis and risk assessments, taking avoiding action and implementing locally appropriate control measure and seeking external assistance and information or advice to better evaluate the risks and potential solutions. Once identified as problems, people, families, communities and local institutions develop strategies to cope with recurring hazards of varying magnitudes.

Considering vulnerability associated with climate change, Working Group II of the IPCC (2007a: 18) noted that 'future vulnerability depends not only on climate change but also on development pathway'. Furthermore, Working Group III (IPCC, 2007b: 34) concluded that 'making development more sustainable by changing development paths can make a major contribution to climate change mitigation'. Although many poor and marginal groups are seemingly at the mercy of climate change impacts, there is a danger that they are cast as hapless and helpless, inflating their apparent vulnerability. It should be acknowledged that their coping strategies will have been developed and refined through bitter experience, formulated based on help and opportunities immediately to hand; they will be nuanced to local environmental, social and institutional settings and not be reliant on unpredictable, inefficient or short-term external assistance. Typical coping strategies are diverse and include stinting, hoarding, preserving and protecting assets, depleting, diversification, claim making and movement and migration (Chambers and Conway, 1991). People, groups and communities, however impoverished, often have a range of assets and capabilities that are easily overlooked but constitute an important lifeline when disaster strikes or times are particularly hard.

Resilience

Early studies on social resilience explored potential parallels between resilience in ecological and social systems. Resilience of ecological systems is often used to describe the functioning of the system, rather than stability of component populations or ability to maintain a steady state. Adger (2000) noted that there was no agreed relationship between diversity of ecosystems and their resilience. Some tropical terrestrial ecosystems with stable and diverse populations can exhibit relatively low resilience. Whereas coastal and estuarine ecosystems frequently exhibit low species diversity owing to physical change and species mobility, they might be considered resilient due to high levels of functional diversity (Costanza et al., 1995) encompassing the range of ecological interactions amongst species; for example, predation, competition, parasitism and mutualism and ecological processes such as processing detritus and nutrient recycling and retention (Temperate Forest Foundation, 2010).

Ecological systems resilience has been described as both the capacity to resist, absorb or cope with disturbances and the rate of recovery from perturbations. Parallels have been drawn between ecological and social resilience, and relationships between ecological and social resilience have been explored in a number of settings. Assuming that ecological and social resilience exhibit common characteristics, however, neglects the fact that there are important differences in behaviour and structure between social groups or institutions and ecological systems (Adger, 2000). This author noted that 'social resilience is an important component of the circumstances under which individuals and social groups adapt to environmental change' (Adger, 2000: 347). Social–ecological systems resilience is interpreted as the ability and capacity to respond and adapt to change, and resilience depends on 'diverse mechanisms for living with, and learning from, change and unexpected shocks' (Adger et al., 2005: 1036). Perceptions of ecological systems, whether benign, balanced or resilient, can affect worldviews held by people and consequently influence the philosophical basis for environmental management (Holling, 1995). Other important aspects of social resilience discussed by Adger (2000) include the degree of reliance of social systems on the environment; the resilience of institutions, ranging from socialised behaviour to structures of governance and legislation; and pressures exerted by market liberalisation, urbanisation and globalisation.

Communities reliant on a narrow resource base, notably single ecosystems, crops or fish stocks, have been said to exhibit resource dependency, with some extending the argument to mineral resources and oil. Such reliance can result in negative social and economic outcomes and constrain investment in social capital maintenance and enhancement. Considering resource-dependent communities in North America and coastal areas, Adger (2000: 352) noted that 'promotion of specialization in economic activities has negative consequences in terms of risks for individuals within communities and for the communities themselves'. An emerging body of evidence, however, suggests that entrepreneurs and small-scale business development can stimulate economic growth, providing employment, delivering knowledge and services beyond the scope and resources of government institutions and revenue to government to fund investment in infrastructure, sanitation, education, schooling, policing and environmental protection.

Adaptation and alternative development pathways

Adaptation to climate change is a priority for many communities, countries and development agencies. It is being researched, promoted and funded in various programmes and initiatives, many with a focus on livelihoods, food production, energy systems, cities, environmental management and biodiversity conservation. Whether it constitutes an efficient or justifiable use of resources to promote adaptation when livelihoods practices are deemed to be unsustainable may be questionable. Investment in relocating particularly vulnerable communities and retaining a focus on poverty alleviation and addressing pressing short-term needs might constitute a more effective and equitable approach to achieving medium- to long-term livelihood improvements (Ahmed et al., 2014). Parallels might be drawn here with coastal management predicaments where tough decisions are required whether to hold and advance the line of coastal defences or adopt a strategy of managed retreat. Where less fundamental change is happening or foreseen, an assessment invoking a sustainable livelihoods framework and focused on interactive stakeholder participation (Pretty, 1995) might be useful to facilitate joint assessment of potential adaptation strategies and likely outcomes. Furthermore, evidence is emerging that responses to environmental change can, instead of dampening impacts, result in continued decline or amplifi cation of change in stocks and flows of ecosystem services (Cinner et al., 2010).

A common strategy adopted within the realms of aquaculture development to counter feed and fertiliser shortages or avoid costs associated with external inputs has been to exploit waste resources to enhance production. Sewage-fed aquaculture practices developed in several countries in Asia, especially in China, West Bengal in India and Vietnam, although most are in decline today (Edwards, 2005). Smaller systems were developed in conjunction with sewage treatment plants in Eastern European countries during the Soviet era (Bunting and Little, 2005). Biogas production from communal waste for use in cooking and lighting has been widely promoted in China and Vietnam, with waste material often being channelled to ponds to enhance fish production. Integrated agriculture–aquaculture, where livestock waste is used to fertilise fishponds, is reported to have been adopted in Argentina in response to the economic crisis of the 1990s (Zajdband, 2009).

Several of the cases described above could be cited as sharing common drivers and constituting adaptations to similar pressures. Governance and market failures often characterised the social–economic settings where these strategies were developed and adopted; furthermore, these failures occurred at regional or national scales. Therefore, ecocultures appear more likely to emerge where policies and prevailing social norms impede free trade and movement of goods and services. Contemplating the establishment of mechanisms to replicate such barriers and market restrictions in functioning democracies and governance regimes might point to taxes and tariffs on external inputs, charges for environmental externalities and drafting of supporting policy and legislation. Horizontal integration or integrated, multi-trophic aquaculture has been proposed as a means to replicate the biophysical principles of waste reuse observed in traditional, resource-efficient systems, internalising negative environmental costs of conventional aquaculture. Assessments have shown, however, that economic gains appear marginal and adoption places additional burdens and demands on workers, managers and authorities. Compelling operators to internalise or pay compensation for environmental costs might establish such approaches as a new paradigm for sustainable aquaculture development.

Considerable forethought would be required, however, to ensure that interventions made to encourage ecocultures do not infringe on other freedoms and rights or disadvantage particular communities. Policies favouring ecocultures could make communities more vulnerable where they are isolated, lack alternative production-enhancing inputs or energy sources or are not compensated for avoiding damage and improving the situation. Considering regimes where freedoms are curtailed and movement of food and goods is restricted or poorly coordinated, there is often market speculation; this combination of factors has been implicated in precipitating food shortages and in extreme cases famines (Sen, 2000). Less severe mechanisms could be deployed to steer livelihoods towards more sustainable pathways drawing on principles derived from historical and contemporary ecocultures to provide incentives for adopting good practices and disincentives to irresponsible practices.

Increased awareness of the issues and problems with unsustainable livelihood practices and outcomes should be promoted so that society comes to view the latter as unacceptable. Pressure from advocates, demonstrators and voters and buying decisions made by retailers and consumers have the potential to transform things away from business as usual. Growth in the organic food sector shows how consumer demand and associated price premiums may incentivise or compensate producers adopting sustainable practices. As voluntary schemes such as organic or freedom foods certification might be considered piecemeal and result in patchy and sporadic change, what are really needed are national standards or international agreements on acceptable production practices. Lessons might be learnt from past environmental or ecocultural revolutions such as global action to curtail use of chlorofluorocarbons, international bans on whaling and trade in ivory,

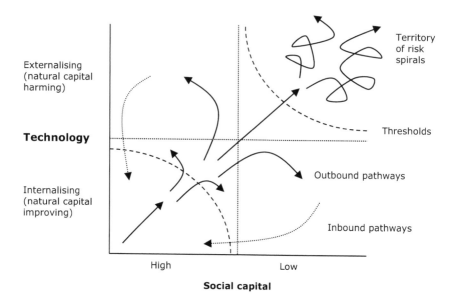

Figure 1.3 Postulated dependence of natural capital on technology type and social capital levels, showing thresholds that may influence development pathways.

Source: Bunting et al. (2010).

carbon-neutral economy commitments by Costa Rica and Iceland, statutory acceptance of sustainable urban drainage approaches throughout Europe and population control policies in China and Vietnam.

Considering the proposition that the interplay between social capital and externalising technology governs natural capital stocks (Figure 1.3), the sustainable livelihoods framework could potentially be used to help explain and account for differences in the actions and principles of ecocultures and other cultures and communities of practice. It might be hypothesised that accumulation of financial assets and access to physical capital lead to a decline of investment in maintaining and enhancing social and natural capital and that until externalities and negative-feedback mechanisms are more explicit in livelihoods frameworks, such representations will be of limited use in understanding individual, household, community and cultural decision-making processes and outcomes.

The driving forces, pressures, state, impacts, responses (DPSIR) framework provides an approach that might be better suited to assess factors catalysing development of ecocultural practices and evaluate associated outcomes. Assessments adopting the DPSIR framework could potentially highlight common or important factors that give rise to ecocultures and might point towards stimuli and tipping-points for ecocultural revolutions that it may be surmised are needed to counter prevailing paradigms of unsustainable lifestyles and development. Assessment of wastewater-fed aquaculture practices in the East Kolkata Wetlands, West Bengal, within the DPSIR framework permitted characterisation of critical aspects of this peri-urban ecocultural system that persists and appears to demonstrate social–ecological resilience (Bunting et al., 2010). Other assessments have focused on urban aquaculture for resilient food systems (Bunting and Little, 2015), the potential contributions of diversified shrimp–rice agroecosystems to social–ecological resilience (Bunting et al., 2017) and innovative biorefinery and intermediary production strategies to capture nutrients before they are lost to the biosphere (see chapter 6, Figure 6.3; Bunting and Edwards, 2018). Comparison across a range of similar assessments might inform typologies for cultures and resilience conditions that could be useful to plan and promote transitions to more desirable, secure and sustainable states.

Frameworks to facilitate systematic and integrated assessments

Appreciation of the role that driving forces play in governing food production strategies, associated value chain configurations, consumer and human behaviour and nutritional and economic outcomes is central to the conceptualisation of food systems formulated by the world-renowned High-Level Panel of Experts (HLPE, 2020). They present a framework that demands comprehensive and systematic assessments and knowledge and understanding of how sub-systems are interconnected and strongly influenced by consumer demand and associated food environments, both external and internal (de Bruyn et al., 2021). Consumers are not homogeneous, and buying behaviours will be influenced by socio-economic status, cultural norms, personal preferences and the spatial and temporal availability and accessibility of different aquatic foods.

It could be argued that risks from shocks and adverse trends may have been overlooked and that, drawing on established sustainable livelihoods frameworks (DFID, 2001), the 'vulnerability context' must be included in food system frameworks and evaluations. The COVID-19 pandemic had a dramatic impact on aquatic food systems globally and resulted in trade disruption and loss of employment and livelihoods (FAO, 2022). Large-scale commercial aquaculture in developing countries is prone to cycles of boom and bust associated with price fluctuation and disease outbreaks (Belton et al., 2017). Consequently, policies, management plans and safety nets may be required to monitor food system functioning and respond effectively to such events.

Disease outbreaks of aquatic animals can cause dramatic and systemic production declines leading to significant production loss, business failures across value chains and threats to food and nutrition security amongst consumers. Environmental degradation can make farmed aquatic animals more susceptible to diseases and external stresses and result in less efficient production and conversion of feed to harvestable biomass and consequently instigate negative feedbacks that exacerbate the situation. Profound inter-connections between the environment and associated ecosystem services, production systems and aquatic animal health and food safety and the food and nutritional security of consumers are exemplified in a One Health framework (Stentiford et al., 2020). Consequently, interdependencies; carrying capacity thresholds; eradication of unethical, unsafe and unstainable practices; and potential trade-offs must be acknowledged and factored into policymaking, food system governance, value chain configurations, farm management decisions and guiding codes of good practice and better management practices.

Perhaps the most important and challenging aspects of pursuing pathways to sustainable aquaculture development globally relate to animal welfare, biosecurity and international trade in feed ingredients. Ensuring that animal welfare is good across all production phases and that when sentient animals are harvested from the wild or cultured to produce feed ingredients (whether directly or indirectly) and when broodstock or juveniles are captured from the wild, this is done responsibly and impacts on by-catch are considered, is necessary. Guaranteeing that inadvertent introductions of invasive species or disease agents, pathogenic organisms and pest species are avoided when animals or plants are imported or moved within countries is vital. Verifying that feed ingredients (e.g. fish meal, fish oil and soyabean meal) are sourced reasonably (including the assessment of historical land-use change, notably deforestation), that nutritious food items are not being diverted away from direct human consumption and exacerbating food and nutrition insecurity (Thiao and Bunting, 2022) and that human rights are respected across seafood value chains (Marschke and Vandergeest, 2016) is of utmost importance.

Sustainable aquaculture development can potentially contribute to enhancing livelihoods across value chains and food systems, contribute to food and nutrition security globally and help alleviate poverty (ASC, 2022; Bunting et al., 2022, 2023). Drawing on the One Health framework analogy, promoting a One Nutrition paradigm from sourcing raw materials, during production (notably for animal welfare), across value chains and food systems, with human health and well-being outcomes being prioritised, could help ensure that practices are equitable (locally and globally) and maximise economic and social development gains. Waste minimisation and recovery and recycling of nutrients (avoiding losses to the biosphere) could be vital to maximising resource-use efficiency and ensuring that healthy diets globally can be sustained in the future (Willett et al., 2019).

Summary

Having reviewed the current status of aquaculture globally, it is apparent the sector has developed rapidly in a number of countries to become a major source of fish and other aquatic animals and plants. Aquaculture can constitute a source of food and nutrition security and generate income and employment, yet production can result in negative environmental, social and economic impacts. Given such concerns, a number of international organisations and environmental NGOs have issued calls to ensure current aquaculture production becomes more responsible and that future development pathways are sustainable. Schemes for organic and ethical aquaculture production have been established, but it is apparent that these initiatives focus on different aspects of production and may not address the full range of potential negative environmental impacts or issues of social and economic development and stakeholder needs. Consequently, a more generic strategy to assessing the sustainability of aquaculture through adopting a sustainable livelihoods approach is presented. Central to such assessments are the assets to which producers, households and communities have access; governing policies and institutions; and aspects of vulnerability. In this context, it is evident that environmental protection, enhanced social capital and avoidance of dependence on externalising technology are necessary to ensure social–ecological resilience.

References

Adger, W.N. (2000) 'Social and ecological resilience are they related?', Progress in Human Geography, vol 24, pp. 347–364.

Adger, W.N., Hughes, T.P., Folke, C., Carpenter, S.R. and Rockstrom, J. (2005) 'Social–ecological resilience to coastal disasters', Science, vol 309, pp. 1036–1039.

Ahmed, N., Bunting, S.W., Rahman, S. and Garforth, C.J. (2014) 'Community-based climate change adaptation strategies for integrated prawn–fish–rice farming in Bangladesh to promote social–ecological resilience', Reviews in Aquaculture, vol 6, pp. 20–35.

Arthington, A.H. and Bluhdorn, D.R. (1996) 'The effect of species introductions resulting from aquaculture operations' in D.J. Baird, M.C.M. Beveridge, L.A. Kelly and J.F. Muir (eds) Aquaculture and Water Resource Management, Blackwell Science, Oxford, UK.

ASC (2022) How Responsible Aquaculture Can Contribute to the UN's Global Sustainable Development Goals (SDGs), Aquaculture Stewardship Council, Utrecht, The Netherlands.

Ashley, C. and Carney, D. (1999) Sustainable Livelihoods: Lessons from Early Experience, Department for International Development, London.

Bardach, J.E. (1997) 'Aquaculture, pollution and biodiversity' in J.E. Bardach (ed) Sustainable Aquaculture, John Wiley & Sons, New York.

Baumann, P. and Subir, S. (2001) Linking Development with Democratic Processes in India: Political Capital and Sustainable Livelihoods Analysis, Overseas Development Institute, Natural Resource Perspectives Number 68, London.

Belton, B. and Azad, A. (2012) 'The characteristics and status of pond aquaculture in Bangladesh', Aquaculture, vol 358–359, pp. 196–204.

Belton, B., Bush, S. and Little, D. (2016) 'Aquaculture: are farmed fish just for the wealthy?', Nature, 538, no 7624, p. 171.

Belton, B., Little, D.C., Zhang, W., Edwards, P., Skladany, M. and Thilsted, S.H. (2020) 'Farming fish in the sea will not nourish the world', Nature Communications, vol 11, pp. 1–8.

Belton, B., Padiyar, A., Ravibabu, G. and Gopal Rao, K. (2017) 'Boom and bust in Andhra Pradesh: development and transformation in India's domestic aquaculture value chain', Aquaculture, vol 470, pp. 196–206.

Beveridge, M.C.M. (2004) Cage Aquaculture, third edition, Blackwell Publishing, Oxford, UK.

Beveridge, M.C.M. and Little, D.C. (2002) 'The history of aquaculture in traditional societies' in B.A. Costa-Pierce (ed) Ecological Aquaculture, Blackwell Science, Oxford, UK.

Beveridge, M.C.M. and Phillips, M.J. (1993) 'Environmental impact of tropical inland aquaculture' in R.S.V. Pullin, H. Rosenthal and J.L. Maclean (eds) Environment and Aquaculture in Developing Countries, ICLARM Conference Proceedings 31, International Centre for Living Aquatic Resources Management, Makati City, Philippines.

Bilio, M. (2007) 'Controlled reproduction and domestication in aquaculture: the current state of the art – part II', Aquaculture Europe, vol 32, no 3, pp. 5–23.

Black, K.D., Ezzi, I.A., Kiemer, M.C.B. and Wallace, A.J. (1994) 'Preliminary evaluation of the effects of long-term periodic sublethal exposure to hydrogen sulphide on the health of Atlantic salmon (*Salmo salar* L)', Journal of Applied Ichthyology, vol 10, pp. 362–367.

Blue Food Assessment (2023) The Blue Food Assessment, https://bluefood.earth/

Bonsdorff, E., Blomqvist, E.M., Mattila, J. and Norkko, A. (1997) 'Coastal eutrophication: causes, consequences and perspectives in the archipelago areas of the northern Baltic Sea', Estuarine, Coastal and Shelf Science, vol 44, supplement A, pp. 63–72.

Boyd, C.E. (1999) 'Aquaculture sustainability and environmental issues', World Aquaculture, vol 30, no 2, pp. 10–13.

Brugere, C., McAndrew, K. and Bulcock, P. (2001) 'Does cage aquaculture address gender goals in development? Results of a case study in Bangladesh', Aquaculture Economics and Management, vol 5, no 3–4, pp. 179–189.

Bunting, S.W. (2001) A Design and Management Approach for Horizontally Integrated Aquaculture Systems, PhD thesis, Institute of Aquaculture, University of Stirling, UK.

Bunting, S.W., Bosma, R.H., van Zwieten, P.A.M. and Sidik, A.S. (2013) 'Bioeconomic modeling of shrimp aquaculture strategies for the Mahakam delta, Indonesia', Aquaculture Economics and Management, vol 17, no 1, pp. 51–70.

Bunting, S.W., Bostock, J., Leschen, W. and Little, D.C. (2023) 'Evaluating the potential of innovations across aquaculture product value chains for poverty alleviation in Bangladesh and India', Frontiers in Aquaculture, vol 2, no 3, pp. 1–23.

Bunting, S.W. and Edwards, P. (2018) 'Global prospects for safe wastewater reuse through aquaculture' in B.B. Jana, R.N. Mandal and P. Jayasankar (eds) Wastewater Management through Aquaculture, Springer, New York.

Bunting, S.W., Kundu, N. and Ahmed, N. (2017) 'Evaluating the contribution of diversified shrimp–rice agroecosystems in Bangladesh and West Bengal, India to social–ecological resilience', Ocean and Coastal Management, vol 148, pp. 63–74.

Bunting, S.W. and Little, D.C. (2005) 'The emergence of urban aquaculture in Europe' in B.A. Costa-Pierce, P. Edwards, D. Baker and A. Desbonnet (eds) Urban Aquaculture, CAB International, Wallingford, UK.

Bunting, S.W. and Little, D.C. (2015) 'Urban aquaculture for resilient food systems' in H. de Zeeuw and P. Drechsel (eds) Cities, Food and Agriculture: Towards Resilient Urban Food Systems, Earthscan, Abingdon, UK.

Bunting, S., Pounds, A., Immink, A., Zacarias, S., Bulcock, P., Murray, F., et al. (2022) The Road to Sustainable Aquaculture: On Current Knowledge and Priorities for Responsible Growth, World Economic Forum, Cologny, Switzerland.

Bunting, S.W., Pretty, J. and Edwards, P. (2010) 'Wastewater-fed aquaculture in the East Kolkata Wetlands: anachronism or archetype for resilient ecocultures?', Reviews in Aquaculture, vol 2, pp. 138–153.

Burbridge, P.R. (1994) 'Integrated planning and management of freshwater habitats, including wetlands', Hydrobiologia, vol 285, pp. 311–322.

Carney, D. (1998) 'Implementing the sustainable rural livelihoods approach' in D. Carney (ed) Sustainable Rural Livelihoods: What Contribution Can We Make?, Department for International Development, UK Government, London.

Chambers, R. and Conway, G. (1991) Sustainable Rural Livelihoods: Practical Concepts for the 21st Century, Institute of Development Studies, IDS Discussion Paper 296, Brighton, UK.

Cinner, J.E., Folke, C., Daw, T. and Hicks, C.C. (2010) 'Responding to change: using scenarios to understand how socioeconomic factors may influence amplifying or dampening exploitation feedbacks among Tanzanian fishers', Global Environmental Change, vol 21, pp. 7–12.

Corea, A., Johnstone, R., Jayasinghe, J., Ekaratne, S. and Jayawardene, K. (1998) 'Self-pollution: a major threat to the prawn farming industry in Sri Lanka', Ambio, vol 27, pp. 662–668.

Costanza, R., Kemp, M. and Boynton, W. (1995) 'Scale and biodiversity in estuarine ecosystems' in C. Perrings, K.G. Maler, C. Folke, C.S. Holling and B.O. Jansson (eds) Biodiversity Loss: Economic and Ecological Issues, Cambridge University Press, Cambridge.

Crona, B.I., Wassénius, E., Jonell, M., Koehn, J.Z., Short, R., Tigchelaar, M., et al. (2023) 'Four ways blue foods can help achieve food system ambitions across nations', Nature, vol 616, pp. 104–112.

Davies, I.M., McHenery, J.G. and Rae, G.H. (1997) 'Environmental risk from dissolved ivermectin to marine organisms', Aquaculture, vol 158, pp. 263–275.

de Bruyn, J., Wesana, J., Bunting, S.W., Thilsted, S.H. and Cohen, P.J. (2021) 'Fish acquisition and consumption in the African Great Lakes Region through a food environment lens: a scoping review', Nutrients, vol 13, no 7, pp. 1–15.

DFID (2001) Sustainable Livelihoods Guidance Sheets, United Kingdom Government's Department for International Development, London.

Edwards, P. (2005) 'Development status of, and prospects for, wastewater-fed aquaculture in urban environments' in B. Costa-Pierce, A. Desbonnet, P. Edwards and D. Baker (eds) Urban Aquaculture, CAB International, Wallingford, UK.

Ellis, F. (1998) 'Household strategies and rural livelihoods diversification', Journal of Development Studies, vol 35, pp. 1–38.

Ellis, F. (2000) Rural Livelihoods and Diversity in Developing Countries, Oxford University Press, Oxford, UK.

European Union (2007) 'Council Regulation (EC) No 834/2007 of 28 June 2007 on organic production and labelling of organic products and repealing Regulation (EEC) No 2092/91', Official Journal of the European Union, L189, pp. 1–23.

FAO (1995a) Aquaculture Production Statistics 1984–1993, FAO Fisheries Circular 815, Rev. 7, FAO, Rome.

FAO (1995b) Code of Conduct for Responsible Fisheries, FAO, Rome.

FAO (2010a) Report of the Regional Workshop on Methods for Aquaculture Policy Analysis, Development and Implementation in Selected Southeast Asian Countries, Bangkok, 9–11 December 2009, FAO Fisheries and Aquaculture Report No. 928, FAO, Rome.

FAO (2010b) The State of World Fisheries and Aquaculture 2010, FAO, Rome.

FAO (2022) The State of World Fisheries and Aquaculture 2022, FAO, Rome.

FAO (2023) 'Fishery and aquaculture statistics. Global aquaculture production 1950–2020', FishStatJ, FAO, Rome.

FAWC (2011) Five Freedoms, Farm Animal Welfare Committee, https://www.gov.uk/government/groups/farm-animal-welfare-committee-fawc#assessment-of-farm-animal-welfare---five-freedoms-and-a-life-worth-living

Folke, C., Kautsky, N. and Troell, M. (1994) 'The costs of eutrophication from salmon farming: implications for policy', Journal of Environmental Management, vol 40, pp. 173–182.

Gephart, J.A., Golden, C.D., Asche, F., Belton, B., Brugere, C., Froehlich, H.E., et al. (2021) 'Scenarios for global aquaculture and its role in human nutrition', Reviews in Fisheries Science and Aquaculture, 29, no 1, pp. 122–138.

Gillibrand, P.A., Turrell, W.R., Moore, D.C. and Adams, R.D. (1996) 'Bottom water stagnation and oxygen depletion in a Scottish sea loch', Estuaries, Coastal and Shelf Science, vol 43, pp. 217–235.

GoI (2005) National Rural Employment Guarantee Act, 2005, Government of India, New Delhi.

Golden, C.D., Koehn, Z.J., Shepon, A., Passarelli, S., Free, C.M., Viana, D.F., et al. (2021) 'Aquatic foods to nourish nations', Nature, vol 598, pp. 315–320.

Hernandez, R., Belton, B., Reardon, T., Hu, C., Zhang, X. and Ahmed, A. (2018) 'The "quiet revolution" in the aquaculture value chain in Bangladesh', Aquaculture, vol 493, pp. 456–468.

Hicks, C.C., Gephart, J.A., Koehn, J.Z., Nakayama, S., Payne, H.J., Allison, E.H., et al. (2022) 'Rights and representation support justice across aquatic food systems', Nature Food, vol 3, pp. 851–861.

HLPE (2020) Food Security and Nutrition: Building a Global Narrative towards 2030, High Level Panel of Experts, Rome.

Holling, C.S. (1995) 'What barriers? What bridges?' in L. Gunderson, C.S. Holling and S.S. Light (eds) Barriers and Bridges to the Renewal of Ecosystems and Institutions, Columbia University Press, New York.

IPCC (2007a) Climate Change 2007: Impacts, Adaptation and Vulnerability. Contribution of Working Group II to the Forth Assessment Report of the Intergovernmental Panel on Climate Change, Summary for Policy Makers, https://archive.ipcc.ch/publications_and_data/ar4/wg2/en/contents.html

IPCC (2007b) Climate Change 2007: Mitigation and Climate Change. Contribution of Working Group III to the Forth Assessment Report of the Intergovernmental Panel on Climate Change, Summary for Policy Makers, https://archive.ipcc.ch/publications_and_data/ar4/wg3/en/contents.html

IPCC (2007c) Climate Change 2007: The Physical Science Basis. Contribution of Working Group I to the Forth Assessment Report of the Intergovernmental Panel on Climate Change, Summary for Policy Makers, https://archive.ipcc.ch/publications_and_data/ar4/wg1/en/contents.html

Kelly, L.A. and Karpinski, A.W. (1994) 'Monitoring BOD outputs from land-based fish farms', Journal of Applied Ichthyology, vol 10, pp. 368–372.

Kibria, G. (2004) 'Gender roles in aquaculture: some findings from aquaculture development on the northern uplands of Viet Nam project', FAO Aquaculture Newsletter, vol 32, pp. 15–18.

Liu, Y., Bunting, S.W., Luo, S., Cai, K. and Yang, Q. (2019) 'Evaluating impacts of fish stock enhancement and biodiversity conservation actions on the livelihoods of small-scale fishers on the Beijiang River, China', Natural Resource Modeling, vol 32, no 1, pp. 1–17.

Loch, D.D., West, J.L. and Perlmutter, D.G. (1996) 'The effect of trout farm effluent on the taxa richness of benthic macroivertebrates', Aquaculture, vol 147, pp. 37–55.

Lumb, C.M. (1989) 'Self-pollution by Scottish fish-farms?', Marine Pollution Bulletin, vol 20, pp. 375–379.

Marschke, M. and Vandergeest, P. (2016) 'Slavery scandals: unpacking labour challenges and policy responses within the off-shore fisheries sector', Marine Policy, vol 68, pp. 39–46.

McAllister, P.E. and Bebak, J. (1997) 'Infectious pancreatic necrosis virus in the environment: relationship to effluents from aquaculture facilities', Journal of Fish Disease, vol 20, pp. 201–207.

Moser, C.O.N. (1989) 'Gender planning in the third world: meeting practical and strategic gender needs', World Development, vol 17, pp. 1799–1825.

Muir, J.F. (2005) 'Managing to harvest? Perspectives on the potential of aquaculture', Philosophical Transactions of the Royal Society B, vol 360, pp. 191–218.

Muir, J.F., Brugere, C., Young, J.A. and Stewart, J.A. (1999) 'The solution to pollution? The value and limitations of environmental economics in guiding aquaculture development', Aquaculture Economics and Management, vol 3, pp. 43–57.

Naylor, R.L., Kishore, A., Sumaila, U.R., Issifu, I., Hunter, B.P., Belton, B., et al. (2021) 'Blue food demand across geographic and temporal scales', Nature Communications, vol 12, pp. 1–14.

NCC (1990) Fish Farming and the Scottish Freshwater Environment, Nature Conservancy Council, Edinburgh, UK.

Newton, R., Zhang, W., Xian, Z., McAdam, B. and Little, D.C. (2021) 'Intensification, regulation and diversification: the changing face of inland aquaculture in China', Ambio, 50, no 9, pp. 1739–1756.

Oberdorff, T. and Porcher, J.P. (1994) 'An index of biotic integrity to assess biological impacts of salmonid farm effluents on receiving waters', Aquaculture, vol 119, pp. 219–235.

Pant, J., Barman, B.K., Murshed-E-Jahan, K., Belton, B. and Beveridge, M. (2014) 'Can aquaculture benefit the extreme poor? A case study of landless and socially marginalized Adivasi (ethnic) communities in Bangladesh', Aquaculture, vol 418–419, pp. 1–10.

Phillips, M.J., Lin, C.K. and Beveridge, M.C.M. (1993) 'Shrimp culture and the environment: lessons from the world's most rapidly expanding warm water aquaculture sector' in R.S.V. Pullin, H. Rosenthal and J.L. Maclean (eds) Environment and Aquaculture in Developing Countries, International Centre for Living Aquatic Resources Management, Makati City, Philippines.

Pilgrim, S., Smith, D. and Pretty, J. (2007) 'A cross-regional assessment of the factor affecting ecoliteracy: implications for policy and practice', Ecological Applications, vol 17, pp. 1742–1751.

Pretty, J.N. (1995) 'Participatory learning for sustainable agriculture', World Development, vol 23, pp. 1247–1263.

Pretty, J.N. (2003) 'Social capital and the collective management of resources', Science, vol 302, pp. 1912–1915.

Pretty, J.N. (2008) 'Agricultural sustainability: concepts, principles and evidence', Philosophical Transactions of the Royal Society B, vol 363, pp. 447–465.

Pretty, J.N., Ball, A.S., Lang, T. and Morison, J.I.L. (2005) 'Farm costs and food miles: an assessment of the full cost of the UK weekly food basket', Food Policy, vol 30, pp. 1–9.

Primavera, J.H. (1997) 'Socio-economic impacts of shrimp culture', Aquaculture Research, vol 28, pp. 815–827.

Robertson, A.I. and Phillips, M.J. (1995) 'Mangroves as filters of shrimp pond effluent: predictions and biogeochemical research needs', Hydrobiologia, vol 295, pp. 311–321.

RSPCA (2010) RSPCA Welfare Standards for Farmed Atlantic Salmon, Royal Society for the Prevention of Cruelty to Animals, Horsham, West Sussex, UK.

SeafoodSource (2016) Massive aquaculture shut-down in central China as government gets tough on pollution, https://www.seafoodsource.com/features/massiveaquaculture-shut-down-in-central-china-as-government-gets-tough-on-pollution

Selong, J.H. and Helfrich, L.A. (1998) 'Impacts of trout culture effluent on water quality and biotic communities in Virginia headwater streams', Progressive Fish-Culturist, vol 60, pp. 247–262.

Sen, A. (2000) Development as Freedom, Oxford University Press, Oxford, UK.

Short, R.E., Gelcich, S., Little, D.C., Micheli, F., Allison, E.H., Basurto, X., et al. (2021) 'Harnessing the diversity of small-scale actors is key to the future of aquatic food systems', Nature Food, vol 2, no 9, pp. 733–741.

Stentiford, G.D., Bateman, I.J., Hinchliffe, S.J., Bass, D., Hartnell, R., Santos, E.M., et al. (2020) 'Sustainable aquaculture through the One Health lens', Nature Food, vol 1, no 8, pp. 468–474.

Stern, N. (2007) The Economics of Climate Change: The Stern Review, Cabinet Office, HM Treasury, London.

Temperate Forest Foundation (2010) Ecological Diversity, www.forestinfo.org/discover/diversity

Thiao, D. and Bunting, S.W. (2022) Socio-Economic and Biological Impacts of the Fish-Based Feed Industry for Sub-Saharan Africa, FAO Fisheries and Aquaculture Circular 1236, FAO, Rome.

Thilsted, S.H., Thorne-Lyman, A., Webb, P., Bogard, J.R., Subasinghe, R., Phillips, M.J., et al. (2016) 'Sustaining healthy diets: the role of capture fisheries and aquaculture for improving nutrition in the post-2015 era', Food Policy, vol 61, pp. 126–131.

Thorpe, J.E. (1980) 'The development of salmon culture towards ranching' in J. Thorpe (ed) Salmon Ranching, Academic Press, London.

Tigchelaar, M., Leape, J., Micheli, F., Allison, E.H., Basurto, X., Bennett, A., et al. (2022) 'The vital roles of blue food in the global food system', Global Food Security, vol 33, 100637.

Tran, T.B., Le, C.D. and Brennan, D. (1999) 'Environmental costs of shrimp culture in the rice-growing regions of the Mekong Delta', Aquaculture Economics and Management, vol 3, pp. 31–42.

Turner, K. (1991) 'Economics and wetland management', Ambio, vol 20, pp. 59–63.

United Nations (1987) Our Common Future, Report of the World Commission on Environment and Development, United Nations, New York.

United Nations (2002) Report of the World Summit on Sustainable Development, United Nations, New York.

United Nations (2015) Transforming Our World: The 2030 Agenda for Sustainable Development, United Nations, New York.

Welcomme, R.L. (1988) International Introductions of Inland Aquatic Species, FAO Fisheries Technical Paper 294, FAO, Rome.

Weston, D.P. (1996) 'Environmental considerations in the use of antibacterial drugs in aquaculture' in D.J. Baird, M.C.M. Beveridge, L.A. Kelly and J.F. Muir (eds) Aquaculture and Water Resource Management, Blackwell Science, Oxford, UK.

Willett, W., Rockström, J., Loken, B., Springmann, M., Lang, T., Vermeulen, S., et al. (2019) 'Food in the Anthropocene: the EAT–Lancet Commission on Healthy Diets from Sustainable Food Systems', Lancet, vol 393, no 10170, pp. 447–492.

Williams, M. (2011) Better Science, Better Fish, Better Life: Overview and Better Science, Plenary Session, 9th Asian Fisheries and Aquaculture Forum, 21–25 April 2011, Shanghai, China.

World Bank (2007) Changing the Face of the Waters: The Promise and Challenge of Sustainable Aquaculture, The World Bank, Washington, DC.

Zajdband, A.D. (2009) Resilience of Integrated Agri-aquaculture Systems in Subtropical NE Argentina, University of Buenos Aires, Buenos Aires, Argentina.

Zou, L. and Huang, S. (2015) 'Chinese aquaculture in light of green growth', Aquaculture Reports, vol 2, pp. 46–49.

2 Resource-conserving and agroecosystem-enhancing aquaculture

Key points

The aims of this chapter are to:

- Review prevailing aquaculture production systems and classification schemes used to describe and differentiate production strategies, notably concerning the intensity of management and production-enhancing inputs.
- Describe the diverse management regimes and culture facilities employed to enable aquaculture in marine and freshwater environments, provide salient examples and highlight animal welfare considerations.
- Describe prevailing value chain configurations and highlight key opportunities and constraints to more sustainable production.
- Discuss prospects for converting shellfish waste to active seafood packaging (de la Caba et al., 2019) as a circular economic strategy and introduce the concept of the 'shell biorefinery' (Yan and Chen, 2015).
- Introduce high-level process mapping and include example for agriculture-based livelihoods and discuss opportunities and constraints to the analysis of financial, social and environmental interactions.
- Review the concept of product units, edible portions and food loss and waste and their potential contribution to better characterising the nature of production and discuss the implications for allocating and managing impacts across value chains.
- Review environmental pressures that can result from poorly planned or managed aquaculture development and discuss the nature of physical, chemical and ecological impacts.
- Introduce the concept of regenerative agriculture and discuss opportunities and constraints to regenerative aquaculture.
- Discuss resource-conserving and agroecosystem-enhancing aquaculture strategies and management options to avoid environmental impacts, notably water management and wastewater treatment and nutrient recovery and reuse options.

DOI: 10.4324/9781003342823-2

• Review development of an ecosystem approach to aquaculture and discuss prospects for polyculture, integrated aquaculture–agriculture, ecological water conditioning (e.g. green-water culture), integrated multi-trophic aquaculture and horizontally integrated aquaculture development.

Prevailing aquaculture production systems

Prevailing aquaculture production systems can be classified employing various systems features or indices, and a number of common constraints and opportunities associated with each category are apparent (Table 2.1a,b). Aquaculture systems are routinely categorised based on the intensity of management inputs, which are often related to area-based production levels. A typology ranging from extensive to semi-extensive to intensive was proposed by Coche (1982), whilst the FAO (2007) described key characteristics of such culture systems and suggested typical yields (Figure 2.1). Further categories of semi-extensive and super-intensive production have been proposed in certain cases, but in most schemes there is a degree of crossover, as with the management of actual culture systems.

Extensive production tends to be adopted where producers are poor and have limited resources to invest, demand from local markets or market networks does not warrant higher production of aquaculture products or risks associated with production deter producers from adopting more intensive production means. Semi-intensive production entails either chemical fertiliser or manure inputs to enhance autotrophic and heterotrophic production and/or supplementary feed inputs to augment natural productivity. Significant areas of land previously under semi-intensive shrimp production throughout the tropics have either been abandoned or reverted to extensive production owing to problems with shrimp disease and concerns over financial returns. Transition to complete dependence on formulated feed is the principal characteristic of intensive production.

Prevailing hydrological conditions or the relative openness of the culture facility constitute further ways of classifying aquaculture systems that are useful for assessing the prospective standing stock that might be maintained given the carrying capacity of the site; risks posed by pests, diseases and pollution; and constraints and opportunities to retaining and reusing nutrients and recycling process water. Where water is recycled within the system, the recirculation rate, expressed as the percentage of water in the system that is recirculated per unit time or proportion of water discharged from the system and requiring replenishment, is a commonly cited measure of management intensity and indicator of energy and water demands.

Classification according to the environment or geographic area where production occurs can provide insights concerning probable salinity, temperature and rainfall regimes or climatic conditions and institutional arrangements and market opportunities. Aquaculture is practiced in a range of

Table 2.1a Marine aquaculture systems, alternative management regimes and examples

Aquaculture system	Management regimes	Examples
Ponds	• extensively managed coastal ponds with supplementary catches of wild shrimp, fish and crabs	• milkfish, shrimp and seaweed in ponds throughout the tropics • rearing brackish water herbivorous fish in *tambak* in Indonesia
	• semi-intensive ponds with water exchange dependent on the tide • semi-intensive ponds with limited water exchange • pellet-fed marine ponds • seasonal marine and brackish water ponds facilitating integrated production	• reservoir-fed earthen ponds in France for seabass integrated with phytoplankton, oysters and clam culture • better management practices for shrimp culture in Ache, Indonesia, for tsunami rehabilitation and enhanced shrimp health • shrimp and marine finfish such as grouper and seabass in Asia • shrimp–rice, fish–rice and prawn–shrimp ponds in Bangladesh and West Bengal, India
Tanks	• intensively managed recirculating systems • flow-through • nursery and hatchery facilities • compartmentalised, three-dimensional and tray-based systems	• seabass, turbot and sole in Europe and Asia • marine ornamentals notably in Asia and Europe • sea urchins and abalone in Asia, North America and Europe • sea urchins, shrimp and *Salicornia* spp. in Israel • juvenile cod and halibut • lobsters in Norway, flatfish in Europe
Raceways	• raceways with paddle-wheels • raceways with airlifts	• micro-algae production in southern Israel • seaweed production as part of IMTA systems in northern Israel • system for seabass and seabream in southern Israel
Intertidal and on-bottom culture	• shore-based • on-bottom culture	• trestles for oysters in France • poles and concrete slabs and iron supports for spat collection • oyster beds in Essex, England • clams near mangroves in Vietnam • Yesso scallops produced by cooperatives in Japan
Coastal wetlands	• silviculture • lagoons	• shrimp ponds integrated with mangroves throughout Asia • pens for crab culture in mangrove stands • southern European lagoons

Embayments		• early trials at Ardtoe, Scotland
		• *loko kuapa* or seawater ponds constructed in Hawaii
Cages and pens	• cage culture for finfish	• salmon cages in Canada, Chile, Norway and Scotland
		• grouper and snapper culture in Vietnam and Indonesia
		• offshore production in the Gulf of Mexico
Open-water suspended culture	• pens	• pens for fish and prawns in West Bengal, India
	• longlines	• shellfish suspended from longlines in North America and Europe
		• seaweed suspended from longlines in Asia and Europe
	• rafts	• shellfish suspended from rafts in North America and Europe
		• seaweed suspended from rafts in Asia and Europe
Integrated open-water marine aquaculture	• IMTA	• salmon and mussels in Norway
		• salmon and seaweed in Canada
Ranching and culture-based operations	• ecological engineering	• cages and blocks forming artificial reefs
		• cod in North America and Norway in 1890s
		• salmon ranching in North America, East Asia and Europe
		• reefs in Europe, North America and Asia
		• 27 countries involved with stocking over 65 marine and brackish water species (Svasand and Moksness, 2004)

Table 2.1b Freshwater aquaculture systems, alternative management regimes and examples

Aquaculture system	Management regimes	Examples
Ponds	• static water ponds with minimal water exchange to conserve nutrients • earthen flow-through ponds for freshwater salmonids • household and farm ponds	• much of Asian aquaculture producing considerable volumes of fish • large fishponds for carp in central Europe • wastewater-fed fishponds in Asia, notably West Bengal, India • used for brown and rainbow trout grow-out for food and re-stocking in Europe and North America • integrated with agriculture and social systems throughout Asia • integrated agriculture–aquaculture systems in Africa (e.g. Cameroon, Egypt and Zambia) and Argentina
Tanks	• intensively managed closed systems totally dependent on formulated feed • flow-through systems for juvenile salmonids • hatcheries • ornamental producers • depuration and holding facilities	• eels in Europe and Asia • tilapia and catfish in Europe and Africa • smolt farms throughout Northwest Europe, North America and Chile • prawn seed throughout the tropics; for example, in Bangladesh and West Bengal, India • numerous species in Asia and Europe • supporting shellfish culture in Europe
Raceways	• temperate water systems for salmonid grow-out	• Arctic charr in Scandinavia • trout in Western and Southern Europe
Integrated freshwater aquaculture	• aquaponics • wastewater management	• lettuce, tomatoes, herbs and tilapia in the USA • social-enterprise and educational facilities in UK, USA and Cambodia • living machines • duckweed lagoons in Bangladesh and USA • constructed wetlands in Europe and North America
Wetlands and embayments	• flooded forests • embayments in lakes and reservoirs	• mellaluca fishponds in Vietnam • chinampas in Mexico, traditional and recent developments in Bangladesh • fingerponds developed on the shores of lakes in Africa • enclosed coves and bays in Vietnam for gobies

Flooded fields	aquaculture with rice cultivation	• prawn–rice and crab–rice ponds in Asia
	aquatic plants cultivated as a main crop	• water mimosa, dropwort, morning-glory cultivated around cities in Asia
		• freshwater upland ponds for taro cultivation in Hawaii
Pens and cages	cages in reservoirs	• hydroelectric and water supply schemes in China, Indonesia and Vietnam
	cages in lakes	• cage-based culture of tilapia species in the African Great Lakes Region
	pens in lakes	• pens for prawns in India
	cages in rivers	• tilapia culture in southern Vietnam
		• small cages for fattening species from capture fisheries in Bangladesh
Ranching and culture-based operations	reservoirs and lakes	• urban lakes leased to companies in Vietnam for carp production
		• self-help groups managing ox-bow lakes in India
	seasonal water bodies	• *beels* in West Bengal and *hoars* in Bangladesh
	wetland agroecosystems	• farmer-managed systems

Extensive

low level of control, minimal costs, high dependence on prevailing climate and environment, large and natural water bodies, natural feed, production <500 kg ha^{-1} y^{-1}

Semi-intensive

water exchange and aeration employed, improved ponds and simple cages, natural food augmented with fertiliser and supplementary feed, production 2-20 t ha^{-1} y^{-1}

Intensive

high degree of control, significant initial costs and technology requirement, operation independent of prevailing climate and water quality, complete reliance of formulated feed, production up to 200 t ha^{-1} y^{-1}

Figure 2.1 Characteristics of aquaculture systems operated at management intensities indicated.

marine (Table 2.1a) and freshwater culture systems (Table 2.1b) and is increasingly integrated with other activities within landscapes or waterscapes and urban and industrial systems and processes.

Discussion concerning aquaculture development usually focuses on the on-growing or grow-out phase, which frequently appropriates the majority of inputs and is often the most visible stage of the production cycle. Three other distinct production phases have been identified with requirements and characteristics that demand that they be considered separately in aquaculture development and planning (Muir, 2005).

Rearing the early stages of aquatic species is critical in closing the life-cycle in captivity. At this stage, animals are acutely sensitive to environmental conditions, encompassing water quality, flow rates and regimes, shelter and disturbance, and display highly specific and rapidly evolving feeding prefer-ences and requirements. Intermediate or part-grown animals, including fish fingerlings, shrimp post-larvae and mollusk seed, exhibit intermediate sensitivity to environmental factors and feed quality and often represent the life-stage in nature that makes a notable migration from freshwater to brackish water and marine environments in the case of catadromous species or from marine to freshwaters for anadromous species or transform from a pelagic to sedentary or sessile existence.

Broodstock are either caught and collected from the wild or derived from on-growing stock, where selection may be based on key attributes or multiple criteria. Refined broodstock management programmes will strive to select desirable traits, avoid negative selection and promote heterozygosity (Bert et al., 2002; Muir, 2005). In practice, this is achieved through developing a broodstock management strategy, maintaining records, identifying criteria upon which to base brood stock selection, potentially using genetic tags to identify desirable genetic alleles or traits and purposefully managing stocks of brood animals and pursuing brood stock from other sources to ensure ongoing stock improvement. Feeding and environmental conditions are often critical to condition animals and stimulate spawning. Animal welfare considerations must be paramount across life-cycles of cultured organisms. Practices such as unilateral eyestalk ablation of crustaceans must be avoided, and alternative spawning strategies can significantly enhance the health and welfare of broodstock and produce progeny that are more robust to prevalent and damaging diseases (Zacarias, 2020; Zacarias et al., 2021).

Prevailing value chain configurations

Characteristics of four broadly representative species-specific value chain configurations for aquatic foods have been described, namely: accessible commodity, accessible niche, luxury commodity and luxury niche (Henriksson et al., 2021). Accessible commodity species (i.e. filter-feeding organisms and omnivorous fish) are produced in large volumes (e.g. 0.5–6 million tonnes per year), deemed relatively affordable, 'costing less than half the global production weighted average' (Henriksson et al., 2021: 1223), and are important for food and nutrition security. Accessible niche species (e.g. gourami, mullets and mussels) are a similar price but currently produced in smaller volumes (e.g. 10,000–100,000 tonnes per year) typically for regional markets. Luxury commodity species (e.g. Atlantic salmon, red swamp crayfish and whiteleg shrimp) are produced at high volume (e.g. >1 million tonnes per year), command a relatively high price (above the global weighted average) and are often consumed by mid- to high-income groups and are traded regionally and international. Luxury niche species (e.g. abalone, bluefin tuna and groupers) are produced principally for international trade and high-income consumers. It was noted that action to close the performance gap (i.e. yields realised versus those achievable in a specific region) for affordable commodity species could have the greatest impacts on production whilst enhancing environmental performance globally (Henriksson et al., 2021).

Prospects for circular economy strategies

Principles of managing waste globally must focus primarily on prevention, reduction, recycling and reuse (United Nations, 2015). Adopting best or better management strategies, optimal feed management and delivery and/or fertiliser applications and ensuring good animal health and welfare can contribute to the efficient conversion of nutrient inputs to harvestable biomass. Cropping multiple species that can utilise different feeding niches in ponds can help maximise production. When small carp or other indigenous species are produced, these could be important sources of food and nutrition for poorer consumers (Roos et al., 2007; Bunting et al., 2015, 2017; Thilsted et al., 2016; Kumar Saha and Kumar, 2020; Karim et al., 2021; Padiyar et al.,

2021), and marketing and consumption of whole small fish can help avoid issues of food loss and waste (Kruijssen et al., 2020; Bunting et al., 2023). Accumulations of nutrients in sediments and wastewater flows can potentially be utilised to cultivate secondary crops and help maximise financial returns and avoid losses to the biosphere; strategies to achieve this are discussed in detail below. Innovative aquaculture practices could potentially provide a viable means to utilising wastewater and other by-products and waste streams to produce intermediate products to feed to other animals or biomass to supply novel biorefinery processes (Bunting and Edwards, 2018).

Appreciation is growing rapidly that aquatic animals and plants must be fully utilised to realise the greatest value across food systems, maximise financial returns to producers and avoid waste. Selected by-products from centralised processing activities can be highly valued and find ready markets internationally. Other components may require more refined strategies to realise their true value, but the transition to productively utilising 100 per cent of fish is underway, with Iceland leading the way (Sigfusson, 2022). As this paradigm takes hold globally, it has potential to benefit both producers with greater financial returns and many thousands of people who could find employment and income-generating opportunities in novel processing sectors and biorefinery operations. There is a danger, however, that existing primary processors or users of processing by-products may be overlooked or priced out of the market, thus possibly disadvantaging poor and marginal groups, notably women (Thiao and Bunting, 2022). Globally, biomass from seafood processing could be processed into fish meal and fish oil to supply the aquaculture sector (Bunting et al., 2022). This could help reduce pressure on overexploited wild stocks and avoid competition for small pelagic fish species with local processors and for direct human consumption (Thiao and Bunting, 2022).

The concept of the 'shell biorefinery' was described by Yan and Chen (2015) and could realise significant economic returns and potentially contribute to social development. Prawn heads are utilised by poorer communities as food in, for example, Bangladesh, and consequently safeguards may be needed to ensure that innovative biorefinery processes are not undermining food and nutrition security and ideally can contribute to poverty alleviation (de la Caba et al., 2019; Bunting et al., 2023). The conversion of crustacean shells and fish gelatine to novel active seafood packaging will demand that risks from allergens to consumers be carefully evaluated and acceptable and prospects for recycling or composting such materials be assessed and optimised (de la Caba et al., 2019). Systematic approaches to the analysis of issues across value chains and food systems are required in developing such innovative strategies, and these are discussed further below.

High-level process mapping for value chain and food system analysis

Visualisation of the phases and processes involved in value chains and food systems through high-level process mapping can be used to explain what is happening, build a common understanding, target subsequent data collection and present results in a clear and compelling manner. The British Standards Institution (2011) advocate this as a first step in carrying out systematic, standardised and reproducible product carbon footprint assessments. Subsequent steps include defining the scope of the study, disregarding processes that make a small contribution to the overall footprint (<5 per cent typically), collecting data on the remaining processes and using independently verified and openly accessible emission conversion factors to arrive at a carbon equivalent (CO_{2e}) assessment value.

High-level process mapping of established and emerging marine aquaculture sectors (i.e. cage-based salmon, continuous longline blue mussel, intertidal oysters on trestles and land-based halibut) in the United Kingdom revealed the most significant processes and value chain phases for greenhouse gas emissions (Bunting, 2014b). An example of a preliminary map was shown to participants in a stakeholder Delphi assessment of opportunities and constraints to better manging greenhouse gas emissions from the sector (Bunting, 2014a). Mapping and the associated evaluation of carbon equivalent footprints highlighted where and when the greatest emissions occurred. On farms using formulated feeds, this is a major source, and efforts to reduce these could yield the greatest benefits.

Comparing the emissions of typical product units of different farmed seafoods produced in the UK demonstrates that finfish that are raised exclusively on formulated feeds have the largest footprints, whilst shellfish that depend purely on natural feeds have the lowest (Bunting, 2014b). Omnivorous fish raised in ponds and lakes with natural productivity that is enhanced using mostly organic fertilisers and using largely plant-based feeds can have intermediate footprints. When comparing emissions, it is important to note that aquatic foods often have much lower carbon footprints than those associated with meat from terrestrial ruminant livestock (Crona et al., 2023). Practices at retailers and consumer choices can have a disproportionate impact too, as relatively large amounts of energy may be expended on storing and cooking individual product units, thus markedly increasing the overall carbon footprint (Bunting, 2014b).

Application of high-level process mapping to small-scale farming livelihoods in Buxa Tiger Reserve, West Bengal, India (see Figure 2.2), helped guide data collection for a bioeconomic modelling assessment and illustrate findings (Bunting et al., 2015). Mapping was able to establish that access to common property resources is important in livelihoods strategies and that people have developed diverse portfolios of activities that contribute to social–ecological resilience in this remote area that is prone to natural disasters and vulnerable to worsening climate change impacts. Study findings demonstrated that if households were to adopt the system of rice intensification (SRI) in conjunction with small-scale fish culture, this could increase incomes by 42 per cent above the typical baseline of US$1257 (Rs. 70,250) to US$1783 (Rs. 99,690) per year. Modest amounts of fish could help contribute

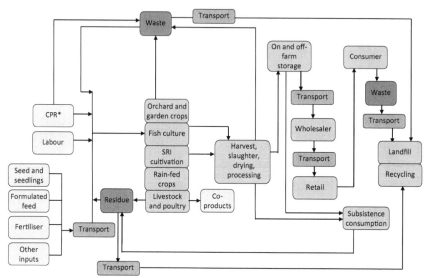

Figure 2.2 High-level process map for typical agriculture-based livelihoods in Buxa, West Bengal, India.

Source: Bunting et al. (2015).

to food and nutrition security in the community and supplement supplies of wild-caught fish. Constraints noted included a lack of dependable fish seed supplies and technical skills locally; physical limitations to transporting live fish seed and products to market; risks of introducing invasive species, disease agents and pests; and threats from impounded water harbouring animal and human disease agents or parasites. In regions where irrigated rice cultivation predominates and prospects for fish culture are good, adoption of SRI principles and integrated fish culture could contribute to higher rice yields and greater availability of nutritious aquatic foods. SRI can facilitate more efficient water and fertiliser use, and intermittent wetting can help avoid methane emissions from permanently flooded fields (Li et al., 2018).

Product units, edible portions and food loss and waste

Standardised carbon footprint assessments demand a focus on the typical product unit as it is presented to consumers, and this encompasses the packaging and labelling and any non-edible parts that may be included. Emissions associated with typical residual management or waste disposal practices are also attributed to the product unit in question. The mode of waste management can result in unexpected outcomes; disposal of plastic packaging generates lower emissions as compared with lightweight wooden cartons when these are composted. Recalculating carbon footprints per unit of edible biomass, protein or essential micronutrients could be useful for optimising potential human nutritional benefits versus climate change impacts.

Production data from academic studies, commercial trials and those collated by international bodies (FAO, 2022) usually refer to the total biomass of aquatic animals and plants produced. This masks a great deal of variation in the actual proportion of this biomass that can be eaten. It could be more instructive to use appropriate conversion factors to arrive at the 'edible portion' of different aquatic organisms that are being cultured (Edwards et al., 2019). Adopting this approach could permit more ready comparison of yields and production efficiencies with poultry and terrestrial livestock. Specifying the edible portion also makes apparent that which is deemed non-edible and could potentially be utilised for other purposes.

Assessments of food loss and waste embody a more comprehensive analysis across value chains and food systems to quantify when and where supplies of edible biomass and nutrients could be bolstered and utilised more efficiently. Studies of typical aquatic food systems globally estimated that 29–50 per cent of production is lost or wasted across the entire value chain (Kruijssen et al., 2020). The highest values were reported for North America and Oceania. By quantifying food loss and waste, it is possible to identify where there is greatest scope to reduce this through better understanding of dynamics in external and internal food environments and typical consumer knowledge, attitudes and practices. Raising awareness of the disproportionate influence that food waste by consumers can have on the overall environmental and social impacts of food systems could help prompt more positive behaviours and outcomes. Further studies are needed, however, to evaluate the impact of different interventions on food loss and waste reduction by value chains actors and consumers (Kruijssen et al., 2020).

Aquaculture-induced environmental pressures

Systematic accounts concerning aggregate environmental impacts directly attributable to specific aquaculture operations, production systems or sectors within regions or countries are scarce. Most assessments are based on individual case studies, and owing to site-specific characteristics, variable farm configuration and husbandry practices and differences in the carrying capacity of receiving environments, it is not generally appropriate to directly extrapolate findings. Examples of environmental impacts and ecological consequences attributed to specific aquaculture facilities and practices are summarised in Table 2.2 and discussed in further detail below. Good site selection, responsible farm management and operating within the carrying capacity of the environment are critical to ensuring that negative environmental and ecological impacts are avoided; planning and management procedures with potential to enshrine and facilitate practical implementation are discussed in chapter 3.

Table 2.2 Ecological impacts associated with wastewater discharges from commercial aquaculture

Genera	Impact	Setting	Reference
Phytoplankton	• modified growth potential for marine diatoms and dinoflagellates	• wastewater from turbot rearing tanks tested in bioassays	Arzul et al. (1996)
	• *Selenastrum capricornutum* and *Oscillatoria redekei* bioassays demonstrated an increase in bioavailable phosphorus	• wastewater from six land-based farms culturing trout and salmon	Massik and Costello (1995)
Periphyton	• enhanced growth on polyethylene sheets	• study conducted downstream of a trout farm on the River Hull, England	Carr (1988; cited in NCC, 1990)
	• increased standing stock on rocks	• observations made in streams directly below trout farms in Virginia, USA	Selong and Helfrich (1998)
	• significantly higher biomass of *Cladophora glomerata* in sheltered bays	• recorded in sheltered bays adjacent to cage farms culturing trout in the northern Baltic Sea	Ruokolahti (1988)
Macrophytes	• increased epiphytic growth implicated in smothering *Ranunculus penicillatus*	• observed downstream from trout farms on the River Hull, England	Carr (1988; cited in NCC, 1990)
	• decline of the rarest British plant *Crassula aquatic*	• attributed to possible effects of increased cage aquaculture in Loch Shiel, Scotland	NCC (1990)
Benthic invertebrates	• decreased diversity of pollution-sensitive species, with taxa richness remaining significantly lower 1.5 km from the farm	• downstream of trout farms in North Carolina, USA	Loch et al. (1996)
	• decreased abundance and diversity of pollution-sensitive species with an increase in pollution-tolerant species	• downstream of trout farms in Virginia, USA	Selong and Helfrich (1998)
	• decreased taxa diversity with increasing sediment organic enrichment under salmon cages	• samples collected from beneath salmon cages in Bliss Harbour, Bay of Fundy, Canada	Duplisea and Hargrave (1996)

Fish	• increase in abundance and biomass with pollution-tolerant *Rutilus rutilus* and exotic *O. mykiss* appearing, whilst the pollution-sensitive goby *Cottus gobio* was excluded	• observations made in streams receiving wastewater from farms in Brittany, France	Oberdorff and Porcher (1994)
	• increased species diversity downstream attributed to increased primary production increasing the trophic base	• recorded in Appalachian headwater streams in Virginia receiving wastewater from trout farms	Selong and Helfrich (1998)

Source: Bunting (2001).

Physical impacts

Water appropriation for aquaculture and wastewater discharges can impact on local hydrological conditions and affect water quality and the chemical composition of sediments in the receiving environment. Reviewing environmental impacts attributed to tropical inland aquaculture, Beveridge and Phillips (1993) noted that appropriation of water resources can result in adverse consequences, notably:

• altering channel morphology and sedimentation patterns;
• reducing access to spawning and nursery areas;
• creating barriers to migratory fish;
• changing thermal regimes;
• modifying biological communities.

Groundwater abstraction for coastal aquaculture was implicated by these authors in causing saline intrusion and subsidence. Globally, several deltas are sinking owing to accelerated compaction promoted by water, gas and oil extraction (Syvitski et al., 2009), and abstraction of water for aquaculture may exacerbate this problem. Saline aquaculture wastewater discharges have been implicated in causing the salinisation of land and surface-water resources (Phillips et al., 1993; Tran et al., 1999). Appropriation of water for aquaculture in sub-tropical and temperate regions poses similar threats to those reported by Beveridge and Phillips (1993) for tropical aquaculture. Abstraction for flow-through trout farming in the UK reduced river flows between the abstraction point and farm discharge downstream so that migratory fish movements were hampered (Jones, 1990).

Deposition of particulate material entrained in aquaculture wastewater can cause a range of problems. Aesthetically it is unsightly; physically it can promote substrate embeddedness, thus reducing interstitial water flows; whilst ecologically it can smother macrophytes and sessile invertebrates and restrict access by certain species to substrates suitable for spawning. Sedimentation is more likely where the dilution capacity of the receiving environment is limited (Jones, 1990), whilst certain habitats are particularly vulnerable to siltation; for example, deeper river stretches and low-energy zones in lakes or along the coast. Excessive sedimentation can rapidly inundate such habitats, alter the sediment composition and result in anoxic conditions that will displace or eliminate pollution sensitive species.

Ammonia and nutrients

Possibly compounding physical impacts described above, a range of prospective pollutants can become entrained in aquaculture wastewater, primarily originating from uneaten feed, faecal matter, excreta and treatments to control and eradicate pests and diseases and maintain sediment and water quality (Beveridge and Phillips, 1993; Brooks et al., 2002). Ammonia arising from the degradation of proteinaceous material can be toxic to fish and invertebrates. Highland streams and rivers are particularly vulnerable as they are generally characterised by having good water quality and being populated by diverse pollution-sensitive fish and benthic invertebrate communities (NCC, 1990).

Nutrients released from aquaculture commonly contribute a small proportion to overall anthropogenic inputs to aquatic ecosystems (Ackefors and Enell, 1990; Foy and Rosell, 1991; Kronvang et al., 1993; Páez-Osuna et al., 1998). Shrimp farming in Mexican coastal states was found to account for 1.5 and 0.9 per cent of nitrogen and phosphorus inputs, respectively to marine environments (Páez-Osuna et al., 1998). Adverse environmental impacts locally may, however, be attributed to such comparatively small nutrient inputs. Whilst nutrient discharges from aquaculture to enclosed water bodies such as Lake Taal, Philippines, and Cirata Reservoir, Indonesia, contributed to massive fish-kills. Fertiliser, manure and other organic matter inputs to semi-intensive, pond-based aquaculture result in elevated nutrient concentrations in the culture water, but the relatively closed nature of the system means that nutrients are not routinely lost to the surrounding environment. Semi-intensive shrimp ponds in Bangladesh, for example, can act as sinks for nutrients from the wider environment (Wahab et al., 2003).

Chemicals and therapeutants

Lime is predominantly applied to ponds following draining to increase hardness, alkalinity, sediment and water pH levels; kill disease organisms; and promote production of phytoplankton and benthic organisms (Beveridge and Phillips, 1993). Application rates of 2–3 t ha^{-1} y^{-1} have been reported, but required rates vary depending on sediment pH and water exchange regimes. Better management practices for shrimp culture recommend application rates to dry ponds of 100 kg ha^{-1} y^{-1} and 1000 kg ha^{-1} y^{-1} where sediment pH levels are 6.5 and 5.0, respectively (ADB et al., 2007). Risks associated with routine use are expected to be minimal. Comprehensive reviews documenting the diversity of chemicals employed in aquaculture and sometimes offering estimates for amounts used annually were conducted for shrimp culture (Phillips et al., 1993), salmonid aquaculture (Burka et al., 1997), tropical aquaculture (Beveridge and Phillips, 1993) and aquaculture generally (Beveridge et al., 1991; Bergheim and Asgard, 1996). Such assessments demand constant updating, however, whilst it is virtually impossible to evaluate the nature and extent of illicit and illegal activity. Evolving legislation, new treatments and preventative measures, notably, probiotics and vaccines and better knowledge of appropriate biosecurity measures, have substantially reduced the number of chemicals licensed for use in several jurisdictions and helped avoid the necessity of resorting to chemical treatments elsewhere (Bunting et al., 2022).

Biocides derived from plants such as nicotine, rotenone and saponin are commonly applied between production cycles in tropical aquaculture to kill pests and predators, including insects, molluscs, amphibians, crabs, snakes and fish (Baird, 1994). Being considered natural in origin and often readily biodegradable, such agents are commonly assumed to be less hazardous than synthetic compounds, but their non-specific nature and health implications for workers preparing and applying these agents constitute serious concerns. Persistent and highly toxic chemicals such as chlorinated hydrocarbons and organotins used as biocides constitute major health hazards, whilst chemicals routinely used as disinfectants (benzalkonium chloride, formalin sodium and hypochlorite), particularly in hatcheries, and copper-based algicides and antifoulants pose environmental, animal and human health risks (Phillips et al., 1993; Howgate et al., 2002).

Other chemicals are used in aquaculture in an attempt to prevent disease outbreaks and treat bacterial and fungal infections and parasite infestations and as anaesthetics to reduce stress and avoid physical damage during handling. Comparing chemotherapeutant use in salmonid aquaculture in Europe and North America with treatments used with mammals, Burka et al. (1997) noted that only a limited number were available to producers. Significant proportions of medications, especially those intended to be delivered orally to infected animals that will probably have suppressed appetites, are not ingested or absorbed and consequently remain in sediments and culture water or are discharged to the receiving environment (Weston, 1996). Therapeutants and their residues can constitute serious environmental, animal and public health hazards, with development of antibiotic-resistant bacteria, particularly amongst human pathogens, representing a major concern.

Chemical use in aquaculture must be regulated to protect the environment; ensure the health and safety of operators, other resource user groups and local community members; and safeguard the quality of products for consumers. Aquaculture sectors developed rapidly in many countries, and in several instances monitoring and regulations were inadequate for detecting and deterring illegal and irresponsible chemical use. Monitoring procedures established to ensure the quality of products destined for export markets and testing of imported products are powerful deterrents to the use of unapproved chemicals, but concern has been expressed that use of prohibited chemicals may still occur during the grow-out phase, whilst errors in sampling and testing can seriously affect the image and prosperity of wrongly implicated aquaculture sectors. Responsible and safe chemical use in aquaculture can effectively be promoted through voluntary adoption of codes of conduct and better management practices, but compliance testing and spot-checks, backed by appropriate sanctions, are still necessary to safeguard confidence in the process and detect and deter illicit activity.

Dissolved oxygen

Wastewater discharged from aquaculture commonly has an elevated bio-chemical oxygen demand (BOD), and reduced dissolved oxygen levels owing to on-farm respiration and can subsequently be depressed further. Dissolved oxygen concentrations 250 m downstream of a trout (*Oncorhynchus mykiss*) farm on the Coura River in northern Portugal producing 500t of fish annually were reportedly 9.6 mg l^{-1}, compared to 10.7 mg l^{-1} above the farm (Boaventura et al., 1997), whilst BOD concentrations upstream of the farm and 1000 m downstream were 1.6 and 5.6 mg l^{-1}, respectively. According to these authors, the maximum permissible BOD level of 3 mg l^{-1} was exceeded in 1 km of the river and concentrations did not return to background levels until 12 km downstream. Oxygen concentrations adjacent to cages containing rainbow trout in Lac du Passage, Quebec, were 6 mg l^{-1} at a depth of 4 m, whilst 400 m away at the same depth a concentration of 12 mg l^{-1} was recorded (Cornel and Whoriskey, 1993). Comparatively high oxygen deple-tion rates in Loch Ailort, Scotland, were attributed by Gillibrand et al. (1996) to microbial breakdown of organic material, whilst the authors estimated that half of the organic carbon input to the loch was from two salmon farms. Elevated sediment oxygen consumption was recorded in the Bay of Fundy, Canada, beneath salmon cages and in Lake Kariba, Zimbabwe, beneath tilapia cages (Hargrave et al., 1997; Troell and Berg, 1997). Direct oxygen consumption and subsequent compensation for biochemical oxygen demand can be compounded by secondary oxygen consumption whereby nutrient discharges stimulate algae and plant production, which consume oxygen when they decompose (Bailey-Watts, 1994).

Ecological consequences

Eutrophication was defined under Council Directive (91/271/EEC) of the Council of the European Communities (1991: 2) as:

> enrichment of water by nutrients, especially compounds of nitrogen and/or phosphorus, causing an accelerated growth of algae and higher forms of plant life to produce an undesirable disturbance to the balance of organisms present in the water and to the quality of the water concerned.

The contribution of aquaculture-derived nutrients to eutrophication will depend on the overall input of nutrients to the system, prevailing ecological status of the receiving environment and external factors governing primary production; for example, temperature, light levels, pH, transparency, micro-nutrient limitation, grazing intensity and inhibitory factors such as tannins, and especially the retention time of the water body.

Both organic and inorganic matter discharged from commercial aquaculture operations have been implicated in stimulating periphyton and phytoplankton production downstream of culture facilities and adjacent to cage farms

(Table 2.2). Changes in periphyton and phytoplankton assemblages and standing stocks often constitute the first evidence of eutrophication; consequently, such communities may be considered important indicators for monitoring aquaculture-induced ecological impacts. A number of confounding factors demand consideration, however, notably:

- production of the dinoflagellate (*Alexandrium minutum*) was inhibited *in vitro* when cultured in diluted turbot (*Psetta maxima*) rearing process water (Arzul et al., 1996);
- periphyton and phytoplankton assemblage responses to aquaculture discharges will be moderated by wastewater composition, original community composition, nutrient status of the receiving environment and ambient conditions;
- growth factors other than nutrients, including carbon, dissolved humic matter, trace elements and vitamins, can limit production (NCC, 1990);
- periphyton standing crops downstream of trout farms showed a positive correlation with feeding rates at farms but a negative correlation with overhead cover (Selong and Helfrich, 1998).

Concern has been voiced that aquaculture wastewater discharges may contribute in some instances to the development of toxic algal blooms, crucially, that formulation of commercial feeds with lower phosphorus levels may increase the nitrogen-to-phosphorus ratio in aquaculture wastewater, thus promoting the proliferation of toxic species (Folke et al., 1994), although the likelihood of this and level of risk were subsequently subject to scientific debate (Black et al., 1997; Folke et al., 1997).

Increased epiphyte growth downstream of trout farms has been implicated in smothering submerged macrophytes, whilst in Loch Shiel, declines in the rarest British plant (*Crassula aquatica*) were attributed to rural development and possible impacts of fish farming in cages (NCC, 1990). Comparison of periphyton growth on artificial substrates positioned underneath cages containing rainbow trout in Loch Fad, Scotland, with a control site 1.6 km to the south-west revealed, however, that the annual geometric mean abundance of only one species from twenty-two recorded was significantly different (Stirling and Dey, 1990). This was attributed to phytoplankton and periphyton production in the loch being light and not nutrient limited owing to high background nutrient levels, although wind-induced turbulence and turbidity resulting from farm waste discharges were also believed to have a role in regulating algae growth.

Shifting trophic status and interactions

Nutrient enrichment can cause a shift in the trophic status of receiving water bodies, characterised by increases in the abundance of pollutant-tolerant species and declines in overall diversity. Impacts of aquaculture development on zooplankton assemblages, benthic invertebrate communities and native fish populations are summarised in Table 2.2 and briefly reviewed below.

Zooplankton assemblages

Species composition, abundance and size structure of zooplankton assemblages can be affected by eutrophication. Changes in phytoplankton community structure resulting from eutrophication may induce a trophic cascade, whereby zooplankton assemblages comprising largely copepods are replaced with smaller herbivorous species. Consequently, grazing pressure on phytoplankton will be affected, leading to modified species composition, size structure and biomass of the phytoplankton assemblage; grazing-resistant species such as cyanobacteria may proliferate. Zooplankton play a crucial role in nutrient cycling within water bodies through migration and faecal pellet deposition, with lower copepod numbers reducing the rate of nutrient transfer to sediments in faecal pellets and higher cladoceran numbers recycling faecal material, maintaining water column nutrient concentrations, potentially promoting eutrophication (Ferrante and Parker, 1977; Lehman, 1980; NCC, 1990). Possible changes in the availability of zooplankton at specific times during the year may negatively impact fish populations, juvenile survival can be seriously affected if suitable zooplankton prey species are not available and fish such as Arctic charr (*Salvelius alpinus*) may not come into breeding condition (NCC, 1990). Biocides are employed within some aquaculture systems to suppress or promote certain zooplankton species, and this is discussed further below.

Benthic invertebrate communities

Monitoring macro-invertebrate communities in streams in western North Carolina, USA, Loch et al. (1996) recorded a lower diversity of pollution sensitive taxa 1.5 km below three trout farms, as compared with control stations upstream. Similarly, sampling in streams in Virginia, USA, showed that the diversity and abundance of sensitive taxa decreased below five trout farms, whilst those tolerant of pollution increased (Selong and Helfrich, 1998). In assessing the composition of meiobenthic taxa at sites in Bliss Harbour, Bay of Fundy, Canada, reduced diversity was recorded in sediments subject to enrichment with organic matter under salmon cages (Duplisea and Hargrave, 1996). Studies have demonstrated similar findings beneath cages being used to culture species including seabass and seabream and under mussel rafts (Karakassis et al., 1999; Stenton-Dozey et al., 1999). Comprehensively reviewing studies on benthic impacts associated with aquaculture, Costa-Pierce (2002: 355) noted that impacts were generally confined to within 100–200 m of the culture facility, although impacts were reported 1 km away from a poorly sited farm employing 'trash fish' as feed.[1] Ecological impacts associated with open-water aquaculture are moderated by prevailing hydrological conditions.

Assessing the impact of oyster (*Crassostrea gigas*) culture on macrobenthic communities in Dungarvan Bay, Ireland, De Grave et al. (1998: 1137) noted that 'no evidence of organic enrichment was found' beneath trestles being used for culture owing to the 'highly dissipative nature' of the site.

Native fish populations

Employing the Index of Biotic Integrity (IBI) that assumes that fish species assemblages change in a predictable manner with changing stream quality, Oberdorff and Porcher (1994) assessed the impact of discharges from nine fish farms on receiving waterways in Brittany, France. Fish species abundance and biomass increased in general below farm sites. Compared with upstream monitoring stations, rudd (*Rutilus rutilus*), classified as pollution tolerant, and introduced rainbow trout (*O. mykiss*) were recorded below farms, whilst the goby (*Cottus gobio*), classified as pollution sensitive, was absent. Low fish species diversity characteristic of temperate oligotrophic waters limited application of the original IBI to assessing the status of such environments. Invoking a modified IBI to ensure that no particular taxa exerted undue influence, Selong and Helfrich (1998) assessed the impact of wastewater discharges from trout farms on fish assemblages in receiving waters as compared with Appalachian headwater streams in Virginia. These authors attributed consistent increases in diversity downstream to enhanced primary production.

Aquaculture development is accompanied by the attendant risk of cultured organisms escaping or being released unintentionally, and occasionally deliberately, from production facilities. Impacts attributed to the introduction of alien or non-native species and strains from different regions from aquaculture have been reviewed extensively (Welcomme, 1988; Beveridge et al., 1994; Arthington and Bluhdorn, 1996; Bardach, 1997; Beardmore et al., 1997). Five distinct impacts from introductions originating from aquaculture were identified by Beveridge and Phillips (1993), namely:

- direct disruption to the receiving environment;
- disruption to the host community, through predation or competition;
- genetic degradation of local stocks;
- introduction of diseases and pathogens;
- socio-economic impacts.

Enhancing prospects for regenerative aquaculture

Regenerative or regenerating agriculture (Pretty, 1995) is garnering a great deal of attention and investment and is set to grow rapidly (Investing in Regenerative Agriculture and Food, 2022). Regenerative aquaculture strategies could build on traditional aquaculture practices devised by farmers, and these are discussed further below and in chapters 4, 5 and 6. These approaches share a fundamental objective, to achieve efficient and optimal resource use within the carrying capacity of supporting ecosystem areas (Bunting, 2007; Bunting et al., 2022). Regenerative farming (e.g. conservation tillage, contour farming, cover crops and green manures, field margins and hedgerows, integrated pest management and water conservation and harvesting) has potential to avoid the use of costly agrochemicals, enhance carbon sequestration and mitigate climate change impacts, promote nutrient-use efficiency, restore soil structure and prevent water runoff (Pretty, 1995; Regeneration International, 2022). Regenerative practices must be developed in harmony with the prevailing ecological, social and political setting.

Integrated agriculture–aquaculture, polyculture and silvofishery (e.g. mangrove–shrimp–fish culture) strategies could sequester blue carbon, bolster agrobiodiversity and social–ecological resilience and contribute to food and nutrition security (Bosma et al., 2012; Bunting et al., 2013, 2017; Ahmed et al., 2014, 2017; IPCC, 2019). Optimising the delivery of multiple benefits will demand considered management and coordination of farming and biodiversity conservation and ecosystem services enhancement activities. This may require incentives to encourage producers to transition away from conventional practices and steps to eliminate subsidies to unsustainable practices. An enabling policy environment is essential to promoting the uptake of regenerative practices (Pretty, 1995). Transparent mechanisms to certify regenerative aquatic foods and trustworthy means to identify them in retail outlets and broader food environments could be important in guiding positive consumer behaviour.

Mitigating and managing environmental impacts

Improved site selection and limits placed on farm size would help avoid problems of excessive water abstraction for aquaculture. Restricting withdrawals to within the carrying capacity of the supporting aquatic resource would ensure that environmental flows are maintained, thus sustaining habitat functioning and ecological processes and safeguarding ecosystem services benefiting other user and social groups.

Production system design and operation largely dictates water-use efficiency (Table 2.3) owing to the need to maintain water quality and safeguard production levels. Broadly speaking, it appears that more intensive modes of production are increasingly efficient in terms of water use. Recycling process water, possible owing to adoption of mechanical and biological filtration and supplementary aeration, can significantly improve the efficiency of water use in aquaculture. Recirculation strategies, where technology and energy are employed to augment ecosystem services are encumbered with inherent risks, however, whilst associated financial and environmental costs, must be considered when assessing the overall performance and sustainability of such systems.

Anticipating that water requirements for pond aquaculture were expected to be higher in more arid climates, Boyd and Gross (2000) proposed a range

Table 2.3 Water-use efficiency in aquaculture systems

Aquaculture system	Water-use efficiency ($m^3 kg^{-1}$ fresh weight)	Water management characteristics
Traditional extensive fishpond culture	45*	Rainwater and drainage water are routinely channelled into fishponds to compensate for seepage and evaporation losses; excessive water exchange is detrimental as it is desirable to retain nutrients within the pond.
Flow-through ponds	30.1*	Water exchange of 20 per cent of the pond volume per day removes waste and replenishes oxygen levels; production of 30 t $ha^{-1} y^{-1}$ is attainable but seepage and evaporation contribute to water loss in the system.
Semi-intensive fishponds	11.5*	Producing two crops annually with complete drainage to facilitate harvest, fishponds fed with supplementary feed can yield 6 t $ha^{-1} y^{-1}$; one-fifth of water consumption is associated with feed inputs.
Wastewater-fed aquaculture	11.4	Wastewater is routinely fed into fishponds in the East Kolkata Wetlands to make up the water to a desirable level; estimates suggest that 550,000 $m^3 d^{-1}$ of wastewater is used to produce 18,000 t y^{-1} of fish in 3900 ha of ponds.
Intensively managed ponds	2.7*	Lined ponds are used to produce 100 t $ha^{-1} y^{-1}$, whilst intensive mixing results in evaporation of 2000 mm y^{-1}.
Super-intensive recirculation systems	0.5–1.4*	Process water is recirculated with pumps and treated with mechanical filters, biofilters and disinfection technology; stocked animals are entirely dependent on high-protein formulated feed inputs.

Notes
* Values derived from Verdegem et al. (2006).

of practical water conservation measures, notably, seepage control achieved by employing good construction practices, minimising water exchange and storing rain and runoff water.

Wastewater management[2]

Concerns regarding aquaculture wastewater impacts, including nutrient enrichment and alterations in the biotic community in the receiving water, have been addressed in several fundamental ways. Effluent loading rates have been significantly reduced through careful feed management, formulation of energy-dense and low-pollution diets and more efficient sludge collectors and mechanical filters. In the UK, the range of developments and initiatives mentioned above resulted in discharge consents for freshwater aquaculture being attained more frequently. Doughty and McPhail (1995: 563) cautioned, however, that this compliance 'must be considered in context of the low sampling frequency employed: large daily variations in losses of total phosphorus and total nitrogen (of which ammonia is the largest fraction) from fish farms have been noted'.

A systematic approach is essential to identify prospective management strategies to reduce and mitigate environmental impacts of aquaculture development. Life cycle assessment, discussed in detail in chapter 7, constitutes a promising approach to assessing the environmental performance of production (Gephart et al., 2021). This approach encompasses everything from the production of raw materials to the disposal of waste that would highlight processes and strategies where improvements could be attainable. Potential measures to enhance the efficiency of production and reduce waste can be envisaged for several aspects of the production process, ranging from feed formulation, wastewater treatment and systems design and engineering (Table 2.4). Considering wastewater management for Atlantic salmon (*Salmo salar*) culture, Alanara et al. (1994) proposed an integrated strategy including demand feeders to enhance feeding efficiency, rotary filters fitted with screens corresponding to the size distribution of particulate waste and disposal of accumulated sludge to infiltration basins. Practical and physical limitations to prevailing strategies for aquaculture waste management (Table 2.4) signify, however, that incremental performance and efficiency gains for conventional approaches may not enable producers to make substantial improvements or necessarily attain increasingly stringent discharge standards.

Settlement ponds and mechanical filters demonstrate limited effectiveness with regards to dissolved organic matter and nutrients and fine particles (Hennessy, 1991; Cripps and Kelly, 1996). Sludge accumulates in settlement ponds and is produced by mechanical and biological filters, including those in systems employing water reuse, and must be managed, ultimately requiring disposal. Risks that sludge disposal will only delay and relocate pollution problems were summed up by Gunther (1997: 159) as 'hampered effluent

Table 2.4 Strategies for reducing waste loadings from commercial aquaculture

Management area	Strategy	Constraints	Reference
Feed formulation	• constituents selected to match nutritional demands; processing techniques used to maximise digestibility; increased energy levels limit protein catabolism	• quality feed constituents are expensive and limited in availability; excess feed lipid levels can affect product quality	Wiesmann et al. (1988); Seymour and Bergheim (1991); Cho and Bureau (1997)
Feeding strategy	• feeding regime tailored to behaviour; feed delivery automated; careful handling limits dust production	• disease can seriously restrict feeding; automation requires monitoring and investment	Kadri et al. (1991); Seymour and Bergheim (1991)
Facility layout and management	• pipes configured to limit particulate break-up by reducing length and number of bends; oxygenation to reduce water volume requiring treatment	• pipe layout constrained by physical and hydrological conditions; oxygenation increases operating costs	NCC (1990)
Modification of the receiving environment	• stream and river channels modified to promote mixing and aeration, limiting sediment deposition and compensating for depressed oxygen levels	• modifying receiving environments may create separate problems; environmental impacts may only move downstream	
Wastewater collection	• preferential collection and treatment of wastewater produced after feeding, cleaning, grading and harvesting and *in situ* sludge separation	• requires increased attention to management and facility design; culture water produced during routine operation may still require treatment	Lystad and Selvik (1991); Selvik and Lystad (1991); Schwartz and Boyd (1994)
Settlement	• settlement provides a simple approach to wastewater treatment; design parameters developed; effective at removing suspended particles and associated waste fractions	• settlement ineffective at removing soluble waste and small particles; prohibitive land area required; sludge requires regular removal and disposal	Henderson and Bromage (1988); Hennessy (1991); Allcock and Buchanan (1994); Teichert-Coddington et al. (1999)

Mechanical treatment	screens configured to wastewater characteristics; treatment efficiency predictable and a range of commercial systems available	frequent monitoring and maintenance required; sludge requires de-watering, storage and disposal; capital and operating costs represent a financial burden	Makinen et al. (1988); Cripps (1991, 1995); Bergheim et al. (1993, 1997)
Water reuse	culture water treated and reused to retain heat energy; optimal conditions maintained for species cultured; facilitates aquaculture where access to water is limited	systems require constant monitoring; sludge produced requires de-watering, storage and disposal; capital and operating costs high	Chen et al. (1993); Muir (1994)

Source: Bunting (2001).

accumulation processes (HEAP) traps, where former point source pollution is ultimately converted into non-point source pollution'.

Nutrient discharges from aquaculture are increasingly covered by guidelines and governed by legislation, with releases in excess of permissible levels incurring pollution taxes and non-compliance leading to court action and fines. Often, however, monitoring and enforcement is constrained in practice where farming operations are in isolated regions or far from suitably staffed and equipped facilities, making sampling and analysis difficult and expensive. Consequently, the onus comes to rest on producers to demonstrate that they are aware of possible environmental impacts and are taking steps to minimise and mitigate discharges of nutrients and other waste fractions.

Filter maintenance constitutes a burden in terms of time and costs for producers, demanding constant monitoring and regular cleaning. Risks from mechanical failure and electricity supply disruption can render filters useless, and shock loadings associated with cleaning and handling stock can seriously impinge on treatment performance. Constraints to removing dissolved nutrients from aquaculture wastewater represent a major factor governing the volume of production possible at land-based sites and significant concern for producers with regards conforming to stricter discharge standards. Environmental management systems (reviewed in chapter 3) and better management practices (reviewed in chapter 8) provide valuable concepts for farm- or producer-level interventions to enhance environmental performance and improve resource-use efficiency.

Resource-conserving and agroecosystem-enhancing practices

Intentional development of resource-conserving and agroecosystem-enhancing aquaculture systems drawing on traditional and artisanal practices has been a constant theme within the realms of aquaculture development and proffered as an antidote to widely perceived unsustainable practices propping-up intensive aquaculture. Acknowledging the potential significance of indigenous, resource-conserving and resilient agricultural practices, the FAO, with support from the Global Environmental Facility, has instigated a programme to identify and designate land use systems with intrinsic cultural and biodiversity values as Globally Important Agricultural Heritage Systems (GIAHS). With explicit objectives of promoting dynamic conservation through leveraging global and national recognition of GIAHS and institutional support for their preservation, building capacity amongst farmers and institutions to manage and conserve the systems and sustainably add economic value to products and promoting enabling institutional environments that support conservation, adaptation and viability (Koohafkan and Altieri, 2011).

Knowledge of these agricultural systems that have generally persisted for many hundreds of years has the potential to inform and influence future agricultural development pathways. Rice–fish culture in Zhejiang Province, China, has been recognised as an agroecosystem worthy of inclusion under this initiative, and brackish water and freshwater aquaculture integrated with rice cultivation are discussed further in chapters 4 and 5, respectively. Complementary to the GIAHS scheme were the FAO (2010) programme on developing an ecosystem approach to aquaculture (EAA) and joint United Nations Development Programme and International Water Management Institute initiative on ecosystems for water and food security (Boelee et al., 2011).

Stocking several species with different feeding behaviours and preferences with the intention of exploiting a wide array of feeding niches is commonly referred to as polyculture and constitutes one of the oldest strategies embodying an EAA-based approach to aquaculture development. Fundamental aspects of polyculture and other promising practices (i.e. ecological water conditioning, integrated aquaculture–agriculture, horizontally integrated and integrated multi-trophic aquaculture) to enhance resource-use efficiency, facilitate nutrient recovery and reuse and avoid negative environmental impacts are reviewed below.

Polyculture[3]

Polyculture practices employing indigenous fish species have developed predominantly in China and India, to optimise yields in static ponds where the production of diverse biota may be stimulated through introducing fertiliser, particularly organic waste. Traditional Chinese approaches are based on stocking at least four to five species with diverse feeding behaviours to exploit the various nutritional resources found in managed ponds. Typically, these include species that feed preferentially on phytoplankton (silver carp), zooplankton (bighead carp), macrophytes (grass carp, *Ctenopharyngodon idella*), detritus (common carp and mud carp, *Cirrhina molitorella*) and snails (black carp, *Mylopharyngodon piceus*; Ruddle and Zhong, 1988).

There may also be synergistic effects of stocking different species together; for example, common carp consuming silver carp faeces, rich in partially digested phytoplankton, may attain a higher yield than in monoculture, whilst nutrient-rich sediments put into suspension by foraging common carp suppress the development of filamentous algae and higher plants and stimulate the production of phytoplankton, upon which the silver carp feed (Milstein, 1992). This author also noted that polyculture may contribute to improved water quality, with appropriate species used to regulate phytoplankton and macrophyte populations, contributing to more stable dissolved oxygen regimes, whilst benthic feeders prevent the accumulation of organic sediments, reducing oxygen consumption and ammonia production. However, despite the potential benefits, poorly managed or inappropriate polyculture practices could result in unstable phytoplankton populations, leading to oxygen depletion, blooms of undesirable algae or competition amongst species for food.

Production rates of 5–10 t ha^{-1} y^{-1} have been reported for ponds managed for polyculture in China (Ruddle and Zhong, 1988; Korn, 1996). Accounts of initiatives where bream (*Megalobrama bramula*) and tilapia (*Oreochromis* spp.) were stocked with six carp species to enhance production reported yields of 7 t ha^{-1} y^{-1} (Ruddle and Christensen, 1993); this is still within the range given for traditional carp polycultures. When compared to the Chinese system, the greater plasticity in feeding behaviour exhibited by fish used in Indian polycultures appears to have constrained productivity (Little and Muir, 1987). The feeding preferences of fish in the Chinese system tend to mean that particular species forage in different spatial zones, minimising competition for nutritional resources, facilitating optimal foraging and limiting energy expenditure during behavioural interactions. Furthermore, the diversity of fish species stocked in polycultures may introduce a degree of resilience to disease and market pressures. However, these potential benefits must be considered with respect to possible constraints, including restricted access to the required quantity and species of fry for stocking, extra handling costs during harvest and poor market value and acceptability of some species (Little and Muir, 1987).

In geographical locations outside Asia, traditional polyculture practices have not developed, possibly due to the limited selection of indigenous fish that are cultured. However, in a number of situations, recent innovations have included culturing chinook salmon (*Oncorhynchus tshawytscha*) with Pacific oysters (T.O. Jones and Iwama, 1991), shrimp (*P. vannamei*) with clams (*Mercenaria mercenaria*) and oysters (*Crassostrea virginica*; Hopkins et al., 1993) and shrimp (*Penaeus chinensis*) with hybrid tilapia (Wang et al., 1998). Culturing bivalves in static water ponds may contribute to maintaining water quality, particularly in shrimp culture ponds where the inefficient feeding of shrimp frequently results in dense phytoplankton blooms.

Hopkins et al. (1993) found that dissolved oxygen concentrations in polyculture ponds were consistently higher than in a control pond stocked only with shrimp. Furthermore, shrimp production in polyculture and control ponds was similar at 10.6 and 11.5 t ha^{-1}, respectively; however, overall production in polyculture ponds, including bivalves, was 24.7 t ha^{-1}. High mortality rates for clams were associated with seeding juveniles directly into pond sediments; in contrast, the survival of oysters supported above the bottom of the pond was significantly higher, although the authors noted that the trays used might interfere with management procedures on commercial farms. Wang et al. (1998) investigated using tilapia to regulate the development of phytoplankton blooms in marine shrimp ponds, and optimal stocking densities for shrimp (60,000 ha^{-1}) and tilapia (400 kg ha^{-1}) were reported. A possible problem with this strategy would be the unregulated reproduction of tilapia; regular netting could be employed to maintain the optimal biomass or a predatory species could be stocked to control the population (Little and Muir, 1987); however, the impact of a predator on shrimp culture must also be considered.

Within the definition of horizontally integrated aquaculture (see below), aquatic plants may be considered as potential candidates for polyculture. Their capacity to assimilate inorganic nutrients also makes them potentially important as *in situ* biological filters. McCord and Loyacano (1978) reported that Chinese water chestnut (*Eleocharis dulcis*) cultivated on ponds stocked with channel catfish (*Ictalurus punctatus*) produced 4660 kg ha^{-1} of corms. The presence of water chestnuts was associated with significantly lower concentrations of phytoplankton, total ammonia nitrogen and nitrate; nutrient removal during the growing season averaged 53.8, 3.4 and 18.6 mg m^{-2} d^{-1} for nitrogen, calcium and magnesium, respectively. The authors proposed that adopting this strategy could allow supplementary feed levels to be increased, enhancing fish production. However, despite its promise, possible disadvantages associated with the polyculture of macrophytes and fish include reduced light penetration into the pond restricting primary production in the water column, which could limit fish growth and restrict dissolved oxygen levels. Other problems include potential interference with routine pond management and the possibility that the macrophytes could harbour fish pathogens or their hosts.

Integrated aquaculture–agriculture[4]

The poor resource base of many small-scale farms in developing countries means that unexploited nutrient sources – for example, crop by-products, terrestrial weeds, aquatic plants and manure – represent important production enhancing inputs to fishponds (Edwards et al., 1996; Newton et al., 2021). Little and Edwards (1999) reviewed the alternative strategies that have evolved to integrate the production of livestock and aquaculture. Manure from cattle, buffalo, sheep, pigs and poultry has been employed to enhance production in aquaculture systems (Teichert-Coddington et al., 1999; Edwards et al., 1994; Maclean et al., 1994; Garg, 1996). Zhu et al. (1990) studied the response of ponds stocked with silver carp (*Hypophthalmichthys molitrix*), bighead carp (*Aristichthys nobilis*), common carp and crucian carp (*Carassius carassius*) to various applications of fermented pig manure. Manure rates of 31–48 kg ha^{-1} d^{-1} produced a net fish yield of 10.2 kg ha^{-1} d^{-1}; control ponds receiving equivalent inputs of inorganic fertiliser produced 4.3 kg ha^{-1} d^{-1}. These authors proposed that higher fish production observed in manured ponds, as compared to ponds receiving equivalent nutrient inputs in the form of inorganic fertiliser, was attributable to increased heterotrophic production. This demonstrates that both autotrophic and heterotrophic pathways are important in optimising fish production in ponds.

Integrated rearing of poultry above fishponds has been widely advocated (Christensen, 1993, 1994; Njoku and Ejiogu, 1999). It has been practised in Asia for several decades and has been shown to confer a number of advantages. Fermentation, evaporation and coagulation are reduced, preserving the nutritional quality of faeces; feed residues are consumed directly by fish; costs associated with collecting, storing and transporting manure are eliminated, and as compared with separate culture units, less land is required (Barash et al., 1982). These authors found that production in fishponds receiving direct inputs of excreta from ducks was equivalent to that in ponds receiving equivalent inputs of dry poultry manure and supplementary feed. However, despite the apparent benefits of integrating poultry farming with aquaculture, constraints have been identified (Little, 1995). These include limited access to inputs – for example, capital, labour and feed – and lack of skills amongst potential operators to successfully manage such systems. Small-scale farmers integrating the production of fish and ducks may also encounter problems obtaining inputs and selling products, especially if competing with large-scale producers (Edwards, 1998).

Within the dike–pond farming system that evolved over several hundred years in the Zhujiang Delta, Guangdong Province, South China, excreta from pigs was used extensively to fertilise ponds (Ruddle and Christensen, 1993). Ruddle and Zhong (1988) suggested that in many cases, manure production was the primary reason for keeping pigs. Concern has also been expressed regarding the role that pigs may play in facilitating contact between avian and human influenza viruses, possibly leading to influenza pandemics. It has therefore been recommended that co-location of pigs and poultry in integrated farming systems be avoided and that the promotion of this management strategy be discontinued (Scholtissek and Naylor, 1988).

Replying to this commentary, Edwards et al. (1988) stated that pigs and ducks had been raised together in traditional Asian farming systems for centuries without apparent risks and that promoting integrated aquaculture would be unlikely to increase the risk posed by new influenza strains. Furthermore, problems with management and marketing had constrained the integration of more than one species of livestock or poultry, making the dangers of wider multi-species transfers less likely. In response, Naylor and Scholtissek (1988: 506) called for a proposal from the Consultative Group on International Agricultural Research advocating 'fully integrated crop/live-stock/fish farming' to be qualified.

Skladany (1996) chronicled the evolution of the resulting debate and highlighted the potential of following the research, publications, projects and controversies of innovators to inform sociologists studying aquaculture development. The discourse described is based largely on a hypothetical risk; however, Muratori et al. (2000) showed that 11.2 per cent of fish sampled ($n = 540$) from ponds fertilised with pig manure harboured *Edwardsiella tarda*. This causes tropical fish septicaemic disease and is considered an emergent foodborne health hazard for humans. Therefore, risks associated with using animal manure in aquaculture require consideration, and precautionary measures such as pre-treatment to kill pathogens and depuration may be advisable.

Ecological water conditioning

Promoting autotrophic and heterotrophic food webs in fishponds was viewed principally as a means to enhance growth and yields of cultured aquatic products; however, an emerging paradigm is centred on exploiting photo-synthetic pathways in 'green-water' systems or bacterial processes in 'brown-water' systems to concomitantly condition culture water, principally through the conversion of toxic un-ionised ammonia, *in situ* nutrient recycling producing fish feed and in green-water systems, oxygen production (Little et al., 2008).

Green-water culture has been refined by farmers in the Philippines and entails shrimp culture in water that has first been conditioned in a reservoir stocked with tilapia (Bosma and Tendencia, 2014). Financial analysis demon-strated that this approach could potentially reduce per unit production costs for shrimp in the Philippines (Bosma and Tendencia, 2014) and if combined with mangrove–shrimp culture could help extensive farmers in the Mahakam Delta, East Kalimantan, Indonesia, to become more productive, whilst maintaining stocks and flows of ecosystem services (Bosma et al., 2012).

Production of fish in green-water systems can be significant, with maximum theoretical yields ranging from 10 to 13 t ha^{-1} y^{-1} having been estimated (Little and Muir, 1987; Mara et al., 1993). Assessment focusing on tilapia, however, indicated that growth rates observed in trial ponds and gut content analysis showing high volumes of algae could not be achieved by feeding on algae suspensions alone (Dempster et al., 1995). These authors concluded that grazing on periphyton and algae-derived detritus and feeding on cyanobacteria-covered particles and algae aggregates must contribute significantly to tilapia growth under normal culture conditions.

Consequently, research has been conducted to assess potential productivity gains associated with adopting attached-growth green-water systems. Bamboo *acadjas* operated in Benin represent an alternative approach to enhancing attached production in nutrient-rich water through the introduc-tion of substrate for periphyton colonisation (Hem et al., 1995). Traditional *acadjas* consisted of bundled sticks positioned in lagoons to attract fish owing to the refuge they offer and periphyton and invertebrate food sources they sustain. Adaptation of the *acadja* concept gave rise to an apparently appropriate aquaculture strategy to enhance production from freshwater fishponds in West Africa. Fish production with *acadjas* was 3.6 t ha^{-1} annually, significantly better than prevailing pond management practices of the time, but lower than production of 4–8 t ha^{-1} y^{-1} obtained in traditional coastal lagoon settings (Hem et al., 1995). A similar development pathway is apparent in Bangladesh where principles of traditional *katha* fisheries, like the *acadja* of West Africa, have been applied to pond-based culture systems (Wahab et al., 1999a, 1999b), and research and development outcomes are discussed further in chapter 5. Traditional *acadja* practices in West Africa reportedly led to overexploitation of local forest resources, necessitating introduction of bamboo as the substrate for *acadja*-based aquaculture.

Concerns that similar problems might ensue from promotion of substrate-based aquaculture development in Bangladesh were addressed in part through surveying prospective adopters, and outcomes suggested that local norms and limited access to common property forest resources would probably curtail negative impacts (Bunting et al., 2005).

Integrated Multi-Trophic Aquaculture

Defined as 'a practice in which the by-products (wastes) from one species are recycled to become inputs (fertiliser, food and energy) for another' (Barrington et al., 2009: 9) integrated multi-trophic aquaculture (IMTA) was described in detail as the 'incorporation of species from different trophic or nutritional levels in the same system' and 'more intensive cultivation of the different species in proximity to each other (but not necessarily right at the same location), connected by nutrient and energy transfer through water'.

Knowledge of traditional aquaculture practices such as polyculture of multiple fish species within ponds to optimise production has been cited as critical for informing the guiding principles underpinning development of IMTA strategies. Notable differences, however, centre on farming primary producers, filter-feeders, detritivores and scavengers to assimilate nutrients released from culture activities receiving formulated feed and integration within open-water systems. Culture of salmon, seaweed and mussels in Canada is reportedly being adopted by a growing number of producers and has garnered a great deal of interest, raising awareness of the potential of such strategies. Prospects for broader adoption of IMTA strategies drawing on experiences with open-water marine culture in Canada are critically assessed in chapter 4.

Horizontally integrated aquaculture development

Building on the comprehensive review of integrated aquaculture practices including 'use of animal and crop wastes as fish feed or fertiliser ... irrigation structures and water supplies to raise fish' and 'production of fish in rice fields' by Little and Muir (1987: xix), a definition was proposed for horizontally integrated aquaculture thus: 'use of unexploited resources derived from primary aquaculture activities to facilitate the integration of secondary aquaculture practices' (Bunting, 2001: 56).[5] Principally, this definition was proposed as a means to highlight the need for a new paradigm of aquaculture development where perceived waste streams should be viewed as potential inputs or contributions to achieving more efficient or diversified aquaculture production. Box 2.1 provides a brief insight concerning types of culture system that exemplify the principles of horizontal integration, whilst Table 2.5 presents a summary of ecologically orientated operational systems and integration strategies.

Box 2.1   Horizontally integrated land-based marine aquaculture

Responding to concerns over possible environmental degradation associated with open-water marine aquaculture and competition for space and resources with other coastal zone users, several bodies have advocated the development of horizontally integrated or multi-trophic land-based marine aquaculture systems. Within such systems, species farmed using formulated feeds – for example, fish, shrimp or abalone – are cultured together with other organisms, notably microalgae, shellfish and seaweed, that convert nutrients released from the fed component to harvestable biomass, either for use as a supplementary food source or to generate additional revenue. In Israel, tank-based culture systems have been developed combining fish or abalone with seaweed; abalone, fish and seaweed; fish and shellfish; fish, microalgae and shellfish; and fish, shellfish, abalone and seaweed (Figure 2.3).

Figure 2.3 Commercial production of sea urchin, fish, bivalves and seaweed in horizontally integrated culture systems in Israel.

Photo credit: Stuart Bunting.

Constructed wetlands planted with samphire (*Salicornia* spp.) that can be harvested for use as a vegetable, forage or biofuel have been evaluated to a limited extent for additional nutrient polishing (Bunting and Shpigel, 2009), but further work is required to assess likely production from commercial-scale systems, labour demands associated with management and harvesting, market perceptions and risks associated with this strategy.

Integrated systems permit the generation of higher revenues and more regular cash flows from water pumped ashore or from underground or available via tidal exchange. A pond-based system combining fish, microalgae and shellfish developed on the Atlantic coast of France received water from a tidally filled reservoir; however, the capacity of the reservoir limited the biomass of fish and consequently integrated production that could be maintained in the systems.

In tropical coastal areas, integrated farming systems combining pond-based fish and shrimp production with shellfish and seaweed production have been developed. However, high suspended solids concentrations can constrain shellfish growth, and high turbidity and grazing can limit algae production. Mangrove stands have been used to condition incoming water and treat aquaculture wastewater. Economies of integration associated with horizontally integrated systems, using the same water, feed inputs, infrastructure, equipment and labour to produce multiple crops, appear to offer a potential advantage over monoculture systems. However, integration places new demands on farmers in terms of skills and knowledge requirements and results in additional risks, in particular related to engineering requirements and pests and disease, and poses new and poorly defined statutory and marketing challenges.

Unexploited resources that might contribute to horizontal integration or diversification are considered here in a broad sense, and an asset- or capital-based assessment derived from the sustainable livelihoods framework presented in chapter 1 appears to constitute a valuable approach in this regard. Transition from a focus on biological and ecological synergy to assessing potential opportunities to exploit underutilised infrastructure, natural capital or ecosystem services, human resources and networks (supply, marketing, professional contacts) and associations encompassing social capital could contribute to identifying the full range of potential integration and diversification options (Bunting et al., 2015, 2017; Short et al., 2021). Some options might contribute to vertical integration within the business or livelihood strategy, whereby inputs previously purchased or sourced elsewhere are produced as part of the broader farm or household activity (Newton et al., 2021; Bunting et al., 2023). Specific strategies that appear to show promise in achieving the goals of horizontal

Table 2.5 Horizontal integration strategies for selected aquaculture systems

Aquaculture system	Horizontally integrated system	Species groups with potential for integration	Constraints	Opportunities
Semi-intensive and intensive ponds, tanks and raceways	Single-species flow-through	• bivalve molluscs • phytophagous feeding fish • sea cucumbers • macroalgae • microalgae • floating macrophytes	• fouling with particulate matter • biofouling with epiphytes and colonial organisms • predation by fish and shrimps of juveniles for stocking • aeration may be needed at night • danger of shock loads of waste being released during harvest	• reduction in the waste load discharged to the receiving environment • low–medium engineering demands
	Multi-species flow-through	• macroalgae/bivalves • macroalgae/herbivorous fish • algae/phytophagous fish • periphyton/fish	• estimating the nutrient dynamics within the system • fouling and bio-fouling • finding suitable combinations of indigenous species	• increased reduction of wastes when compared with single-species systems • medium–high engineering demands
	Constructed wetlands	• reeds • mangrove ferns • mangroves	• extensive area required • few accounts regarding the use of wetlands to treat the wastewater from aquaculture • potentially limited capacity for nutrient retention • limited opportunities for producing valuable crops • sediment accumulation	• appropriate for both freshwater and marine environments • limited level of maintenance required

(Continued)

Table 2.5 (Continued)

Aquaculture system	Horizontally integrated system	Species groups with potential for integration	Constraints	Opportunities
Semi-intensive and intensive cages and pens	Open	• macroalgae • bivalves	• difficulty in predicting the dispersion pattern for nutrients in dynamic environments • herbivory and predation • loss due to storm damage • seasonal production in temperate climates	• compensates for the lack of waste treatment systems developed for open aquaculture systems • produces oxygen that can compensate for the respiratory demands of the aquaculture system
	Artificial reefs	• macrophytes • bivalves • fish • lobsters • crabs	• physical interference with access to the aquaculture system • loss due to predation and poaching • securing sole access to the fishery	• nutrient removal from the system • foraging around the reef contributes to the assimilation capacity of the system

Source: Bunting (2001).

integration in marine, freshwater or peri-urban and urban settings are critically assessed in chapters 4, 5 and 6, respectively.

Ecosystem approach to aquaculture development

An EAA was defined as a 'strategy for the integration of the activity within the wider ecosystem in such a way that it promotes sustainable development, equity, and resilience of interlinked social and ecological systems' (Soto et al., 2008: iv), whilst the purpose of adopting such an approach was summarised as being 'to plan, develop and manage the sector in a manner that addresses the multiple needs and desires of societies, without jeopardizing the options for future generations to benefit from the full range of goods and services provided by aquatic ecosystems' (p. 1). Jointly organised by the FAO and the Universitat de les Illes Balears, an expert workshop was convened in Palma de Mallorca Spain, 7–11 May 2007, to elaborate the EAA concept and outline principles to guide the EAA strategy and consider implications for EAA adoption at farm, waterbody and associated watershed or aquaculture zone and global scales. Aquaculture practices that exemplified EAA for promotion by policymakers were identified by workshop participants (Soto et al., 2008) and focused on:

- integrated aquaculture, particularly integrated multi-trophic aquaculture;
- ecosystem-based approaches to mitigating environmental impacts of aquaculture;
- broad stakeholder participation and local and relevant knowledge use;
- integration across sectors and deployment of appropriate incentives;
- EAA-specific research promotion.

Recommendations originating from the workshop included a focus on guidelines and promoting an enabling environment promoting dialogue amongst different institutions and sectors to coordinate initiatives on legislation and management and awareness-raising concerning EAA, particularly amongst consumers.

Summary

Aquaculture production systems have been classified according to various schemes and attributes, but it is important to reflect on the diversity of production systems adapted to local environmental conditions. Specific culture practices and management regimes have also evolved in response to prevailing social–economic conditions and potential hazards and risk levels deemed manageable by producers and acceptable by authorities. A range of environmental problems can result from poorly planned and managed aquaculture development. Various production principles including the optimisation of resource-use efficiency and productive use of waste resources could potentially minimise such negative impacts, and an array of strategies for resource-conserving and agroecosystem-enhancing aquaculture have been presented. Adoption of the EAA and production practices including polyculture, integrated aquaculture–agriculture, ecological water conditioning, IMTA and horizontally integrated aquaculture has the potential to make more efficient use of appropriated resources and mitigate environmental impacts.

Notes

1 'Trash fish' is a generic term used widely in the past to describe small, locally sourced fish used with minimal processing for aquaculture feed, but given the parlous state of most fish stocks and subsidies to unsustainable fisheries and recognising that such fish would probably have been processed and consumed by people from local communities and that there is a critical need to better appreciate the value of ecosystems services, including the provisioning of fish, sustaining society (Thiao and Bunting, 2022), this term is now generally regarded as inappropriate.
2 Section adapted from Bunting (2001) with contemporary references and edits to conform to publishing house style and conventions and ensure coherence with previous text.
3 Section adapted from Bunting (2001) with contemporary references and edits to conform to publishing house style and conventions and ensure coherence with previous text.
4 Section adapted from Bunting (2001) with contemporary references and edits to conform to publishing house style and conventions and ensure coherence with previous text.
5 Horizontal integration from a business management perspective is deemed to occur 'when a firm acquires or merges with a major competitor, or at least another firm operating at the same stage in the added value chain' (Thompson, 1993).

References

Ackefors, H. and Enell, M. (1990) 'Discharge of nutrients from Swedish fish farming to adjacent sea areas', Ambio, vol 119, pp. 28–35.
ADB, ACIAR, AwF, BRR, DKP, FAO, GTZ, IFC, MMAF, NACA and WWF (2007) Practical Manual on Better Management Practices for Tambak Farming in Aceh, Asian Development Bank ETESP, Australian Centre for International Agriculture Research, Food and Agriculture Organization of the United Nations, International Finance Corporation of the World Bank Group, Banda Aceh, Indonesia.
Ahmed, N., Bunting, S.W., Glaser, M., Flaherty, M.S. and Diana, J.S. (2017) 'Can greening of aquaculture sequester blue carbon?', Ambio, vol 46, pp. 468–477.
Ahmed, N., Bunting, S.W., Rahman, S. and Garforth, C.J. (2014) 'Community-based climate change adaptation strategies for integrated prawn–fish–rice farming in Bangladesh to promote social–ecological resilience', Reviews in Aquaculture, vol 6, pp. 20–35.
Alanara, A., Bergheim, A., Cripps, S.J., Eliassen, R. and Kristiansen, R. (1994) 'An integrated approach to aquaculture wastewater management', Journal of Applied Ichthyology, vol 10, p. 389.

Allcock, R. and Buchanan, D. (1994) 'Agriculture and fish farming' in P.S. Maitland, P.J. Boon and D.S. McLusky (eds) The Fresh Waters of Scotland, John Wiley & Sons, New York.

Arthington, A.H. and Bluhdorn, D.R. (1996) 'The effect of species introductions resulting from aquaculture operations' in D.J. Baird, M.C.M. Beveridge, L.A. Kelly and J.F. Muir (eds) Aquaculture and Water Resource Management, Blackwell Science, Oxford, UK.

Arzul, G., Clément, A. and Pinier, A. (1996) 'Effects on phytoplankton growth of dissolved substances produced by fish farming', Aquatic Living Resources, vol 9, pp. 95–102.

Bailey-Watts, A.E. (1994) 'Eutrophication' in P.S. Maitland, P.J. Boon and D.S. McLusky (eds) The Fresh Waters of Scotland, John Wiley & Sons, New York.

Baird, D.J. (1994) 'Pest control in tropical aquaculture: an ecological hazard assessment of natural and synthetic control agents', Mitteilungen Internationale Vereinigung fur Theoretische und Angewandte Limnologie, vol 24, pp. 285–292.

Barash, H., Plavnik, I. and Moav, R. (1982) 'Integration of duck and fish farming: experimental results', Aquaculture, vol 27, pp. 129–140.

Bardach, J.E. (1997) 'Aquaculture, pollution and biodiversity' in J.E. Bardach (ed) Sustainable Aquaculture, John Wiley & Sons, New York.

Barrington, K., Chopin, T. and Robinson, S. (2009) 'Integrated multi-trophic aquaculture (IMTA) in marine temperate waters', in D. Soto (ed) Integrated Aquaculture: A Global Review, FAO, FAO Fisheries and Aquaculture Technical Paper No 529, Rome.

Beardmore, J.A., Mair, G.C. and Lewis, R.I. (1997) 'Biodiversity in aquatic systems in relation to aquaculture', Aquaculture Research, vol 28, pp. 829–839.

Bergheim, A. and Asgard, T. (1996) 'Waste production from aquaculture' in D.J. Baird, M.C.M. Beveridge, L.A. Kelly and J.F. Muir (eds) Aquaculture and Water Resource Management, Blackwell Science, Oxford, UK.

Bergheim, A., Rønhovde, J. and Mundal, H. (1997) 'Efficient sludge treatment for land-based fish farms', Fish Farming International, vol 24, no 4, pp. 30–32.

Bergheim, A., Sanni, S., Indrevik, G. and Holland, P. (1993) 'Sludge removal from salmonid tank effluent using rotating microsieves', Aquacultural Engineering, vol 12, pp 97–109.

Bert, T.M., Tringali, M.D. and Seyoum, S. (2002) 'Development and application of genetic tags for ecological aquaculture' in B.A. Costa-Pierce (ed) Ecological Aquaculture, Blackwell Publishing, Oxford, UK.

Beveridge, M.C.M. and Phillips, M.J. (1993) 'Environmental impact of tropical inland aquaculture' in R.S.V. Pullin, H. Rosenthal and J.L. Maclean (eds) Environment and Aquaculture in Developing Countries, ICLARM Conference Proceedings 31, International Centre for Living Aquatic Resources Management, Makati City, Philippines.

Beveridge, M.C.M., Phillips, M.J. and Clarke, R.M. (1991) 'A quantitative and qualitative assessment of wastes from aquatic animal production' in D.E. Brune and J.R. Tomasso (eds) Aquaculture and Water Quality, Advances in World Aquaculture, vol 3, World Aquaculture Society, Baton Rouge, LA.

Beveridge, M.C.M., Ross, L.G. and Kelly, L.A. (1994) 'Aquaculture and biodiversity', Ambio, vol 23, pp. 497–502.

Black, E., Gowen, R., Rosenthal, H., Roth, E., Stechy, D. and Taylor, F.J.R. (1997) 'The costs of eutrophication from salmon farming: implications for policy – a comment', Journal of Environmental Management, vol 50, pp. 105–109.

Boaventura, R., Pedro, A.M., Coimbra, J. and Lencastre, E. (1997) 'Trout farm effluents: characterization and impact on the receiving streams', Environmental Pollution, vol 95, pp. 379–387.

Boelee, E. (ed) Atapattu, S., Barron, J., Bindraban, P., Bunting, S.W., Coates, D., et al. (main authors) (2011) Ecosystems for Water and Food Security, United Nations Environment Programme, Nairobi and International Water Management Institute, Colombo.

Bosma, R.H. and Tendencia, E.A. (2014) 'Comparing profits from shrimp aquaculture with and without green-water technology in the Philippines', Journal of Applied Aquaculture, vol 26, no 3, pp. 263–270.

Bosma, R.H., Tendencia, E.A. and Bunting, S.W. (2012) 'Financial feasibility of green-water shrimp farming associated with mangrove compared to extensive shrimp culture in the Mahakam Delta, Indonesia', Asian Fisheries Science, vol 25, no 3, pp. 258–269.

Boyd, C.E. and Gross, A. (2000) 'Water use and conservation for inland aquaculture ponds', Fisheries Management and Ecology, vol 7, pp. 55–63.

British Standards Institution (2011) Newly Revised PAS 2050 Poised to Boost International Efforts to Carbon Footprint Products, British Standards Institution, London.

Brooks, K.M., Mahnken, C. and Nash, C. (2002) 'Environmental effects associated with marine netpen waste with emphasis on salmon farming in the Pacific northwest' in R.R. Stickney and J.P. McVey (eds) Responsible Marine Aquaculture, CAB International, Wallingford, UK.

Bunting, S.W. (2001) A Design and Management Approach for Horizontally Integrated Aquaculture Systems, PhD thesis, Institute of Aquaculture, University of Stirling, UK.

Bunting, S.W. (2007) 'Regenerating aquaculture: enhancing aquatic resources management, livelihoods and conservation' in J. Pretty, A. Ball, T. Benton, J. Guivant, D. Lee, D. Orr, M. Pfeffer and H. Ward (eds) The SAGE Handbook of Environment and Society, SAGE Publications, London.

Bunting, S.W. (2014a) Stakeholder Delphi Assessment of UK Marine Aquaculture Sector Carbon Footprint Management, Final Report for The Crown Estate, University of Essex, Colchester, UK.

Bunting, S.W. (2014b) UK Marine Aquaculture Sector Carbon Budgeting, Final Report for The Crown Estate, University of Essex, Colchester, UK.

Bunting, S.W., Bosma, R.H., van Zwieten, P.A.M. and Sidik, A.S. (2013) 'Bioeconomic modeling of shrimp aquaculture strategies for the Mahakam Delta, Indonesia', Aquaculture Economics and Management, vol 17, no 1, pp. 51–70.

Bunting, S.W., Bostock, J., Leschen, W. and Little, D.C. (2023) 'Evaluating the potential of innovations across aquaculture product value chains for poverty alleviation in Bangladesh and India', Frontiers in Aquaculture, vol 2, no 3, pp. 1–23.

Bunting, S.W. and Edwards, P. (2018) 'Global prospects for safe wastewater reuse through aquaculture' in B.B. Jana, R.N. Mandal and P. Jayasankar (eds) Wastewater Management through Aquaculture, Springer, New York.

Bunting, S.W., Karim, M. and Wahab, M.A. (2005) 'Periphyton-based aquaculture in Asia: livelihoods and sustainability' in M.E. Azim, M.C.J. Verdegem, A.A. van Dam and M.C.M. Beveridge (eds) Periphyton: Ecology, Exploitation and Management, CAB International, Wallingford, UK.

Bunting, S.W., Kundu, N. and Ahmed, N. (2017) 'Evaluating the contribution of diversified shrimp–rice agroecosystems in Bangladesh and West Bengal, India to social–ecological resilience', Ocean and Coastal Management, vol 148, pp. 63–74.

Bunting, S.W., Mishra, R., Smith, K.G. and Ray, D. (2015) 'Evaluating sustainable intensification and diversification options for agriculture-based livelihoods within an aquatic biodiversity conservation context in Buxa, West Bengal, India', International Journal of Agricultural Sustainability, vol 13, pp. 275–294.

Bunting, S., Pounds, A., Immink, A., Zacarias, S., Bulcock, P., Murray, F. and Auchterlonie, N. (2022) The Road to Sustainable Aquaculture: On Current Knowledge and Priorities for Responsible Growth, World Economic Forum, Cologny, Switzerland.

Bunting, S.W. and Shpigel, M. (2009) 'Evaluating the economic potential of horizontally integrated land-based marine aquaculture', Aquaculture, vol 294, pp. 43–51.

Burka, J.F., Hammell, K.L., Horsberg, T.E., Johnson, G.R., Rainnie, D.J. and Speare, D.J. (1997) 'Drugs in salmonid aquaculture: a review', Journal of Veterinary Pharmacology and Therapeutics, vol 20, pp. 333–349.

Chen, S., Timmons, M.B., Aneshansley, D.J. and Bisogni, J.J., Jr. (1993) 'Suspended solids characteristics from recirculating aquacultural systems and design implications', Aquaculture, vol 112, pp. 143–155.

Cho, C.Y. and Bureau, D.P. (1997) 'Reduction of waste output from salmonid aquaculture through feeds and feeding', Progressive Fish-Culturist, vol 59, pp. 155–160.

Christensen, M.S. (1993) 'An economic analysis of floating cage culture of tinfoil barb, *Puntius schwanenfeldii*, in East Kalimantan, Indonesia, using chicken manure and other fresh feeds', Asian Fisheries Science, vol 6, pp. 271–281.

Christensen, M.S. (1994) 'Growth of tinfoil barb, *Puntius schwanenfeldii*, fed various feeds, including fresh chicken manure, in floating cages', Asian Fisheries Science, vol 7, pp. 29–34.

Coche, A.G. (1982) 'Cage culture of tilapias', in R.S.V. Pullin and R.H. Lowe-McConnell (eds) Biology and Culture of Tilapias, International Centre for Living Aquatic Resource Management, Metro Manila, Philippines.

Cornel, G.E. and Whoriskey, F.G. (1993) 'The effects of rainbow trout (*Oncorhynchus mykiss*) cage culture on the water quality, zooplankton, benthos and sediments of Lac du Passage, Quebec', Aquaculture vol 109, pp. 101–117.

Costa-Pierce, B.A. (2002) 'Ecology as the paradigm for the future of aquaculture' in B.A. Costa-Pierce (ed) Ecological Aquaculture, Blackwell Publishing, Oxford, UK.

Council of the European Communities (1991) 'Council Directive of 21 May 1991 Concerning Urban Waste Water Treatment', Official Journal of the European Communities, vol. 135, pp. 40–52.

Cripps, S.J. (1991) 'Comparison of methods for the removal of suspended particles from aquaculture effluents', in N. De Pauw and J. Joyce (eds) Aquaculture and the Environment, European Aquaculture Society, Special Publication 14, Bredene, Belgium.

Cripps, S.J. (1995) 'Serial particle size fractionation and characterisation of an aquacultural effluent', Aquaculture, vol 133, pp. 323–339.

Cripps, S.J. and Kelly, L.A. (1996) 'Reductions in wastes from aquaculture' in D.J. Baird, M.C.M. Beveridge, L.A. Kelly and J.F. Muir (eds) Aquaculture and Water Resource Management, Blackwell Science, Oxford, UK.

Crona, B.I., Wassénius, E., Jonell, M., Koehn, J.Z., Short, R., Tigchelaar, M., et al. (2023) 'Four ways blue foods can help achieve food system ambitions across nations', Nature, vol 616, pp. 104–112.

De Grave, S., Moore, S.J. and Burnell, G. (1998) 'Changes in benthic macrofauna associated with intertidal oyster, *Crassostrea gigas* (Thunberg) culture', Journal of Shellfish Research, vol 17, pp. 1137–1142.

de la Caba, K., Guerrero, P., Trung, T.S., Cruz-Romero, M., Kerry, J.P., Fluhr, J., et al. (2019) 'From seafood waste to active seafood packaging: an emerging opportunity of the circular economy', Journal of Cleaner Production, vol 208, pp. 86–98.

Dempster, P., Baird, D.J. and Beveridge, M.C.M. (1995) 'Can fish survive by filterfeeding on microparticles? Energy balance in tilapia grazing on algal suspensions', Journal of Fish Biology, vol 47, pp. 7–17.

Doughty, C.R. and McPhail, C.D. (1995) 'Monitoring the environmental impacts and consent compliance of freshwater fish farms', Aquaculture Research, vol 26, pp. 557–565.

Duplisea, D.E. and Hargrave, B.T. (1996) 'Response of meiobenthic size-structure, biomass and respiration to sediment organic enrichment', Hydrobiologia, vol 339, pp. 161–170.

Edwards, P. (1998) 'A systems approach for the production of integrated aquaculture', Aquaculture Economics and Management, vol 2, pp. 1–12.

Edwards, P., Demaine, H., Innes-Taylor, N. and Turongruang, D. (1996) 'Sustainable aquaculture for small-scale farmers: need for a balanced model', Outlook on Agriculture, vol 25, pp. 19–26.

Edwards, P., Lin, C.K., Macintosh, D.J., Leong Wee, K., Little, D. and Innes-Taylor, N.L. (1988) 'Fish farming and aquaculture', Nature, vol 333, pp. 505–506.

Edwards, P., Pacharaprakiti, C. and Yomjinda, M. (1994) 'An assessment of the role of buffalo manure for pond culture of tilapia. I. On-station experiment', Aquaculture, vol 126, pp. 83–95.

Edwards, P., Zhang, W., Belton, B. and Little, D.C. (2019) 'Misunderstandings, myths and mantras in aquaculture: its contribution to world food supplies has been systematically over reported', Marine Policy, vol 106, 103547.

FAO (2007) The State of World Fisheries and Aquaculture 2006, FAO, Rome.

FAO (2010) Aquaculture Development 4, Ecosystem Approach to Aquaculture, FAO Technical Guidelines for Responsible Fisheries, FAO, Rome.

FAO (2022) Fishery and Aquaculture Statistics. Global Aquaculture Production 1950–1920, FishStatJ, FAO, Rome.

Ferrante, J.G. and Parker, J.I. (1977) 'Transport of diatom frustules by copepod fecal pellets to the sediments of Lake Michigan', Limnology and Oceanography, vol 22, pp. 92–98.

Folke, C., Kautsky, N. and Troell, M. (1994) 'The costs of eutrophication from salmon farming: implications for policy', Journal of Environmental Management, vol 40, pp. 173–182.

Folke, C., Kautsky, N. and Troell, M. (1997) 'Salmon farming in context: response to Black et al.', Journal of Environmental Management, vol 50, pp. 95–103.

Foy, R.H. and Rosell, R. (1991) 'Loadings of nitrogen and phosphorus from a Northern Ireland fish farm', Aquaculture, vol 96, pp. 17–30.

Garg, S.K. (1996) 'Brackishwater carp culture in potentially waterlogged areas using animal waste as pond fertilisers', Aquaculture International, vol 4, pp. 143–155.

Gephart, J.A., Henriksson, P.J.G., Parker, R.W.R., Shepon, A., Gorospe, K.D., Bergman, K., et al. (2021) 'Environmental performance of blue foods', Nature, vol 597, no 7876, pp. 360–365.

Gillibrand, P.A., Turrell, W.R., Moore, D.C. and Adams, R.D. (1996) 'Bottom water stagnation and oxygen depletion in a Scottish sea loch', Estuaries, Coastal and Shelf Science, vol 43, pp. 217–235.

Gunther, F. (1997) 'Hampered effluent accumulation process: phosphorous management and societal structure', Ecological Economics, vol 21, pp. 159–174.

Hargrave, B.T., Phillips, G.A., Doucette, L.I., White, M.J., Milligan, T.G., Wildish, D.J. and Cranston, R.E. (1997) 'Assessing benthic impacts of organic enrichment from marine aquaculture', Water, Air and Soil Pollution, vol 99, pp. 641–650.

Hem, S., Avit, J.B.L.F. and Cisse, A. (1995) 'Acadja as a system for improving fishery production' in J.J. Symoens and J.C. Micha (eds) The Management of Integrated Freshwater Agro-piscicultural Ecosystems in Tropical Areas, Seminar Proceedings, 16–19 May 1994, Technical Centre for Agricultural and Rural Co-operation (CTA), Royal Academy of Overseas Sciences, Brussels.

Henderson, J.P. and Bromage, N.R. (1988) 'Optimising the removal of suspended solids from aquaculture effluents in settlement lakes', Aquacultural Engineering, vol 7, pp. 167–181.

Hennessy, M. (1991) 'The efficiency of two aquacultural effluent treatment systems in use in Scotland' in N. De Pauw and J. Joyce (eds) Aquaculture and the Environment, European Aquaculture Society, Special Publication 14, Bredene, Belgium.

Henriksson, P.J.G., Troell, M., Banks, L.K., Belton, B., Beveridge, M.C.M., Klinger, D.H., et al. (2021) 'Interventions for improving the productivity and environmental performance of global aquaculture for future food security', One Earth, vol 4, pp. 1220–1232.

Hopkins, J.S., Hamilton, R.D., Sandifer, P.A. and Browdy, C.L. (1993) 'The production of bivalve mollusks in intensive shrimp ponds and their effect on shrimp production and water quality', World Aquaculture, vol 24, no 2, pp. 74–77.

Howgate, P., Bunting, S., Beveridge, M. and Reilly, A. (2002) 'Aquaculture associated public, animal, and environmental health issues in nonindustrialized countries', in M. Jahncke, S. Garrett, R. Martin, E. Cole and A. Reilly (eds) Public, Animal and Environmental Aquaculture Health, John Wiley & Sons, New York.

Investing in Regenerative Agriculture and Food (2022) Why These 16 Funds Could Transform Agriculture, Investing in Regenerative Agriculture and Food, https://investinginregenerativeagriculture.com/2022/02/02/why-these-15-funds-could-transform-agriculture-updated-april-2020/

IPCC (2019) The Ocean and Cryosphere in a Changing Climate, A Special Report of the Intergovernmental Panel on Climate Change, Intergovernmental Panel on Climate Change.

Jones, J.G. (1990) 'Pollution from fish farms', Water and Environment Journal, vol 4, pp. 14–18.

Jones, T.O. and Iwama, G.K. (1991) 'Polyculture of the Pacific oyster, Crassostrea gigas (Thunberg), with Chinook salmon, Oncorhynchus tshawytscha', Aquaculture, vol 92, pp. 313–322.

Kadri, S., Metcalfe, N.B., Huntingford, F.A. and Thorpe, J.E. (1991) 'Daily feeding rhythms in Atlantic salmon in sea cages', Aquaculture, vol 92, pp. 219–224.

Karakassis, I., Hatziyanni, E., Tsapakis, M. and Plaiti, W. (1999) 'Benthic recovery following cessation of fish farming: a series of successes and catastrophes', Marine Ecology Progress Series, vol 184, pp. 205–218.

Karim, A.T., Baten, M., Sarker, A.K. and Ullah, H. (2021) Suchana: Ending the Cycle of Undernutrition in Bangladesh, Homestead Food Production (HFP) Aquaculture Carp–Tilapia–Mola Polyculture, WorldFish, Malaysia.

Koohafkan, P. and Altieri, M.A. (2011) Globally Important Agricultural Heritage Systems, FAO, Rome.

Korn, M. (1996) 'The dike–pond concept: sustainable agriculture and nutrient recycling in China', Ambio, vol 25, pp. 6–13.

Kronvang, B., Aertebjerg, G., Grant, R., Kristensen, P., Hovmand, M. and Kirkegaard, J. (1993) 'Nationwide monitoring of nutrients and their ecological effects: state of the Danish aquatic environment', Ambio, vol 22, pp. 176–186.

Kruijssen, F., Tedesco, I., Ward, A., Pincus, L., Love, D. and Thorne-Lyman, A.L. (2020) 'Loss and waste in fish value chains: a review of the evidence from low and middle-income countries', Global Food Security, vol 26, 100434.

Kumar Saha, M. and Kumar, B. (2020) A Strategy on Increase Production and Marketing of Mola and Other Small Indigenous Species of Fish (SIS) in Bangladesh, The WorldFish Center Working Papers, WorldFish, Malaysia.

Lehman, J.T. (1980) 'Release and cycling of nutrients between planktonic algae and herbivores', Limnology and Oceanography, vol 25, pp. 620–632.

Li, J., Wan, Y., Wang, B., Waqas, M.A., Cai, W., Guo, C., et al. (2018) 'Combination of modified nitrogen fertilizers and water saving irrigation can reduce greenhouse gas emissions and increase rice yield', Geoderma, vol 315, pp. 1–10.

Little, D.C. (1995) 'The development of small-scale poultry–fish integration in northeast Thailand: potential and constraints' in J.J. Symoens and J.C. Micha (eds) The Management of Integrated Freshwater Agro-piscicultural Ecosystems in Tropical Areas, Seminar Proceedings, 16–19 May 1994, Technical Centre for Agricultural and Rural Co-operation (CTA), Royal Academy of Overseas Sciences, Brussels.

Little, D.C. and Edwards, P. (1999) 'Alternative strategies for livestock–fish integration with emphasis on Asia', Ambio, vol 28, pp. 118–124.

Little, D.C. and Muir, J. (1987) A Guide to Integrated Warm Water Aquaculture, Institute of Aquaculture, University of Stirling, UK.

Little, D.C., Murray, F.J., Azim, E., Leschen, W., Boyd, K., Watterson, A. and Young, J.A. (2008) 'Options for producing warm-water fish in the UK: limits to "Green Growth"?' Trends in Food Science and Technology, vol 19, pp. 255–264.

Loch, D.D., West, J.L. and Perlmutter, D.G. (1996) 'The effect of trout farm effluent on the taxa richness of benthic macroivertebrates', Aquaculture, vol 147, pp. 37–55.

Lystad, E. and Selvik, J.R. (1991) 'Reducing environmental impact through sludge control in land-based fish farming' in N. De Pauw and J. Joyce (eds) Aquaculture and the Environment, European Aquaculture Society, Special Publication 14, Bredene, Belgium.

Maclean, M.H., Brown, J.H., Ang, K.J. and Jauncey, K. (1994) 'Effects of manure fertilization frequency on pond culture of the freshwater prawn, *Macrobrachium rosenbergii* (de Man)', Aquaculture and Fisheries Management, vol 25, pp. 601–611.

Makinen, T., Lindgren, S. and Eskelinen, P. (1988) 'Sieving as an effluent treatment method for aquaculture', Aquacultural Engineering, vol 7, pp. 367–377.

Mara, D.D., Edwards, P., Clark, D. and Mills, S.W. (1993) 'A rational approach to the design of wastewater-fed fishponds', Water Research, vol 27, pp. 1797–1799.

Massik, Z. and Costello, M.J. (1995) 'Bioavailability of phosphorus in fish farm effluents to freshwater phytoplankton', Aquaculture Research, vol 26, pp. 607–616.

McCord, C.L. and Loyacano, H.A. (1978) 'Removal and utilization of nutrients by Chinese water chestnut in catfish ponds', Aquaculture, vol 13, pp. 143–155.

Milstein, A. (1992) 'Ecological aspects of fish species interactions in polyculture ponds', Hydrobiologia, vol 231, pp. 177–186.

Muir, J.F. (1994) 'Water reuse in aquaculture systems', Infofish International, vol 6, pp. 40–46.

Muir, J.F. (2005) 'Managing to harvest? Perspectives on the potential of aquaculture', Philosophical Transactions of the Royal Society B, vol 360, pp. 191–218.

Muratori, M.C.S., de Oliveira, A.L., Ribeiro, L.P., Leite, R.C., Costa, A.P.R. and da Silva, M.C.C. (2000) '*Edwardsiella tarda* isolated in integrated fish farming', Aquaculture Research, vol 31, pp. 481–483.

Naylor, E. and Scholtissek, C. (1988) 'Fish farming and aquaculture: Naylor and Scholtissek reply', Nature, vol 333, p. 506.

NCC (1990) Fish Farming and the Scottish Freshwater Environment, Nature Conservancy Council, Edinburgh, UK.

Newton, R., Zhang, W., Xian, Z., McAdam, B. and Little, D.C. (2021) 'Intensification, regulation and diversification: the changing face of inland aquaculture in China', Ambio, vol 50, pp. 1739–1756.

Njoku, D.C. and Ejiogu, C.O. (1999) 'On-farm trials of an integrated fish-cum-poultry farming system using indigenous chickens', Aquaculture Research, vol 30, pp. 399–408.

Oberdorff, T. and Porcher, J.P. (1994) 'An index of biotic integrity to assess biological impacts of salmonid farm effluents on receiving waters', Aquaculture, vol 119, pp. 219–235.

Padiyar, A.P., Dubey, S.K., Shenoy, N., Mohanty, B., Baliarsingh, B.K., Gaikwad, A., et al. (2021) Scientific Fish Farming in Gram Panchayat Tanks by Women Self Help Groups in Odisha, India: Crop Outcome Survey Report 2018–2019 and 2019–2020, WorldFish, Program Report: 2021–15, Malaysia.

Páez-Osuna, F., Guerrero-Galvan, S.R. and Ruiz-Fernandez, A.C. (1998) 'The environmental impact of shrimp aquaculture and the coastal pollution in Mexico', Marine Pollution Bulletin, vol 36, no 1, pp. 65–75.

Phillips, M.J., Lin, C.K. and Beveridge, M.C.M. (1993) 'Shrimp culture and the environment: lessons from the world's most rapidly expanding warm water aquaculture sector' in R.S.V. Pullin, H. Rosenthal and J.L. Maclean (eds) Environment and Aquaculture in Developing Countries, International Centre for Living Aquatic Resources Management, Makati City, Philippines.

Pretty, J.N. (1995) Regenerating Agriculture: Policies and Practice for Sustainability and Self-Reliance, Earthscan, London.

Regeneration International (2022) Why Regenerative Agriculture?, Regeneration International, https://regenerationinternational.org/why-regenerative-agriculture

Roos, N., Wahab, M.A., Hossain, M.A.R. and Thilsted, S.H. (2007) 'Linking human nutrition and fisheries: incorporating micronutrient-dense, small indigenous fish species in carp polyculture production in Bangladesh', Food and Nutrition Bulletin, vol 28, no 2, pp. 280–293.

Ruddle, K. and Christensen, V. (1993) 'An energy flow model of the mulberry dike–carp pond farming system of the Zhujiang delta, Guangdong Province, China', in V. Christensen and D. Pauly (eds) Trophic Models of Aquatic

Ecosystems, International Centre for Living Aquatic Resources Management, Makati City, Philippines.

Ruddle, K. and Zhong, G. (1988) Integrated Agriculture–Aquaculture in South China: The Dike–Pond System of the Zhujiang Delta, Cambridge University Press, Cambridge, UK.

Ruokolahti, C. (1988) 'Effects of fish farming on growth and chlorophyll *a* content of *Cladophora*', Marine Pollution Bulletin, vol 4, pp. 166–169.

Scholtissek, C. and Naylor, E. (1988) 'Fish farming and influenza pandemics', Nature, vol 331, p. 215.

Schwartz, M.F. and Boyd, C.E. (1994) 'Effluent quality during harvest of channel catfish from watershed ponds', Progressive Fish-Culturist, vol 56, pp. 25–32.

Selong, J.H. and Helfrich, L.A. (1998) 'Impacts of trout culture effluent on water quality and biotic communities in Virginia headwater streams', Progressive Fish-Culturist, vol 60, pp. 247–262.

Selvik, J.R. and Lystad, E. (1991) 'Removal of solid wastes from fishfarm effluents' in N. De Pauw and J. Joyce (eds) Aquaculture and the Environment, European Aquaculture Society, Special Publication 14, Bredene, Belgium.

Seymour, E.A. and Bergheim, A. (1991) 'Towards a reduction of pollution from intensive aquaculture with reference to the farming of salmonids in Norway', Aquacultural Engineering, vol 10, pp. 73–88.

Short, R.E., Gelcich, S., Little, D.C., Micheli, F., Allison, E.H., Basurto, X., et al. (2021) 'Harnessing the diversity of small-scale actors is key to the future of aquatic food systems', Nature Food, vol 2, no 9, pp. 733–741.

Sigfusson, T. (2022) 'Promising opportunities for fish by-products', New Food Magazine, 31 August.

Skladany, M. (1996) 'Fish, pigs, poultry, and Pandora's Box: integrated aquaculture and human influenza', in C. Bailey, S. Jentoft and P. Sinclair (eds) Aquaculture Development: Social Dimensions of an Emerging Industry, Westview Press, Boulder, CO.

Soto, D., Aguilar-Manjarrez, J. and Hishamunda, N. (eds) (2008) Building an Ecosystem Approach to Aquaculture, FAO, FAO Fisheries and Aquaculture Proceedings No 14, Rome.

Stenton-Dozey, J.M.E., Jackson, L.F. and Busby, A.J. (1999) 'Impact of mussel culture on macrobenthic community structure in Saldanha Bay, South Africa', Marine Pollution Bulletin, vol 39, pp. 357–366.

Stirling, H.P. and Dey, T. (1990) 'Impact of intensive cage fish farming on the phytoplankton and periphyton of a Scottish freshwater loch', Hydrobiologia, vol 190, pp. 193–214.

Svasand, T. and Moksness, E. (2004) 'Marine stock enhancement and sea-ranching' in E. Moksness, E. Kjorsvik and Y. Olsen (eds) Culture of Cold-Water Marine Fish, Blackwell Publishing, Oxford, UK.

Syvitski, J.P.M., Kettner, A.J., Overeem, I., Hutton, E.W.H., Hannon, M.T., Brakenridge, G.R., et al. (2009) 'Sinking deltas due to human activities', Nature Geoscience, vol 2, pp. 681–686.

Teichert-Coddington, D.R., Rouse, D.B., Potts, A. and Boyd, C.E. (1999) 'Treatment of harvest discharge from intensive shrimp ponds by settling', Aquaculture Engineering, vol 19, pp. 147–161.

Thiao, D. and Bunting, S.W. (2022) Socio-economic and Biological Impacts of the Fish-Based Feed Industry for Sub-Saharan Africa, FAO Fisheries and Aquaculture Circular 1236, FAO, Rome.

Thilsted, S.H., Thorne-Lyman, A., Webb, P., Bogard, J.R., Subasinghe, R., Phillips, M.J., et al. (2016) 'Sustaining healthy diets: the role of capture fisheries and aquaculture for improving nutrition in the post-2015 era', Food Policy, vol 61, pp. 126–131.

Thompson, J.L. (1993) Strategic Management: Awareness and Change, Chapman and Hall, London.

Tran, T.B., Le, C.D. and Brennan, D. (1999) 'Environmental costs of shrimp culture in the rice-growing regions of the Mekong Delta', Aquaculture Economics and Management, vol 3, pp. 31–42.

Troell, M. and Berg, H. (1997) 'Cage fish farming in the tropical Lake Kariba, Zimbabwe: impact and biogeochemical changes in sediment', Aquaculture Research, vol 28, pp. 527–544.

United Nations (2015) Transforming Our World: The 2030 Agenda for Sustainable Development, United Nations, New York.

Verdegem, M.C.J., Bosma, R.H. and Verreth, J.A.J. (2006) 'Reducing water use for animal production through aquaculture', Water Resources Development, vol 22, pp. 101–113.

Wahab, M.A., Azim, M.E., Ali, M.H., Beveridge, M.C.M. and Khan, S. (1999a) 'The potential of periphyton-based culture of the native major carp calbaush, *Labeo calbasu* (Hamilton)', Aquaculture Research, vol 30, pp. 409–419.

Wahab, M.A., Bergheim, A. and Braaten, B. (2003) 'Water quality and partial mass budget in extensive shrimp ponds in Bangladesh', Aquaculture, vol 218, pp. 413–423.

Wahab, M.A., Mannan, M.A., Huda, M.A., Azim, M.E., Tollervey, A.G. and Beveridge, M.C.M. (1999b) 'Effects of periphyton grown on bamboo substrates on growth and production of Indian major carp, rohu (*Labeo rohita* Ham.)', Bangladesh Journal of Fisheries Research, vol 3, no 1, pp. 1–10.

Wang, J.-Q., Li, D., Dong, S., Wang, K. and Tian, X. (1998) 'Experimental studies on polyculture in closed shrimp ponds I. Intensive polyculture of Chinese shrimp (*Penaeus chinensis*) with tilapia hybrids', Aquaculture, vol 163, pp. 11–27.

Welcomme, R.L. (1988) International Introductions of Inland Aquatic Species, FAO Fisheries Technical Paper 294, FAO, Rome.

Weston, D.P. (1996) 'Environmental considerations in the use of antibacterial drugs in aquaculture' in D.J. Baird, M.C.M. Beveridge, L.A. Kelly and J.F. Muir (eds) Aquaculture and Water Resource Management, Blackwell Science, Oxford, UK.

Wiesmann, D., Scheid, H. and Pfeffer, E. (1988) 'Water pollution with phosphorus of dietary origin by intensively fed rainbow trout (*Salmo gairdneri* Rich.)', Aquaculture, vol 69, pp. 263–270.

Yan, N. and Chen, X. (2015) 'Don't waste seafood waste', Nature, vol 524, pp. 155–157.

Zacarias, S. (2020) Use of Non-ablated Shrimp (*L. vannamei*) in Commercial Scale Hatcheries, PhD thesis, University of Stirling, UK.

Zacarias, S., Fegan, D., Wangsoontorn, S., Yamuen, N., Limakom, T., Carboni, S., et al. (2021) 'Increased robustness of postlarvae and juveniles from non-ablated Pacific whiteleg shrimp, *Penaeus vannamei*, broodstock post-challenged with pathogenic isolates of *Vibrio parahaemolyticus* (VpAHPND) and white spot disease (WSD)', Aquaculture, vol 532, 736033.

Zhu, Y., Yang, Y., Wan, J., Hua, D. and Mathias, J.A. (1990) 'The effect of manure application rate and frequency upon fish yield in integrated fish farm ponds', Aquaculture, vol 91, pp. 233–251.

3 Equitable aquaculture development

Key points

The aims of this chapter are to:

- Review the potential role of aquaculture optimising the use of aquatic resources and social consequences of poorly planned and managed development.
- Review options to reconcile multiple demands associated with aquaculture development and mitigate social impacts.
- Describe the rationale for environmental management systems, discuss barriers to widespread adoption and consider alternatives suited to producers with limited capacity and resources.
- Describe the principles of responsible aquaculture planning and consider how practices and policies for sustainable development may be conceived and evaluated.
- Consider prospects for social enterprise and private sector initiatives in promoting and facilitating sustainable aquaculture development.
- Introduce third- and second-party standards setting bodies and associated certification and labelling and broad objectives.
- Introduce the concept of the social licence to operate (SLO) and discuss prospects for achieving this in different contexts.
- Outline calls for stakeholder participation in development planning and practice and critically evaluate practical means to achieve interactive stakeholder participation, thereby ensuring that the process is equitable, transparent and trustworthy.
- Discuss the concept of aquatic ecosystem services supporting aquaculture development and critically review the application of the ecological footprint approach to assess aquaculture systems.
- Review the concept of environmental carrying capacity and the need to balance the appropriation and supply of ecosystem services whilst maintaining environmental stocks and flows.
- Review the role of aquaculture in contributing to poverty alleviation, through income and employment and provisioning affordable aquatic food products, and its potential to facilitate ecological restoration.

DOI: 10.4324/9781003342823-3

- Consider the opportunities to integrate aquaculture in multifunctional landscapes, waterscapes and seascapes and discuss the need for coherent planning and better communication (see Bunting et al., 2016; Lynch et al., 2019) and highlight emerging limitations in this regard.
- Describe the potential role of aquaculture development in climate change mitigation and carbon sequestration (see Ahmed et al., 2017) and community-based adaptation strategies (see Ahmed et al., 2014).

Sustainable aquaculture development to optimise aquatic resources use

Sustainable aquaculture development can contribute to the optimal use of shared aquatic resources or accessible ecosystem services (Bunting et al., 2022). Culturing filter-feeding fish species in large waterbodies and bivalves and seaweed in coastal areas can extract nutrients from the biosphere and return them to human food systems or make them available for animal feed and plant fertiliser production (Bunting and Edwards, 2018; Bunting et al., 2022). Aquaculture can significantly increase the availability and accessibility of commodity aquatic species in domestic and regional markets that can contribute to food and nutrition security (Henriksson et al., 2021; Crona et al., 2023). Exploitation of accessible aquatic ecosystems through appropriate aquaculture can generate employment and diverse livelihoods opportunities across value chains and food systems (Short et al., 2021). There is a danger, however, that when contemplating the potential economic and social benefits aquaculture could generate, environmental considerations, notably the cumulative impacts of many apparently successful individual operations, can be overlooked (Taskov et al., 2021; Bunting et al., 2022).

When considering the environmental impacts associated with poorly sited and managed aquaculture systems, a lack of value ascribed to ecosystem services means that the true costs of these are often only considered when the health or well-being of people is affected or aquaculture operators and other resource-user groups are inconvenienced. Self-pollution, where environmental degradation from aquaculture results in negative feedback, can depress productivity and promote disease outbreaks (Muluk and Bailey, 1996; Corea et al., 1998; Bunting, 2001b). This is especially so if aquaculture disrupts the supporting ecosystem services with consequent declines in the quality of available water supplies. Land-based aquaculture operations frequently discharge wastewater to streams and rivers that supply other producers downstream. Currents and tides may convey waste emanating from cages and pens to other aquaculture sites in coastal and inland settings. Waste from both shore-based facilities and cages and pens may contaminate water subsequently used to supply the culture facility.

Respiration by microbial communities flourishing in sediments underlying cages can result in anoxic conditions and lead to potentially harmful releases of hydrogen sulphide and methane, implicated by Black et al. (1994) in causing gill disease outbreaks at salmon farms. Dissolved oxygen concentrations in the water column may also be depressed owing to the respiration of microbial communities and benthic invertebrate assemblages stimulated by organic enrichment originating from cages and pens. Reduced oxygen concentrations were recorded beneath salmon cages in Scotland (Lumb, 1989; Gillibrand et al., 1996), and upwelling of this anoxic water constitutes a health hazard to cultured fish that has been implicated in fish-kills (Beveridge, 2004).

Restricted amenity[1]

Discharging aquaculture wastewater can affect users of the aquatic resource other than aquaculture operators. Deteriorating water quality in the Dutch Canal Mundel lagoon system attributed to the increase in shrimp farming has been blamed for the observed decline in the capture fisheries of the lagoon. This caused resentment in local fishing communities and led to poaching (Corea et al., 1998). A survey of local community members revealed that several people complained of skin diseases attributed to poor water quality. Although no direct link has been established, concerns of this type may strengthen the opposition to shrimp farming in the local population.

Saline wastewater discharged from shrimp farms in Songkhla, Thailand, has been implicated in causing the death of livestock drinking from canals (Primavera, 1997). The author further reported that following the salinisation of surface water caused by the discharge of wastewater from shrimp farms in Nellore district, India, women were forced to spend longer collecting drinking water from distant sources. Yields on farms cultivating rice have also been affected by saline water released from neighbouring shrimp farms (Tran et al., 1999); this situation arose from leaching during the dry season and the pond dikes being breached during the growing season. Shrimp farming may increase the sediment load to local canals and rivers (Tran et al., 1999), and sediment in wastewater from shrimp farms in Thailand and Sri Lanka caused irrigation canals to become silted (Phillips et al., 1993).

Aquaculture wastewater may encourage nuisance growths of aquatic macrophytes that interfere with recreational activities such as angling, boating and swimming. Possible disease transmission from cage farms to native fish stocks has been blamed for causing a decline in the number of Atlantic salmon returning to recreational capture fisheries in Scotland (NCC, 1990). Declines in recreational fisheries may have severe implications for rural economies that receive income from anglers. The introduction of disease with the signal crayfish (*Pacifasticus leniusculus*) reduced native crayfish (*Austropotamobius pallipes*) populations in Europe and caused the wild crayfish capture fishery to decline. Furthermore, the reduced indigenous crayfish populations allowed aquatic macrophytes to proliferate, resulting in the elimination of habitat for game fish (Thompson, 1990).

Reduced functionality[2]

Natural wetland functions produce a wide array of environmental goods and services that sustain economic activities and societal systems (Burbridge, 1994). Wastewater discharged from commercial aquaculture operations may, however, have a detrimental effect on the functional integrity of wetlands, disrupting the supply of environmental goods and services. The assimilative capacity of mangrove ecosystems may be exceeded by the discharge of wastewater from shrimp farms and excessive loads of ammonia and organic matter could lead to anaerobic sediments, resulting in tree mortality (Robertson and Phillips, 1995).

Elimination of mangrove trees may result in the loss of a number of functions. Reduced production of mangrove biomass will be accompanied by a decrease in the assimilation and cycling of nutrients, potentially leading to nutrient export from the mangrove ecosystem, with consequent adverse impacts on the receiving environment. Decline of the mangrove root system could decrease sediment stability, leading to erosion and increased risk of saline water intruding inland. Loss of mangrove habitat may also reduce nursery areas for juvenile fish and shrimp and contribute to a general decrease in biodiversity.

Impacts on option and non-use values[3]

Reduction in the quality of the aquatic environment may influence the value an individual attributes to preserving the resource to allow the individual, other individuals and future generations the option of using the resource at a later date (Muir et al., 1999). The impact of an activity on this option value may be estimated by assessing the willingness-to-pay of an individual to preserve the environment. Folke et al. (1994) extrapolated marginal costs of SEK[4] 50–100 and SEK 20–30 kg^{-1} for nitrogen and phosphorus removal, respectively, from sewage in Sweden, to represent the willingness-to-pay of Swedish society to limit nutrient discharges from salmon aquaculture. Based on a comparison of waste production presented as person equivalents, it was estimated in 1994 that the cost to society of eliminating nitrogen and phosphorus discharges originating from salmon aquaculture equated to SEK 4–4.5 kg^{-1} of production. These authors also calculated that internalising this cost increased production costs for salmon to SEK 31–31.5 kg^{-1}, with possible severely reduced profits and a threat to the viability of salmon farming.

Environments also have non-use values. The intrinsic or existence value of environments is unrelated to humans and their present or potential, direct or indirect, use of the resource (Turner, 1991; Muir et al., 1999). For example, people who are unlikely ever to visit a region may attribute value to its existence and feel a sense of loss if the ecosystem was degraded through the discharge of aquaculture wastewater. Degradation of the environment would also reduce the value ascribed to passing the asset onto future generations, termed the 'bequest value'. Therefore, although changes in non-use values of environments due to aquaculture wastewater discharges have not been described, they may be expected to be negative.

Reconciling multiple demands

Rapid expansion of intensive aquaculture resulting in environmental degradation caused considerable alarm at frequent intervals over the past five decades since the 1960s. Given the rapid development of the sector and novelty of production systems employed, governments and agencies charged with environmental monitoring and enforcing standards were poorly equipped to plan, manage and regulate aquaculture development. Anticipating and managing the cumulative effects of multiple operators in specific geographic areas have constituted particular challenges. Proposed aquaculture developments often result in concerns amongst local communities and other resource user groups, notably fishers, existing aquaculture operators and stakeholders affected by changes in access patterns and restrictions on navigation.

Heightened awareness and sensitivity concerning possible environmental impacts and stakeholder conflicts attributed to poorly planned and managed aquaculture development led to the drafting of sector-specific legislation and statutory wastewater discharge monitoring in many countries. Consequently, aquaculture producers and managers must now comply with regulations and parameters specified in discharge consents or risk incurring punitive taxation or face court action, fines and ultimately closure. Operators of aquaculture production systems are increasingly obliged to strive for continued improvement in environmental performance. Consumer perceptions are growing in influence, and where there is concern over production standards, product quality or safety, demand for and the value of implicated products will decline abruptly.

Appreciable aquaculture development has arguably been market driven and provides more reliable supplies and quality of aquatic products than capture fisheries. Domestic markets have been stimulated in developing countries by enhanced communication infrastructure and burgeoning urban populations. An emerging quandary, however, is the extent to which aquaculture products and practices employed in their culture will meet the demands, expectations and aspirations of consumers and other stakeholder groups. Environmental advocates and campaigners for social justice constitute important stakeholder groups that monitor aquaculture development and highlight damaging and inequitable practices and policies. Due consideration of less vociferous and well-organised stakeholder groups with divergent perspectives and agendas encompassing environmental, social, economic and political motivations is increasingly regarded as critical to successful and sustainable aquaculture development.

Although they may not be regarded as experts or recognised as warranting inclusion in statutory consultation or planning processes, many stakeholder groups possess knowledge and experience with the potential to inform and guide the development process. Stakeholder Delphi outcomes, for example, demonstrated that groups representing producers, consultants, regulators and researchers predominantly from developed countries came up with a range of strategies to reduce impacts associated with aquaculture wastewater (Table 3.1). Key individual stakeholders may be in a position to formulate a policy or draft legislation that ultimately governs prospects for aquaculture development in a particular region or jurisdiction. Equally, influential

Table 3.1 Strategies proposed by stakeholder Delphi participants to reduce negative impacts of aquaculture wastewater; frequency of occurrence in Round 1 (n) and mean rate (x) and ordinal rank following Round 3

Factor	n	x	Rank
Managerial			
Management procedures that improve water quality; e.g. careful feed management, de-sludging lagoons, aeration and harvesting strategies that minimise discharges	5	8.1	1.5
Good planning prior to developing aquaculture facilities and improved site selection	3	8.1	1.5
Adoption of a more holistic/systematic paradigm for aquaculture	2	6.7	12
Adopt extensive as opposed to intensive management practices	1	3.8	18
Institutional			
Encourage collaboration between researchers and commercial enterprises	3	7.9	3
Better education of farmers regarding water quality and environmental management	3	7.1	6
Provide information and direction to government regarding opportunities for the innovative management of aquaculture wastewater	2	7.1	7
Increase and enforce discharge standards for wastewater or implement a pollution tax	3	6.9	10
Open some commercial operations for public tours	1	6.1	15
Technological			
Increase research and development into improved treatment technologies	7	7.5	4
Improved feed quality and lower feed conversion ratios	5	7.3	5
Develop systems for water reuse	6	6.6	13
Develop new vaccines and improve disease control	3	5.2	17
Socioeconomic			
Improved evaluation of benefits associated with management practices that reduce environmental impact	2	7.0	8
The need to portray a positive image will necessitate improved waste management	1	6.9	9
Look at the energy costs of typical intensive production systems	1	6.6	11
Educate the public and managers regarding recycling systems	1	6.5	14
Government funding; e.g. subsidies, grants and tax relief to encourage research and development	2	6.1	16

Source: Bunting (2010).

stakeholder groups representing political and ideological positions, workers' rights or issue-specific interests may ultimately dictate what is acceptable.

There is often little incentive to protect the environment where regulation is absent or poorly enforced, although greater awareness amongst producers and direct action by local communities and pressure groups can lead to improvements. A potential price premium associated with quality assured or organic production constitutes a strong incentive to assume more responsible behaviour and practices. Although transaction costs combined with limited capacity may discourage smaller producers from adopting a conventional environmental management–based approach, greater awareness of damage being caused and knowledge of practical ways to better manage their production systems may assist them in adopting appropriate mitigation measures. Better management practice (BMP) guidelines formulated for *tambak* farming in Aceh, Indonesia, constitute a promising approach in this regard and is discussed further in chapter 8.

Selected strategies to reconcile multiple demands on exploited ecosystems are reviewed below, with a focus on approaches including environmental management systems, responsible aquaculture development planning, cost–benefit analysis, ecological footprint and environmental carrying capacity assessments. Prospects for joint assessment and decision-making, notably participatory action planning and implementation, are critically reviewed in chapter 7. Aspirations and motivations of aquaculture producers are often neglected in the assessment of the most appropriate, equitable and sustainable production strategies and practices. Consequently, participatory approaches to joint assessment and decision-making are reviewed to highlight tools and processes with potential to identify and promote options for feasible and equitable sustainable aquaculture development.

Environmental management systems

Aquaculture development can cause negative environmental impacts (see chapter 2), lead to social tension and conflict and impinge on option and non-use values described above. Action is generally taken to avoid and limit such impacts, but often this is not planned and implemented in a systematic manner. Whether prompted by heightened environmental awareness, image-consciousness or regulatory requirements, increasing numbers of commercial aquaculture producers have adopted the established environmental management system (EMS) approach to help minimise environmental impacts and achieve continued performance gains. Senior-level commitment to improved environmental performance is critical to successful establishment of an effective EMS. This is often made manifest with public disclosure of an environmental policy statement (Figure 3.1). Essential elements of this commitment include pollution prevention, full compliance with relevant environmental legislation and demonstrating continued environmental improvement.

Figure 3.1 Decisions and actions central to EMS implementation.

Prevailing EMS strategies evolved from total quality management. This was applied initially to improve the efficiency and quality of industrial processes during World War II and was founded on key principles including teamwork, commitment, communication, organisation, control and monitoring, planning and inventory control systems. Commitment of senior management towards environmental management was seen as critical, and it was expected that responsibility for design and implementation of the environmental policy of a company should rest with someone at board level. Establishment of an environmental committee was intended to organise implementation of the EMS, monitor progress towards environmental improvement and ensure proper lines of communication. Environmental action teams were convened under the direction of the environmental committee and were tasked with examining particular activities in depth; for example, transport, office procedures, paper usage, packaging, storage and waste management. Ultimately, process improvement teams were formed based on the assumption that everybody working on a particular process had responsibility to improve it and to ensure that waste and energy usage were minimised.

Global standards for EMS development and implementation are overseen by the International Organization for Standardization (ISO) with ISO 14001:2004 and ISO 14004:2004 providing requirements and guidelines, respectively. ISO 14001:2004 establishes 'a framework for a holistic, strategic approach' with regards to the 'environmental policy, plans and actions' of an organisation (ISO, 2011). ISO 14000 standards cover a wide range of environmental management topics, including environmental performance evaluation, life-cycle analysis, environmental auditing and communication. An EMS conforming to ISO 14001:2004 should constitute an effective management tool for organisations to identify and control environmental impacts associated with activities, services and products; continually improve environmental performance; and establish a systematic approach to environmental target setting and demonstrating attainment (ISO, 2011).

Whilst larger companies have the resources and administrative support to implement conventional EMS strategies, smaller producers will generally not have such capacity but will endeavour to meet regulatory requirements to safeguard the environment and avoid penalties that may include fines and court action. Producers may well be concerned that action by regulators or evidence of severe impacts may attract damaging media coverage, but negative publicity may tarnish perceptions of the whole sector and affect not only the farm concerned. Essential elements of an EMS scheme for finfish aquaculture start with a preparatory review covering legislative and regulatory requirements; and codes of practice, legislation at international, supranational, national and state levels; and possibly local by-laws concerned with environmental protection and safeguards (Gavine et al., 1996). Key stages in a cyclical EMS process to reduce the impacts of solid waste originating from a marine cage site are depicted in Figure 3.1. Adoption of such an EMS constitutes a major commitment by the organisation, bringing with it additional costs and risks associated with under-performance, as measured by self-imposed objectives and targets that surpass regulatory requirements.

Preparatory measures to implement an EMS include assessing the strengths and weaknesses of procedures for monitoring environmental performance and dealing with non-compliance. Once the relevant environmental policy has been formulated stating the philosophy and aims of the organisation with respect to the environment, it is necessary to address administrative and personnel requirements, most importantly assigning overall responsibility for environmental management to a senior staff member. Objectives and targets should be stated for all issues noted in the environmental policy and an environmental management programme developed specifying the means by which objectives and targets will be met (Figure 3.1).

Operations with possible environmental consequences should have corresponding procedures that limit these impacts detailed in a management manual. Environmental management records must be kept documenting progress and compliance levels and arguably the success of EMS implementation depends on the record-keeping system. As an independent and credible assessment of EMS implementation and performance, environmental management audits will be required, whilst any internal audit should be subject to external auditing. Critical issues to evaluate include compliance with legislation, achievement of targets (e.g. feed conversion ratios), EMS review and revision at specific intervals and compliance with procedures specified in the environmental management manual. Hazards such as severe weather, criminal damage and accidents could result in unintentional breaches in environmental management procedures of aquaculture operations, but measures should be in place to try and restrict such events and minimise their impact.

Responsible planning and management regimes

Aquaculture development planning features prominently in Article 9 – Aquaculture Development of the Code of Conduct for Responsible Fisheries produced by the United Nations' Food and Agriculture Organisation (FAO, 1995). Notably, provision 9.1.3 specifies that 'states should produce and regularly update aquaculture development strategies and plans, as required, to ensure that aquaculture development is ecologically sustainable and to allow the rational use of resources shared by aquaculture and other activities' (FAO, 1995: 23). Given apparent sporadic and inconsistent implementation, the third session of the FAO Committee on Fisheries, Sub-Committee on Aquaculture, held in New Delhi, India, in September 2006, recommended that an expert consultation be held to address the need for further guidance to support implementation. The need for improved 'aquaculture planning and polity development at national and regional levels' was reiterated at the twenty-seventh session of the Committee on Fisheries in Rome, Italy in March 2007 (FAO, 2008: 10).

Consequently, a meeting on 'improving planning and policy development in aquaculture' was convened at the request of the Director General of the FAO in Rome, 26–29 February 2008. Specific aims identified for the consultation were guidance on the 'most appropriate planning methods and policy contents to support the sustainable development of aquaculture' (FAO, 2008: 1). Outcomes of the meeting included outline FAO technical guidelines for improved planning and policymaking and implementation in support of aquaculture development addressing three connected themes: policy formulation processes, policy implementation processes and supporting policy implementation. Specific guidelines elaborated for each theme are presented in Table 3.2.

Recommendations for further work originating from the meeting included 'gathering and compilation of case study information' and 'best practices in policy formulation and implementation' (FAO, 2008: 4). Lessons that might be drawn from the allocation of salmon farming rights and area management plans in Norwegian waters were deliberated during informal discussions. Furthermore, zoning for aquaculture, fisheries and other maritime activities in New Zealand appears to provide an exemplar for responsible approaches to comprehensive planning for marine aquaculture development.

Accepting that aquaculture producers must meet regulatory requirements, they must also be aware of, and responsive to, concerns of food safety and

Table 3.2 Guidelines for improved planning and policy formulation and implementation for aquaculture development

Theme	Guideline
Theme 1: Policy formulation process	1.1: Aquaculture policy should reflect relevant national, regional and international development goals and agreements
	1.2: The aquaculture sector should be enabled to develop optimally and sustainably
	1.3: A legitimate and competent authority should lead the policy development process
	1.4: General policy formulation approaches from other relevant sectors could be adopted and adapted for aquaculture purposes
	1.5: Consultation with stakeholders should be as extensive as possible
	1.6: Policy development based on consensus is desirable
Theme 2: Policy implementation process	2.1: Implementation of policy should be operationalised through a set of well-defined strategies and action plans
Theme 3: Supporting policy implementation	3.1: Effective implementation of aquaculture policy requires systematic coordination, communication and cooperation between institutions, tiers of government, producers and other stakeholders
	3.2: Where possible, decisions should be taken by the lowest level of competent authority according to the principle of subsidiary
	3.3: The development of human and institutional capacity should reflect sectoral needs (e.g. producer, research, management, trade development, regulatory and associated societal levels)
	3.4: In order to effectively implement policy, adequate resources need to be identified and allocated
	3.5: Policy development and implementation should be supported by a suitable legal framework
	3.6: Incentives, where appropriate, should be used to encourage good practice throughout the sector
	3.7: Aquaculture policy implementation should be supported by appropriate research
	3.8: The impact of policy implementation should be monitored and evaluated to ensure that future policy development remains relevant and effective

Source: Developed from FAO (2008).

environmental and social impacts of production expressed by both consumers and the general public. Producer associations, predominantly in developed countries, have been in the vanguard in formulating codes of conduct to encourage responsible production practices to ensure regulatory compliance, to reassure consumers and buyers and to assist farmers in attainment of their goals and aspirations.

A great deal of sentiment casts agricultural producers as villains. In contrast, most farmers consider themselves as custodians and are uniquely sensitised to the interplay and interdependence of production, climate, weather, environment, society and economy. Such inter-connectedness with nature may be questioned, however, as production becomes increasingly intensive, removed and apparently isolated from the natural environment. Regulatory and economic tools have been introduced or proposed that make explicit dependence on the natural environment, notably, the inclusion of carrying capacity evaluations in environmental impact assessments prior to aquaculture development, water rates, pollution taxes, carbon credits and levies on feed ingredients.

Several interrelated issues appear to warrant investment in developing more sustainable aquaculture production systems and practices. These include predictions of stricter discharge consents, greater financial costs being incurred by producers causing negative environmental externalities and stronger buyer and consumer preferences for products demonstrating environmentally sound and socially responsible credentials. Construction and management enhancements that could optimise water-use efficiency, thus potentially reducing water extraction and movement costs and charges associated with waste and nutrient discharges, are presented in Table 3.3. Recognition that nutrients in wastewater constitute a resource could trigger a paradigm shift to productive use and adoption of integrated aquaculture systems, resulting in greater financial returns and a better spread of risks for producers. Where regulation is poorly developed or enforcement lacking, operators may still be willing to adopt such strategies of their own volition.

Market perceptions and environmental attributes constitute important factors that will become increasingly influential in directing future aquaculture industry growth (Young et al., 1999). Effective wastewater management strategies achieved through productive reuse should therefore be of interest to commercial aquaculture producers in more highly regulated settings. Focus group discussions, for example, indicated that existing seafood consumers in the United Kingdom would be willing to pay a premium of 20–30 per cent for products from integrated production systems (Ferguson et al., 2005). Regulators, policymakers and planners could intensify sanctions and moderate incentive schemes to favour more ecologically sound production practices. Consequently, proactive operators employing more efficient production practices and strategies and managing and mitigating environmental problems would find themselves at an advantage. Developed country producers striving for improved environmental performance may be well

Table 3.3 Practices and policies for optimising water use efficiency in aquaculture

Practice and policy area	Measures to optimise water use efficiency
Construction	• optimal pond design and surface area–to-volume ratios can help minimise evaporative losses • employing appropriate pond bottom sealing techniques and liners on free-draining soils can minimise seepage losses • good maintenance of water control structures can reduce leakage • inclusion of simple treatment systems such as sedimentation ponds, constructed wetlands and mechanical filter screens can significantly improve farm discharge water quality
Systems design and operation	• employing aeration and oxygenation can allow higher stocking densities to be maintained and reduce water use per unit biomass production • incorporation of mechanical filters, biofilters and disinfection technology can permit the recirculation of process water, reducing water exchange and use • culturing species with less exacting environmental requirements could help reduce water demand
Optimising production efficiency	• careful stock or broodstock selection and breeding programme management can enhance production efficiency and contribute to optimal water use • optimal feeding strategies, appropriate grading and generally good husbandry and animal welfare can enhance feed conversion rates, thus reducing waste loadings and water exchange rates required to maintain water quality
Water management practices	• draining fishponds can be avoided where, for example, teaseed cake has been applied to partially anaesthetise tilapia prior to netting • sequential partial draining of fishponds, where 25 cm of water is left in ponds and fish are harvested by netting, reduces the discharge of potential pollutants • adopting drop-fill pond water management strategies, where the water level is permitted to fall to a drop point before being made up to a fill level, can potentially reduce groundwater use and effluent release
Horizontally integrated aquaculture systems	• incorporating primary producers to convert waste nutrients discharged from fed culture organisms can improve water quality and make nutrients available to animals in other integrated culture units • integration of different species within a culture system can increase biomass production per unit water appropriated

Water rates and pollution taxes	• statutory limits to the amount of water that may be abstracted or charges per unit volume would help maximise water-use efficiency
	• statutory discharge standards and taxes on the amount of waste fractions discharged from aquaculture would promote more careful husbandry and considered systems design and operation
Policy and planning	• remove subsidies for fuel that make it easier for producers to abstract and discharge larger water volumes
	• facilitate joint assessment of water resources management schemes by all stakeholders and include consideration of full range of ecosystem services in decision-making
	• consider inclusion of water footprints in product labelling and certification schemes to permit buyers and consumers to make educated choices favouring more water-efficient products and production strategies

placed to access marketing benefits and price premiums associated with quality assurance or organic certification. Potential benefits for producers adopting sustainable production practices appear tangible, but there are associated costs and risks that must be accounted for in assessments and decision-making.

Cost–benefit analysis

Rational appraisal approaches are needed to evaluate potential benefits, such as premiums associated with product certification, increased financial returns from product diversification and more efficient resource use, against expenditure on modifying production systems and transaction costs incurred in developing skills and markets and gaining continued certification. Cost–benefit analysis (CBA) is a standard approach to systematically evaluate the economic costs of a proposed programme or initiative and contrast this with the economic benefits associated with anticipated outcomes. Benefit–cost analysis can be used interchangeably to describe the process and might be more acceptable to some stakeholders in selected cases as it could be seen as accentuating the positives.

Calls for greater use of CBA in guiding aquaculture policymaking and development have been widespread. Recommendations emanating from the meeting on improving planning and policy development in aquaculture included 'analyses (including cost/benefit) of the efficiency of policy imple-mentation instruments in specific contexts' (FAO, 2008: 4). CBA theory and practice evolved across a range of sectors, and unfortunately its application is not standardised and is often hampered by the absence of data concerning broader environmental and social costs and benefits that can readily be converted to monetary equivalents.

Limitations to applying CBA to valuation of alternative management options for mangroves in the Philippines, including aquaculture development, have been highlighted (see original paper by Gilbert and Janssen (1998); critique of the paper by Ronnback and Primavera (2000); and original authors' response, Janssen et al. (2000)). Criticisms levelled at the original assessment were a failure to adequately value existing capture fisheries dependent on the mangroves and a perceived overestimate of the value and sustainability of alternative management options involving aquaculture. Consequently, it was concluded that 'if applied to decision-making, the erroneous results from this partial cost–benefit analysis may have dire consequences for the mangroves and coastal communities of Pagbilao' (Ronnback and Primavera, 2000: 135).

Replying to this critique, Janssen et al. (2000: 141) reiterated the original paper's intended focus of 'communication of values, use of valuation in a decision context and the limitations of valuation to prove that mangroves should be preserved'. The intended purpose of the paper was also restated as 'to show the need for ecological information as basis for valuation', and it was noted that often such data are not available. Other problems agreed on included problems of attribution of benefits and costs, limited availability of economic data and selection and interpretation of measures of sustainability for alternative management options. Janssen et al. (2000: 143) concluded by stating that the central point being made was 'not that mangroves should be converted to fishponds, the point is that valuation may not be the most appropriate weapon to prevent this'.

Comparing CBA with environmental impact assessments, Bateman (2007: 156) stated that CBA has 'grown out of the subjective social science of economics' and it may be regarded as 'overtly anthropocentric and produces assessments which reflect a human rather than physical view of a project' as it entails assigning values to the 'socio-economic interpretation' of environmental impacts. Reducing assessments to a single, monetised unit of measurement discounted over time presents particular dangers. Potential long-term benefits of preserving ecosystems may be foregone for short-term financial gains, whilst assessing the total value of alternative management options may mask the uneven distribution of benefits, with poor and marginal groups probably being disadvantaged, especially in developing countries.

Social enterprise and private sector initiatives

Community-based and social enterprise initiatives focused on aquaculture systems have emerged as an interesting and potentially influential counterpart to mainstream commercial aquaculture production. These include a means to produce food locally, generate employment, regenerate rundown urban neighbourhoods and abandoned industrial sites and provide an educational resource for learning and teaching. A pioneering example was the Stensund wastewater aquaculture system commissioned in Sweden as an 'intensive indoor technology built for demonstration, research, and development of recycling through natural processes using constructed food webs' incorporating recycling of 'wastewater resources, including nutrients for aquacultural production' (Guterstam, 1997: 93). Prospects for contemporary urban aquaculture systems designed as multifunctional facilities are critically reviewed in chapter 6.

Trans-national corporations and larger companies and organisations are increasingly reporting on their environmental and social impacts and mitigation measures. Typically, this encompasses ethical conduct; respect for the rights and dignity of employees; relationships, notably engagement with key stakeholders; health and safety; and environmental management. Companies listed on stock markets, including some engaged in aquaculture production, are often obliged by regulatory authorities to make such disclosures. Other private companies and public bodies are beginning to volunteer information on environmental, social and ethical performance to reassure consumers and the general public and to demonstrate their commitment to continued improvement.

Triple bottom line accounting systems and corporate social responsibility reports have been implemented in many multinational companies and international organisations, the former to monitor and assess economic, environmental and social performance and the latter to cover environmental and social performance, routinely published in conjunction with traditional financial accounts. To guide social and environmental assessments, the AA1000APS 2008 standard has been devised to provide organisations with 'an internationally accepted, freely available set of principles to frame and structure the way in which they understand, govern, administer, implement, evaluate and communicate their accountability' (AccountAbility, 2008: 8). Three principles are specified concerning the attainment of accountability:

- inclusivity (foundation principle): include stakeholders to formulate and implement an accountable sustainability strategy;
- materiality: identify and account for issues of significance that influence decision-making, actions and sustainability performance of an organisation and stakeholders;
- responsiveness: respond to stakeholders concerning sustainability performance through decision-making, actions, performance and communication.

Transaction costs and administrative demands associated with quality assurance scheme participation, organic or sustainability certification and formal EMS implementation constitute barriers to adoption for most producers, especially in the developing world. Support could be provided by the state to assist producers, or there might be opportunities for entrepreneurs and local businesses to facilitate efficient scheme implementation and record-keeping and accounting procedures, benefitting both producers and the economy. One approach to spreading costs and risks of entry would be to advocate group formation to share the burden. Similarly, formal organisation and self-help group, cooperative, cluster and resource-user group formation are recommended for small-scale farmers and community members contemplating adopting BMPs, regional aquaculture improvement projects and collective management of natural resources (Bunting et al., 2022; SFP, 2023b). Limitations tempering perceived advantages have been identified, however, and these are reviewed for cooperative shellfish management in Japan (chapter 4) and cooperatives and self-help groups in the East Kolkata Wetlands, India (chapter 6).

Securing the social licence to operate

Societal perception of, and reaction to, aquaculture development can be mixed, and views and opinions can evolve over time to become more supportive or sceptical. Broad acceptance by civil society of the legitimacy of a production facility to function of a firm to continue to trade can be termed the 'social licence to operate' (SLO). Ensuring that day-to-day operations are considerate of the rights and reasonable expectations of neighbours, other resource users and local communities can help secure the SLO. Community-supporting activities (e.g. maintaining and enhancing shared infrastructure and communal areas, contributing time and resources to local charities and fundraising campaigns or providing teaching and learning opportunities for local schools) can help raise awareness of shared environmental and social goals and help build trust and mutual understanding.

Aquaculture development in remote and marginal communities can help boost local economies and help justify continued or enhanced transport links that could benefit residents across a region (Alexander et al., 2014). Production and processing can create employment opportunities in geographies with relatively low population densities yet competitively high unemployment rates. To maximise benefits for local communities, it may be desirable to raise awareness of potential opportunities amongst the resident population and relevant education providers and employment agencies. Large-scale aquaculture development can necessitate the migration of skilled workers, and it may be necessary to conceive and deploy appropriate support measures (Bunting et al., 2023).

It is important to recognise that some groups have legitimate concerns or may oppose aquaculture development on ethical grounds and to consider this in decision-making and in developing contingency plans. Where there is concern that protest groups, for example, could disrupt or try to interfere with legitimate operations, this must be accounted for with appropriate risk assessments and employees informed of their specific roles and responsibilities in this regard. Online profiles and resources could potentially be targeted, and this may demand the specific assessment of cyber-related risks and the development of appropriate data management and protection protocols and safeguards for employees.

Aquaculture development in support of poverty alleviation must promote inclusive business models and gender-sensitive value chains (Kruijssen et al., 2018; Kaminski et al., 2020) and nutrition-sensitive production and food systems that could benefit people in poor and marginal communities, especially women and young children, who stand to gain the most from enhanced food and nutrition security (1,000 days, 2023; Bunting et al., 2023). Functioning and transparent leasehold markets can potentially enable poor and marginal groups to benefit indirectly from aquaculture development (Padiyar et al., 2014; Belton et al., 2017). When there is a risk that poor and marginal groups may be disadvantaged by aquaculture development (e.g. community-based floodplain aquaculture in Bangladesh), appropriate mitigation and benefit-sharing mechanisms can be developed (Toufique and Gregory, 2008). When consolidation or innovation is proposed in supply chains (e.g. diversion of small pelagic fish species and fish processing by-products away from artisanal feed manufactures) or value chains (e.g. utilisation of crustacean shells in biorefinery processes), safeguards may be needed to ensure that poor and marginal groups are not disadvantaged (de la Caba et al., 2019; Thiao and Bunting, 2022).

Participatory and interdisciplinary principles

Attempts to enhance aquatic resources management and aquaculture sustainability are destined to fail unless the prevailing situation is better understood and, crucially, action is guided by joint assessment founded on the perspectives of stakeholders who invariably possess different worldviews. Transition from understanding complex aquatic systems to developing improvements in an iterative, participatory way involving all concerned stakeholders constitutes a major challenge and much-needed innovation. Livelihood implications of modifying access to, or the management of, aquatic resources are difficult to assess, but invoking a sustainable livelihoods framework (see chapter 1) has the potential to guide analysis and facilitate interventions that benefit the poor. Livelihood impacts associated with aquaculture development, or alternative aquatic resources management strategies aimed at reconciling conservation needs with enhanced outcomes for poor and vulnerable groups (notably women and children), for example, need to be jointly assessed with stakeholders using appropriate measures.

Understanding the role of institutions and organisations, both formal and informal, in mediating change in livelihoods is critical (Carney, 1998). Elucidating his influential book, *Ideas for Development*, Robert Chambers (2005: 94) noted 'the tendency for local elites to capture projects and programmes and use them for their own benefit should indeed by recognized as a fact of life' and that 'there are benefits as well as costs in this'. Leaders are potentially valuable allies in development because they find themselves in such positions through demonstrated ability and garnered support within the community. Leaders of political groups and producer associations engaging in participatory planning may need to report back to their members to seek a mandate to proceed with discussions and negotiations, possibly requiring appropriately planned pauses in planning and implementation. Being accountable to voters and members places greater onus on representatives and leaders to ensure that they engage in development to secure benefits and distribute them widely.

Enhancing natural resources management practices demands elements of participatory action research involving households, user groups, communities and local institutions. Researchers may act as facilitators or catalysts and communicate invaluable external knowledge to the process (Chambers et al., 1989; Pretty et al., 1995). Appropriate communication strategies and upscaling activities are, however, needed to promote wider adoption and appropriate adaptation of promising strategies (see chapter 8). Adaptive learning could be invoked as it has been proven to represent an effective approach to enhance productivity of community managed resources (Garaway et al., 2002). Facilitating change in aquatic resources use patterns with complex access rules risks increasing conflicts, whilst needs and assets of poor and vulnerable groups must be understood and accounted for if they are not to be marginalised further (Murray and Little, 2000).

Adopting an interdisciplinary and integrated assessment approach is essential to promote sustainable aquaculture development and wise use of aquatic resources, whilst safeguarding biodiversity and sustaining stocks and flows of ecosystem services benefitting user groups and society. Integration of people from traditionally separate disciplines such as natural scientists, social scientists, aquaculture specialists, conservationists, marketing and institutional experts, lawyers and economists holds promise for more holistic and comprehensive joint assessments. Challenges persist, however, concerning methodological differences in the approaches respective disciplines favour; for example, qualitative versus quantitative approaches, and, most taxing perhaps, the different timescales for assessment, possibly years for repeated household interviews and longitudinal social surveys, compared with a matter of weeks, potentially, for policy and market assessments. Other limitations include discipline-specific terminology and measures that may take time and effort for mutual comprehension and the not inconsiderable practicalities of assembling as well as managing interdisciplinary teams to facilitate fieldwork and review outcomes. Moreover, a major challenge remains: the identification of the most effective and efficient means to analyse and synthesise outcomes arising from interdisciplinary assessments.

International treaties and agreements have called on states to promote and ensure participation, whilst development practitioners and academics have argued for stakeholder participation in development to be considered alongside other human rights. Participation is routinely espoused as the means to achieve equitable, sustainable and widely supported development, but how it is defined and consequently pursued frequently varies between programmes and projects. Engaging people in joint assessment and decision-making is central to interactive participation with the participation of those with an interest or stake in what is being proposed regarded as being critical (Pretty, 1995). Professor Amartya Sen, recipient of the 1998 Nobel Memorial Prize for Economic Science, expounded his theory of 'Development as Freedom' founded on development 'as a process of expanding the real freedoms that people enjoy' including 'political and civil rights' encompassing 'liberty to participate in public discussion and scrutiny' (Sen, 2000: 3). Commitments to guarantee citizens' freedom to participate in joint assessment and decision-making, in turn, necessitate action to enable and empower poor, disadvantaged, marginalised and powerless groups to enjoy this endowment.

Interactive participation and joint assessment

Preliminary approaches devised to facilitate joint assessment within rural community and agricultural development spheres centred on rapid rural appraisal. This encompasses a variety of approaches ranging from individual and household interviews to structured observation and visualisation techniques such as resource mapping and Venn diagrams (Townsley, 1996). Participatory rural appraisal (PRA) has subsequently been adopted as an umbrella term for approaches enshrining interactive stakeholder participation for joint assessment, and it emphasises the need to avoid extractive data collection to serve project purposes. Four categories of participatory approaches to facilitate 'alternative systems of learning and action' were identified by Pretty (1995), specifically addressing sampling methods, interviewing and dialogue, visualisation and diagramming and group and team dynamics. Approaches suited to PRA assessments are presented in Table 3.4 together with examples and a summary of desirable aspects.

The proper identification of stakeholders and measures to ensure their full and fair representation is critical to conducting PRA assessments. Consequently, a range of approaches may be needed to achieve this, and appropriate safeguards and verification steps should be conceived to ensure that groups, in particular poor and marginal groups, are not overlooked or barred from participation, either intentionally or tacitly. Bringing together local stakeholders to jointly assess a situation may be challenging, yet it is feasible; more problematic may be facilitating joint assessment vertically within hierarchical power and institutional structures. As noted by Pretty (1995: 1255), 'A more sustainable agriculture, with all its uncertainties and complexities, cannot be envisaged without a wide range of actors being involved in continuing processes of learning'. Ensuring that internationally accepted and respected guidelines and good practice enshrine the right of stakeholders to participate in joint assessment and decision-making puts the onus on policy and decision-makers to ensure that local assessment outcomes and plans are given credence and find expression. Moreover, innovative approaches to facilitate joint assessments and decision-making vertically with key and primary stakeholders could make a significant contribution to planning and implementing sustainable aquaculture. Prospects for participatory action planning and stakeholder Delphi assessments are critically reviewed in chapter 7.

Gender assessment framework

Growing recognition that agricultural innovation should be directed by farmers themselves led in the 1980s to what has been termed the 'Farmer First' movement. Contemporary advances in the field of development planning saw practitioners being urged to incorporate consideration of the triple role of women in assessments regarding potential impacts of policies and projects (Moser, 1989). Productive, reproductive and community managing work were presented as the three roles women play in development. Consequently, practical and strategic gender needs should be assessed in this context. Arguably, Farmer First theory had a transforming effect on agricultural development and influenced practice in other sectors, including

Table 3.4 Approaches to facilitate interactive participation for sustainable aquaculture development

Approach	Examples	Benefits and opportunities
Structured observation	• transect walks, field walks	• informal approach that is sensitive to practical and social limitations on people attending meetings and provides opportunities for exploration of inter-household dynamics and organisation
Mapping techniques	• resource mapping, activity mapping, historical mapping	• mapping process permits participants to prioritise features and issues, whilst associated discussion adds context and enables deeper understanding
Process and change assessment	• seasonal calendars, timelines, process diagrams, historical maps, oral histories	• effective means to assess change in the absence of written historical accounts; opportunities to assess responses to past shocks; avoid introducing shifting-baselines and develop understanding of prevailing vulnerability context
Visualisation and diagrams	• problem and decision trees, impact diagrams, flowcharts, bar charts, Venn diagrams	• inter-relationships can be made explicit and followed up accordingly with detailed discussion; problems experienced can be traced back to underlying causes, institutional arrangements can be visualised with Venn diagrams and gaps between state infrastructure and grassroots organisations highlighted
Ranking, scoring and classification	• pair-wise ranking, matrix ranking, scoring options and priorities, local classification schemes	• ranking can be undertaken irrespective of numeracy levels; different rankings for groups disaggregated by gender, age, class, education, ethnicity and wealth can be derived easily; priorities identified can inform planning and decision-making
Interviews and group discussion	• semi-structured household and key informant interviews, focus group discussions	• open questioning and discussion guided by interview topic checklists can facilitate in-depth assessment and permit participants to highlight issues and raise concerns; focus groups with small numbers can help overcome barriers to discussions with individuals, facilitate joint assessment with poor and marginal groups and prompt internal verification and validation
Workshops and meetings	• community meetings and workshops, joint stakeholder meetings, participatory action planning workshops	• opportunities for joint assessment from multiple perspectives and feedback and comment on positions, perceptions and plans; shared experiences should lead to greater mutual understanding and reflection on entrenched views and reformulation of obsolete or unfeasible plans

aquaculture development. Despite calls for a paradigm shift coming to the fore in academic circles at a similar time, consideration of gender issues in agricultural and aquaculture development lagged.

Advancing a strategy to support implementation of participatory and integrated policy formulation for the small-scale fisheries sector, Campbell and Townsley (1996) recommended adoption of a gender analysis framework to ensure that gender issues were included in fisheries governance and management decisions. Moser (1989) presented a conceptual gender-planning framework and reviewed the different policy approaches to including women in development planning, but further elaboration and capacity-building would probably be required to implement such an assessment.

A range of possible approaches to jointly assess the needs of different stakeholder groups, disaggregated by gender and age, employing appropriate participatory approaches can be foreseen (see Table 3.4). Focus groups with women and children might be conceived to help participants feel comfortable in discussing issues that may be sensitive. Use of focus groups would also avoid common problems of men and older people coming to dominate discussions in public spaces and joint meetings. Facilitators should be cognisant, however, of possible barriers to interactive participation. They should manage the participatory process and adopt safeguards to ensure that perspectives of missing and marginal groups are actively sought. As Pretty (1995: 1254) noted, 'There are always differences between women and men, between poor and wealthy, between young and old', but those groups most often missed 'are usually the socially marginalized'. Consequently, 'rigorous sampling' is essential to achieve representative and trustworthy participatory approaches (Pretty, 1995: 1255).

Barriers to interactive participation

Despite efforts made to facilitate the interactive participation of stakeholders in joint assessment and decision-making for sustainable aquaculture development, possible constraints are apparent. These include prevailing institutional arrangements, practical problems in implementing appropriate approaches and personal and social barriers (see Table 3.5). Greater awareness of constraints to effective and interactive participation is required. This would lead to the selection of approaches and processes of enquiry, joint assessment and decision-making to ensure that permitted aquaculture development meets the demands and expectation of a broader array of stakeholders. Where policy-makers and natural resources managers foresee tough decisions and hard choices, engaging in interactive participation could constitute a potential means to communicate constructively with interest groups and protagonists. Furthermore, knowledge emanating from the process could be critical in ensuring that appropriate safeguards are implemented to ensure that sensitive habitats and vulnerable groups are adequately protected (de la Caba et al., 2019; Thiao and Bunting, 2022).

Table 3.5 Constraints to interactive participation and possible mitigation strategies

Issue	Constraint	Mitigation strategy
Development programme or project organisation and implementation	• sectoral organisation (education, health, water and sanitation, natural resources, etc.) • predefined focus, agenda, objectives or targets • authorities fear interactive participation will slow planning and implementation, resulting in token participation that undermines the process and engenders greater distrust and alienation	• support cross-sectoral and multidisciplinary activities that can respond to poor people's most pressing needs; conduct a comprehensive needs analysis with communities and use resources targeted by sector to address issues prioritised by selected poor stakeholders • ensure that activities are demand-led and focused on identifiable groups of the poor or target institutions; incorporate a consultative inception phase to define local needs and explain focus and limitations of studies or projects, thus avoiding false expectations • reassure authorities that interactive participation constitutes a more effective approach than more manipulative types; promote transparency and accountability; ensure strong facilitation and full representation (see below); test trustworthiness of findings through extended interaction, critical observation, triangulation and participant checking
Representation and engagement	• people unable to attend meetings due to logistical and social constraints • stakeholders inadvertently missed in the process • views and opinions of poor, powerless and marginal groups subjugated, ignored or not finding expression	• select conveniently situated, politically and institutionally neutral and socially accessible venues or meeting places; schedule activities at convenient times and give sufficient advance warning • conduct preliminary stakeholder analysis to ensure that the process is as inclusive as possible; encourage newly identified stakeholders to engage with ongoing process on equal terms • seek out views of poor, vulnerable and socially excluded or politically marginalised groups using appropriate participatory tools; avoid situations where participants are unduly influenced by social pressures, peers or the powerful
Barriers to participation	• stakeholders barred from participation due to intra-household and social norms or tacit exclusion • prospective participants unwilling to engage due to a lack of trust, fear of conflict or losing control, scepticism or an alternative personal or political agenda	• adopt methods that negate such barriers (e.g. focus groups and household interviews), ensure transparency and strive for full representation (see above) • establish constructive dialogue, raise awareness of process and objectives, provide reassurance, safeguard anonymity, ensure that findings are trustworthy and promote transparency and accountability

Source: Bunting (2010).

Aquatic ecosystem services to society

Ecosystem services derived from wetlands were comprehensively reviewed during the Millennium Ecosystem Assessment (MEA, 2005). The assessment included the relative magnitude of ecosystem services ranging from low to medium to high per unit area from different coastal wetland types, including coral reefs, lagoons and mangroves. Ecosystem services were defined as:

> benefits people obtain from ecosystems. These include provisioning services such as food and water; regulating services such as regulation of floods, drought, land degradation, and disease; supporting services such as soil formation and nutrient cycling; and cultural services such as recreational, spiritual, religious and other non-material benefits.
>
> (MEA, 2005: v)

Ecosystem services result in benefits to society and contribute to the livelihoods and well-being of individuals. Consequently, comprehensive assessment of the range and magnitude of ecosystem services derived from specific habitats and geographical areas will only be possible through drawing on the principles of participation outlined above.

With reference to the comparative assessment, there is little, if any, associated discussion on the influence of variation within wetland type, area, coverage or status on the supply of ecosystem services. Depending on location, the value of mangroves in providing shoreline protection or spiritual renewal, for example, would vary significantly. Ecosystem services supported by degraded and replanted mangroves might be expected to be less than those from a pristine area.

Ecosystem services associated with mangroves and their significance to particular stakeholder groups in Indonesia, Thailand and Vietnam are summarised in Table 3.6. Nuanced assessments of ecosystems services derived from particular ecosystem support areas by specific communities or user groups, and society more generally, are needed to identify critical areas that demand enhanced protection and conservation given the heterogeneous nature of ecological functioning. Ecosystem services assessments across watersheds in upland areas of Asia were conducted to identify where services are produced or stored as well as where they are extracted and the benefits are realised (Bunting et al., 2013b). Water appropriated to support aquaculture production, for example, may originate from sub-catchments much higher in the river basin, and consequently protection of both sub-catchments and connecting waterways is critical to ensure continued supplies (Lund et al., 2014; Bunting et al., 2016).

Ecological footprints and aquaculture

Assessing approaches to the valuation of goods and services derived from natural and semi-natural ecosystems, de Groot et al. (2002) grouped ecosystem functions into four categories (regulating, habitat, production

Table 3.6 Ecosystem services associated with mangroves and significance for stakeholders and aquaculture in Indonesia, Thailand and Vietnam

Ecosystem services	Significance to stakeholders	References		
		Indonesia	*Thailand*	*Vietnam*
Provisioning Food	• fish, shellfish and crustaceans caught and collected *in situ* for food; wild game, fruits and grains; support in-shore and off-shore fisheries; broodstock and juveniles supply aquaculture; honey production	Bosma et al. (2007)[*]	Paphavasit et al. (2004); Dulyapurk et al. (2007)[*]; SEI and KU (2008)[*]	Hong and Tuan (1997); Tang and Hong (1999); Tho and Tri (1999)
Water	• storage and retention of water; supplying marine and brackish water aquaculture	van Zwieten et al. (2006)[*]	Robertson and Phillips (1995)	Xan and Khuong (1999)
Fibre and fuel	• wood production for fuel; logs and fibrous material for construction, thatching and handicrafts; grazing and fodder; detritus supporting aquaculture production	Bosma et al. (2007)[*]	Dulyapurk et al. (2007)[*]; SEI and KU (2008)[*]	Tho and Tri (1999); Hong et al. (2007)
Biochemical	• medicinal plants; extraction of medicines and other materials from biota; fertiliser for agriculture	Bosma et al. (2007)[*]	Dulyapurk et al. (2007)[*]	Hong and Tuan (1997);
Genetic materials	• genes promoting resistance to plant pathogens and diseases in aquaculture; broodstock and individuals with desirable traits to		Panapitukkul et al. (1998); Macintosh et al. (2002); Dulyapurk et al. (2007)	Hong and Dao (2005)[*]; Hong et al. (2007)

(*Continued*)

Table 3.6 (Continued)

Ecosystem services		Significance to stakeholders	References		
			Indonesia	Thailand	Vietnam
		• enhance aquaculture and address emerging needs; new culture species; bio-prospecting opportunities; ornamental species; propagules for rehabilitating degraded areas and establishing new mangrove areas			
Regulatory	Climate regulation	• influence local and regional temperature, precipitation and other climatic processes; source of and sink for greenhouse gases	Bunting and Pretty (2007)*	Bunting and Pretty (2007)*; SEI and KU (2008)	Hong and Tuan (1997); Bunting and Pretty (2007)*
	Water regulation – hydrological flows	• regulation of groundwater recharge and discharge; buffer to flow of saline water inland		SEI and KU (2008)*	
	Water purification and waste treatment	• retention, recovery and removal of excess nutrients and pollutants from domestic, agricultural and aquaculture wastewater; compensation site for biochemical oxygen demand; entrapment and retention of sediments and reduction in turbidity	Bosma et al. (2007)*	Robertson and Phillips (1995); Dulyapurk et al. (2007)*; SEI and KU (2008)*	Xan and Khuong (1999); Wosten et al. (2003)
	Erosion regulation	• sediment deposition and retention; soil formation; shoreline stabilisation; reduced wave heights and current speeds	Bosma et al. (2007)*	Panapitukkul et al. (1998); Dulyapurk et al. (2007)*; SEI and KU (2008)*	Hong and Tuan (1997); Hong (1999)

	Natural hazard regulation	• flood control; storm protection; tsunami mitigation; protection for dykes and embankments inland	Bosma et al. (2007)*	Dulyapurk et al. (2007)*; SEI and KU (2008)*	Hong and Tuan (1997); Hong (1999); Tho and Tri (1999)
	Pollination	• habitat for pollinators; honey production		Dulyapurk et al. (2007)*	Hong and Tuan (1997); Hong et al. (2007)*; Tri (2005)*
Cultural	Spiritual and inspirational	• source of inspiration; religions and ethnic groups attach spiritual and religious values to aspects of wetlands			
	Recreational	• opportunities for recreational activities; e.g. bird watching, fishing, photography, boating; on-site ecotourism; off-site eco-tourism; e.g. watching migratory birds		Dulyapurk et al. (2007)*; SEI and KU (2008)*	Shinji and Haruyoshi (1999)
	Aesthetic	• many people find beauty or aesthetic value in aspects of wetlands		Dulyapurk et al. (2007)*; SEI and KU (2008)*	Tang and Hong (1999)
	Educational	• opportunities for formal and informal education and training; scientific research and hypothesis testing			Hong and Dao (2005)*; Phuong et al. (2007)*
Supporting	Soil formation	• sediment retention and accumulation of organic matter		Wattayakorn (2004); Wattayakorn and Saramul (2004)	
	Nutrient cycling	• storage, recycling, processing and acquisition of nutrients			Wosten et al. (2003)

Source: Adapted from Bunting (2008).

Notes

* EC MANGROVE project outputs.

and information), identified twenty-three ecosystem functions for natural and semi-natural ecosystems and elaborated a wide range of associated environmental goods and services that benefit society. Further elaborating a 'framework for integrated assessment and valuation of ecosystem functions, goods and service', the authors stated that 'translation of ecological complexity (structures and processes) into a more limited number of ecosystem functions' constitutes the 'first step towards a comprehensive assessment of ecosystem goods and services' (de Groot et al., 2002: 394).

Aquaculture appropriates a range of environmental goods and services from supporting ecosystem areas (Beveridge et al., 1997), whilst dependence of a farming practice, business, community or state on environmental goods and services can be expressed as an ecological footprint (Folke et al., 1997; Wackernagel and Rees, 1997). Application of the ecological footprint to aquaculture production assessment was reviewed by Roth et al. (2000: 461), who stated that the 'concept may provide a reasonable visioning tool to demonstrate natural resource dependence of human activities to politicians and the public at large' but owing to 'inherent weaknesses', it 'fails to provide a cohesive analytical tool for management'.

Conceptual and practical challenges are apparent concerning the application of ecological footprint assessment to aquaculture. A brief review concerning the development of ecological footprints for aquaculture is presented below to illustrate an emerging dichotomy in application and interpretation of findings. Ecological footprint assessments for aquaculture have generally focused on the physical area required by the culture facility the capacity of the environment to assimilate excess nitrogen and phosphorus and the ecosystem area required to compensate for oxygen consumption and facilitate carbon sequestration. Application of the ecological footprint methodology to conventional intensive aquaculture has indicated that environmental goods and services are appropriated from relatively large ecosystem areas when compared to the physical area occupied by the culture facility (Larsson et al., 1994; Berg et al., 1996; Folke et al., 1998). Assessing the extent of ecosystem areas supplying feed inputs and clean water, acting as nurseries and necessary to sequester carbon released from shrimp farming in Colombia, Larsson et al. (1994) estimated that an area 35–190 times that of the semi-intensive ponds was required.

Comparisons between intensive and semi-intensive systems based on production per unit area of the physical production system have implied that semi-intensive systems have smaller ecological footprints (Berg et al., 1996; Kautsky et al., 1997; Folke et al., 1998). Application of the ecological footprint methodology in this manner to assess management strategies for aquaculture fails to account, however, for the efficiency of natural resources use. While interpretation on an areal basis helps visualise the ecosystem area appropriated, it does not provide a comparative measure regarding the degree of natural resource use per unit production under different management strategies and between production systems.

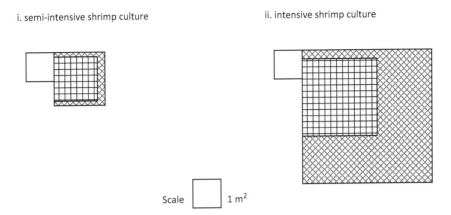

i. semi-intensive shrimp culture   ii. intensive shrimp culture

Scale   1 m^2

Figure 3.2a Ecosystem areas required to assimilate the nitrogen and phosphorus discharged from 1m^2 of (i) semi-intensive and (ii) intensive shrimp culture.

Source: Bunting (2001a).

Although not couched in terms of ecological footprints, Robertson and Phillips (1995) estimated the mangrove area necessary to assimilate excess nitrogen and phosphorus in wastewater and sediments from semi-intensive shrimp farming in Tra Vinh Province, Vietnam, and intensive farming in Chantaburi Province, Thailand. Mangrove areas of 2.5 ha and 3.4 ha were required for nitrogen and phosphorus assimilation from 1 ha of semi-intensive ponds, respectively, increasing to 7.2 ha and 21.7 ha of mangrove forest for nitrogen and phosphorus assimilation from intensive ponds, respectively (Figure 3.2a). Reassessing estimates from Robertson and Phillips (1995) showed that assimilation of nitrogen (547 g kg^{-1}) and phosphorus (69 g kg^{-1}) from semi-intensive shrimp culture would require 25 and 34m^2 per kilogram of shrimp produced, respectively, whereas nitrogen (116 g kg^{-1}) and phosphorus (32 g kg^{-1}) released during production of 1 kg of shrimp from intensively managed ponds would require 5.3 and 16m^2, respectively (Figure 3.2b). Production of 1 kg of shrimp annually in semi-intensive and intensive systems was estimated to require physical areas of 10.2 and 0.7m^2, respectively.

Reinterpretation of outcomes from single assessments of ecological footprints for shrimp farming in Thailand and tilapia culture in Zimbabwe appears to demonstrate that land and water area and nutrient assimilation capacity are used more efficiently during intensive than semi-intensive production. Although the type of system varied with intensity and production, oxygen production capacity was apparently used more efficiently in semi-intensive pond culture of tilapia compared to intensive cage-based culture. Caution is required when drawing lessons from such a small number of comparative assessments, and further work is warranted to evaluate the

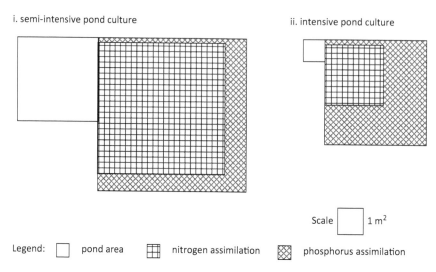

i. semi-intensive pond culture

ii. intensive pond culture

Scale [ ] 1 m²

Legend: [ ] pond area ⊞ nitrogen assimilation ▨ phosphorus assimilation

Figure 3.2b Ecosystem areas to assimilate nitrogen and phosphorus discharged during production of 1 kg of shrimp in (i) semi-intensive and (ii) intensive ponds.

Source: Bunting (2001a).

relative appropriation of environmental goods and services across a broader spectrum of production systems. Moreover, although these assessments are dated, the principle of expressing ecological footprints relative to appropriate product units offers further insights concerning potential husbandry practices (e.g. stocking densities, water exchange rates and feeding regimes) that might make more efficient use of appropriated ecosystem services.

Standards formulated through a series of aquaculture dialogues facilitated by the World Wildlife Fund (WWF) were intended to support the establishment of the Aquaculture Stewardship Council with subsequent product certification and labelling planned to promote responsible production. Similar schemes established for forestry products under the Forestry Stewardship Council and the Marine Stewardship Council for capture fishery landings are widely known. Forestry Stewardship Council certification has been adopted as an industry-wide standard, whereas Marine Stewardship Council certification is less prevalent and generally covers higher-value products and landings from smaller, species-specific fisheries. Standards produced through the WWF dialogues, such as the Pangasius Aquaculture Dialogue Standards, note that 'efficient nutrient utilization may result in less negative impacts on the receiving water bodies' (WWF, 2010: 15) and specify standards for total nitrogen and total phosphorus discharges of 27.5 and 7.2 kg t^{-1} of fish produced, respectively, as opposed to permissible nutrient amounts per unit culture area.

It is often impossible to estimate the extent of ecological footprints without resorting to using data derived from markedly different ecosystems. This is due to an absence of representative data concerning ecological functioning for many ecosystems in specific geographic areas and difficulty in measuring the productivity of ecosystems and their capacity to assimilate waste. Furthermore, inherent variation within ecosystems concerning ecological functioning owing to environmental gradients and patch dynamics combined with seasonal differences make generalisations concerning the capacity for ecosystem services provision meaningless from both scientific and regulatory standpoints. Ecological footprint assessments in isolation are relatively limited in terms of impacts assessed, so other indicators must be considered to evaluate the sustainability of aquaculture systems. Outcomes from ecological footprint assessments might point to potential mismatches between production levels and the carrying capacity of the culture system or ecosystem support area, but monitoring environmental and ecological parameters will be necessary to detect specific impacts on water quality parameters and biodiversity.

Ecological footprints appear, however, to constitute a promising approach to assist stakeholders in visualising the dependence of aquaculture development on environmental goods and services derived from supporting ecosystem areas. Thus, this potentially engenders greater accountability amongst resource users and helps to increase awareness and recognition concerning the need to reconcile multiple demands, constrain notional overlapping footprints and restrict production activities and development and regional aquaculture development to within the capacity of the environment to sustain stocks and flows of ecosystem services.

Environmental carrying capacity

Environmental carrying capacity relates to the amount of aquaculture production that can be sustained by an environment within certain defined criteria (Telfor and Robinson, 2003). Irrespective of production system or environment, Telfor and Robinson noted that estimates of carrying capacity must account for factors determining the productivity of the environment, what the farmed organisms consume and produce, how the environment responds to waste loadings and how much change is permissible. When assessing the potential environmental impact of aquaculture, the degree to which material is concentrated is important (Folke, 1988). Estimated nutrient discharges, for example, may be within the theoretical carrying capacity of the receiving environment but exceed the capacity for nutrient assimilation close to the point of discharge, resulting in localised eutrophication.

Balancing the ecological footprint of an activity with the carrying capacity of the ecosystem is essential to avoid undesirable environmental impacts. A key factor when invoking the principle of ecological footprints is the degree of connectivity that exists between the activity appropriating environmental goods and services and the supporting ecosystem (Bunting, 2001a). Traded goods and services (e.g. feed and fry) derived from other ecosystems can be purchased to augment local supplies, but the sustainability of this will depend on the management and practices of the suppliers. Certain fisheries producing fish meal and oil for aquaculture feed have been certified as sustainable but currently account for a small proportion of the industry. Environmental goods and services that cannot be supplemented will limit the carrying capacity. Therefore, continued dependence of aquaculture on the natural supply of environmental services, particularly the assimilation of waste by local ecosystems, means that appropriation can sometimes exceed the carrying capacity.

Intervention to harness or enhance the capacity of ecosystems to supply environmental goods and services is often referred to as ecological engineering. Marine constructed wetlands planted with mangrove and halophyte species and freshwater ones planted with emergent macrophytes, principally common reed (*Phragmites australis*), have been commissioned in varied settings to treat aquaculture wastewater. Assimilation of waste nutrients by aquatic species cultured in horizontally integrated or integrated multi-trophic aquaculture systems constitutes a further approach to mitigating aquacultural impacts invoking ecological engineering principles (Bunting, 2001b; Troell et al., 2009). Given apparent financial and social benefits and potential reductions in ecological footprints attributed to such strategies, they would seem to constitute a resilient and sustainable aquaculture strategy. Production volumes from such horizontally integrated systems remain insignificant, however, suggesting that constraints to integration remain. If potential environmental, economic and social costs and benefits were assessed more widely in aquaculture development planning and operators brought to account for negative externalities, these may favour integrated production.

Safeguarding environmental stocks and flows

Growing recognition that livelihoods, human well-being and societal systems are dependent on ecosystem services derived from natural and semi-natural ecosystems and agroecological systems must be welcomed. Factoring in the full array of benefits derived from ecosystems in assessments concerning management and development priorities is expected to favour protecting and conserving ecosystems, compared with conversion for other purposes. Dangers are apparent, however, if planning and decision-making were to focus exclusively on the current or potential future values of ecosystem services. Less exploited ecosystems that do not support many livelihoods and generate poor revenues may be left vulnerable, but these may harbour endemic species or populations of global conservation concern.

Exploitation of multiple ecosystem services is not necessarily mutually compatible, and different groups may compete for the same ecosystem services. Means of extraction – for example, large dams impounding irrigation water supplies and for hydroelectric power generation – severely impact stocks and flows of other ecosystem services and devastate both aquatic and riparian ecosystems that are inundated. Acceptance of the concept of ecosystem services without critical reflection threatens to enshrine a dogma that ecosystems generate numerous benefits and that these are readily accessible and should be exploited to maximise benefits to society and meet social and economic development needs. Capture fisheries may be considered sentinel in this regard, as until recently they have been viewed largely as a limitless resource and open to widespread and unregulated exploitation for centuries. Only around 25 per cent of fish-stock groups assessed by the Food and Agriculture Organisation in 2005 were underexploited (3 per cent) or moderately exploited (20 per cent), whilst the majority (52 per cent) were fully exploited, producing catches at or close to maximum sustainable limits with no scope for expansion; the remainder were either overexploited (17 per cent), depleted (7 per cent) or recovering from depletion (1 per cent; FAO, 2007).

Restrictions on the extent of ecosystem services appropriation are critical to ensure that exploitation is sustainable and does not result in disruption to ecological processes that may weaken ecosystem resilience. Flows of ecosystem goods and services necessary to ensure continued ecological functioning are commonly referred to as environmental flows. Similarly, environmental stocks may be regarded as stores, sinks or repositories necessary to sustain and buffer ecosystem functions and maintain resilience.

Aquaculture for poverty alleviation and ecological restoration

Appropriate aquaculture development has the potential to confer a range of health and social–economic benefits on poor and food and nutrition insecure communities (Edwards, 2000; Belton and Azad, 2012; Toufique and Belton, 2014; Belton et al., 2016; Thilsted et al., 2016; Bunting et al., 2023). Large-scale, export-orientated operations can provide livelihoods and income-generating opportunities. For example, commercial shrimp farming development in Bangladesh was credited with empowering women through employment in the processing sector (Shirajee et al., 2013). Ensuring that community members retain access to natural resources on which they may depend for their livelihoods can help avoid disputes and conflicts arising (Bunting et al., 2013a). Well planned and managed aquaculture in low- and middle-income countries can enhance poor livelihoods, promote social–economic resilience and make more efficient use of accessible resources.

Yield gaps, the difference between yields achieved and those that could potentially be obtained for a given production system in a given area, have been noted for several aquaculture sectors (Henriksson et al., 2021). Initiatives to close these could potentially realise modest benefits for many millions of people in countries such as Bangladesh (Bunting et al., 2023). There must, however, be a note of caution as the reasons for apparent yield gaps may be multifaceted, with some producers more concerned with minimising financial and production risks that may be associated with more intensive production modes (Bunting et al., 2013a). Issues of over-intensification have been witnessed in China

(Newton et al., 2021), and it is vital that production by individuals and collectively across regions be maintained with the carrying capacity of supporting ecosystem areas (Bunting, 2001a; SFP, 2023c).

Aquaculture has been promoted in Europe in remote and rural areas to provide employment and compensate for employment loss owing to declines in land-based industries and capture fisheries. Productivity enhancement and mechanisation in sectors such as the Scottish salmon farming sector, for example, have resulted in significant efficiency gains and consequently reduced labour requirements (Muir, 2005). Despite this, the sector is still deemed to make a significant contribution to the Scottish economy and supports a good number of skilled jobs (Alexander et al., 2014).

Waterscape and landscape restoration

Options for wastewater management are severely limited with open culture systems such as cages and pens. Consequently, strategies have been proposed that aim to manage and mitigate nutrient discharges within receiving water bodies. Secondary enclosures stocked with fish and positioned beneath or around cage facilities have been recommended to ameliorate environmental impacts and sequester nutrients. Foraging by grey mullet (*Mugil cephalus*) held in open-bottom cages reportedly improved sediment quality beneath commercial cage-based farms situated in the Gulf of Aqaba just off Eilat, Israel (Porter et al., 1996). Organic matter content and redox potentials returned to levels like those recorded at an undisturbed reference site within seven weeks, whilst the benthic invertebrate assemblage changed from a population comprisin principally small nematodes to a mixed assemblage of nematodes and polychaetes.

Ecosystem enclosures containing unfed bait fish have been proposed as a potential means to intercept fine particles and sequester phosphorus released from trout cages in lakes, whilst the bait fish produced could be harvested for sale or released to enhance local fisheries (Costa-Pierce, 1996). Studies conducted employing indicator particles in feed given to rainbow trout on a farm in Lake Southern Bullaren, Sweden, demonstrated that wild fish, notably bream (*Alburnus alburnus*) and roach (*Rutilus rutilus*), consumed a substantial proportion of the fish faeces emanating from the cages (Johansson et al., 1998). Consequently, it might be envisaged that fish could be stocked in water bodies with the purpose of intercepting particulate waste matter and foraging on primary production that may be stimulated by nutrient discharges. Artificial reefs have been proposed to fulfil a similar role, acting as an enhanced substrate area for colonisation by filter-feeding organisms to sequester nutrients released from cage culture (see chapter 4).

Stocking animals in water bodies to grow largely independently prior to recapture is commonly referred to as ranching and is increasingly being adopted as a strategy to bolster production or appease capture fisheries' interests. Several factors are known to constrain ranching, and notable ecological, environmental and social–economic hazards can be foreseen where it is implemented. Animals stocked for ranching may be predated or caught by fishers, whilst unrestricted movement may result in free-moving animals ranging widely and thus not mitigating local aquaculture development impacts or being accessible for harvest to permanently remove nutrients and contribute to cost recovery and income generation. Invoking optimal foraging theory, it might be concluded, however, that freedom to move would help maximise nutrient assimilation from the ecosystem, and consequently such theoretical concepts should be tested and where appropriate integrated with management strategies.

Containment mechanisms ranging from physical barriers and bubble curtains to behavioural conditioning might assist in restricting animals being ranched to within a reasonable distance from the main culture facility. Management demands and costs associated with such approaches may, however, be prohibitive, and environmental conditions and social and institutional arrangements needed for implementation may severely limit the number of suitable sites. Limitations to curtailing the movement of aquatic animals suggest that species that are broadly regarded as sessile, in particular, shellfish, constitute the best candidates for ranching in association with cage aquaculture to mitigate ecological impacts and sequester nutrients and carbon. Disobligingly, animals stocked to counter benthic impacts and realise potential improvements alluded to by Porter et al. (1996) may, however, avoid heavily impacted sites.

Precautions and safeguards are required to ensure that animals introduced for ranching do not pose unacceptable risks. Where such assessments are constrained by data deficiencies and uncertainty, the precautionary principle should dictate that such interventions be deemed unacceptable. Madeleine Bunting (2009) discussed the various ways in which the English countryside has been conceptualised as a landscape at different times and by different people. This notion can be translated to aquaculture, from which it could be considered that an 'ego-centred' perspective prevails when planners and natural resources managers contemplate waterscapes and landscapes sustaining aquaculture development. Consequently, there is a risk that policies and plans to mitigate the adverse environmental impacts of aquaculture development envisage more ambitious programmes of ecological intervention and manipulation, whilst generating employment and thus revitalising riparian and coastal communities. Allied with technocentric sentiments where human ingenuity is expected to provide solutions to environmental problems and social development needs, there is a danger that efforts will focus on piecemeal management interventions and technology deployment as opposed to fundamental reform of damaging policies and practices.

Where ecosystems are already severely degraded or aquaculture operates in non-natural water bodies, engineered solutions and biomanipulation strategies (see Box 3.1) may be deemed acceptable to rehabilitate and enhance ecosystem services. Replanting mangrove stands to rehabilitate ecosystems services,

---

Box 3.1   Biomanipulation for water quality enhancement

Deliberately stocking species in aquatic ecosystems to induce changes in ecological processes and thus ameliorate negative environmental impacts is a key element of biomanipulation. Approaches to biomanipulation include introducing predatory fish or suppressing zooplanktivores to increase phytoplankton grazing by zooplankton, encouraging filter-feeders to reduce the seston concentration, introducing herbivores to increase the conversion efficiency of primary production and promoting populations of organisms which are readily consumed by fish (Klapper, 1991). Despite several examples where the biomanipulation of fish populations has produced significant improvements in water quality, several constraints to this approach have been identified. Suppressing populations of fish, particularly small species and juveniles, is labour intensive and difficult to maintain (Moss, 1992), and grazing pressure from the increased population of zooplankton may lead to the selection of undesirable phytoplankton species. Furthermore, unless nutrient levels in the water body are permanently reduced, any improvement in water quality will be unsustainable and the ecosystem will tend to revert to its original trophic state.

Source: Bunting (2001b).

---

including services to sustain and support coastal aquaculture, is being widely contemplated. Other strategies being considered for waterscape and landscape regeneration involving aquaculture include culture-based fisheries enhancement programmes, open-water integrated multi-trophic aquaculture, geo-engineering through large-scale algae culture, enhanced aquatic production in farmer-managed systems and integrated peri-urban aquaculture development, but such interventions should be subject to robust and trustworthy environmental and social impact assessments to ensure that negative and unintended consequences are avoided.

An alternative solution to mitigate environmental impacts of cage-based facilities whilst enhancing degraded terrestrial ecosystems involves pumping ashore waste collected under trout cages for the purposes of creating wetland habitats and rehabilitating eroded hillsides (Costa-Pierce, 1996). Potential problems include energy provision for pumping, maintenance demands, capital costs and agreements necessary to permit application of reclaimed waste material to terrestrial habitats, without giving rise to diffuse pollution. Problems have been reported concerning the maintenance of interception devices beneath cages, and consequently such a strategy may be restricted to sheltered sites close to the shore.

Wastewater from traditional land-based marine aquaculture was routinely discharged to coastal wetlands in the belief that ecological processes would treat the water. Nutrient budgets within mangroves and salt marshes, for example, are, however, often finely balanced. Discharging untreated wastewater into natural wetlands can result in elevated nutrient out-welling and disrupt the hydrodynamics of the ecosystem and consequently is generally prohibited. Knowledge is accumulating, however, concerning the design and management of constructed wetlands and mangrove stands planted specifically to treat wastewater and sometimes condition incoming water. Progress in integrating aquaculture in freshwater environments with constructed wetlands is reviewed in chapter 5.

Productive management of sludge, derived from mechanical or biological filters and settlement ponds, through application to agricultural land has been evaluated in various settings. Whilst the practicalities of this pose particular challenges and financial costs may not necessarily be lower than disposal to landfill, additional benefits in terms of enhanced environmental credentials and carbon footprint reductions may assist producers to command premium prices. Application of sediments accumulated within ponds situated in farming landscapes and watersheds to upland areas could contribute to enhanced nutrient and carbon levels in impoverished soils and result in higher yields. Further strategies to better manage stocks and flows of carbon associated with aquaculture are described below.

Aquaculture development within landscapes, seascapes and foodscapes

Freshwater resources globally are under extreme pressure and are fully exploited in many regions, with climate change impacts severely disrupting hydrological regimes and pollution denying people access to safe water sources and negatively impacting biodiversity (United Nations, 2015; Rockstrom et al., 2023). Marine ecosystems face similar problems with pollution and nutrient runoff impacting coastal habitats and many fish-stocks overfished by the large-scale capture fisheries sector. Against this backdrop, Sustainable Development Goal 14 endorsed by the United Nations calls on states to 'conserve and sustainable use the oceans, seas and marine resources for sustainable development' (United Nations, 2015). Appropriate aquaculture development in this context could potentially help restore aquatic ecosystems and fish populations and relieve pressure on overexploited capture fisheries. Globally, increased production from selected aquaculture systems could contribute to enhanced supplies of nutritious and low-impact aquatic foods to contribute to meeting the rapidly growing need of the human population for healthy and sustainable diets (Willett et al., 2019).

Governance structures and regulations will need to be responsive to these pressing needs and enable the optimal use of accessible aquatic resources. Demarcation of areas or zones suited to aquaculture development can assist producers in identifying available locations where supportive planning procedures and regulations exist and help authorities target investments for support infrastructure and services. Development of regional aquaculture improved initiatives, such as those for shrimp producers in Indonesia (SFP, 2023a), and disease management plans with real-time monitoring (Bostock et al., 2010) could enable sectors elsewhere to grow in a responsible and sustainable manner (Bunting et al., 2022). Where aquaculture development occurs spontaneously over large areas or in shared waterbodies, this can lead to undesirable and potentially damaging cumulative impacts. In response to, and ideally in anticipation of this, it may be necessary to determine the potential carrying capacity of the supporting ecosystem areas and restrict sector growth accordingly. When it is not possible to reconcile the needs of aquaculture with other resource users, it may be necessary to outlaw some practices and pursue alternative production strategies (Newton et al., 2021).

Considering aquaculture development within landscapes (see Ezban (2020) for an excellent exploration of this topic), waterscapes or seascapes, it may be necessary to account for and reconcile the demands of multiple users. Agricultural irrigation, capture fisheries, hydroelectric power generation, mining of aggregates, navigation and recreation might typically need to coexist with aquaculture development. The confluence of competing demands or interests concerning natural resources is often termed a nexus. Effective communication and governance arrangements are fundamental requirements in managing what can be highly challenging situations (Lynch et al., 2019). A common understanding of the resource use dilemmas in question, clear and jointly developed management plans and culturally appropriate mechanisms to resolve conflicts in a transparent and timely manner are crucial (Bunting et al., 2016; Islam et al., 2016; Mohammed et al., 2018).

Aquaculture consists of highly diverse production systems, including those such as culture-based fisheries and stock enhancement programmes, that result in a continuum of practice from closed life-cycle culture to inland and marine capture fisheries (see chapter 1). When contemplating the release of aquatic animals or algae to the wild or shared waterways, it is vital that local laws and regulations be respected and international guidelines and best practices followed. When uncertainty exists, the precautionary principle must be respected, with no releases made. Challenges concerning the establishment of ownership of such stocks are addressed in chapter 4.

Climate change adaptation and mitigation

Management decisions and production processes employed can significantly influence the carbon footprint of produce from different aquaculture systems (Bunting and Pretty, 2007; Bunting, 2014a, 2014b). Potential emission reduction strategies for farm-level intervention encompass capital expenditure on more efficient machinery, instilling behaviour change in staff and workers, optimising production efficiency and investment in on-site renewable energy generation. These are summarised in Table 3.7. By deciding to eat aquaculture

Table 3.7 Farm-centred strategies to reduce carbon emissions and enhance sinks

Strategy	Options	Practical actions
Reduced emissions	• reduce energy use and conserve fuel	• reduce machinery use; utilise energy-efficient lighting, pumps and machinery; avoid waste through raised awareness amongst workforce and staff training; better command and control; fit insulation
		• ensure good site selection; reduce stocking densities; improve stock fitness; optimise feeding; minimise disease, pest and predator problems
	• adopt less energy-demanding, more resource-efficient culture practices	• solar, wind, geothermal, water, tide, wave, biomass
	• invest in on-site power generation from renewable sources	
	• switch to renewable energy supplies	• convert machinery, boats, vehicles to run on renewable energy sources; e.g. bio-diesel; switch to renewable energy supplier
	• cultivate biomass crops using waste resources for on-site substitution	• constructed wetlands to treat and reuse wastewater to cultivate biomass crops; cultivate additional crops as bio-fuels
	• reduce inorganic fertiliser inputs	• use waste resources and green manures; optimise fertiliser use efficiency
	• enhanced waste management to avoid greenhouse gas emissions	• digest waste, using methane produced for energy production
	• manage water and soil to reduce carbon and greenhouse gas emissions	• treat wastewater to retain carbon, manage soils to reduce mineralisation, capture carbon in integrated farming systems
Enhanced sinks	• enhance soil organic matter conservation	• reduce organic matter mineralisation; utilise accumulated sediments to improve degraded or poor soils
	• cultivate green manures, leguminous and oilseed crops on-site for fertiliser and feed substitutes	• cultivate *Azolla* spp., duckweed or fodder plants to enhance carbon and nitrogen fixation and use as fertiliser and feed inputs

(Continued)

Table 3.7 (Continued)

Strategy	Options	Practical actions
	• promote *in situ* primary production to stimulate production, waste processing and carbon sequestration • increase tree cover and cultivate biomass crops • restore and protect wetlands • adopt horizontally integrated production systems • convert aquatic farming areas to wetlands and woodlands	• optimise management regimes to stimulate and sustain phytoplankton and periphyton production • plant trees or cultivate short-rotation coppice on embankments and bare land; cultivate biomass crops (reed, willow, mangrove, halophytes) in constructed wetlands • avoid unnecessary development in natural wetlands; mitigate unavoidable wetland loss; restore degraded wetlands • enhance carbon assimilation through adopting horizontally integrated production practices • convert unproductive aquatic farming areas back to wetlands (mangroves, salt marshes, floodplains, riparian wetlands) or woodlands, supporting conservation, mitigation and sustainable livelihoods and production

Source: Bunting and Pretty (2007).

products as opposed to foods with higher greenhouse gas emissions (e.g. meat from terrestrial ruminant livestock), people could reduce their carbon footprint (Crona et al., 2023). With a focus on aquaculture, transitioning to greater culture of omnivorous fish species and filter-feeding organisms could help the sector reduce its carbon footprint (Gephart et al., 2021).

Aquaculture processes and systems have the potential to sequester carbon. Mangrove–shrimp culture, seaweed cultivation and integrated multi-trophic aquaculture systems can sequester blue carbon (Ahmed et al., 2017). Mangrove stands associated with aquaculture systems can provide an array of ecosystem services, notably protection of coastal communities from high winds and storm surges, and this could become increasingly important as predicted climate change impacts worsen (Bosma et al., 2012; Bunting et al., 2013a). Rice–prawn–fish and rice–shrimp–fish agroecosystems developed by farmers in coastal zones produce both cash crops, which are important to economic development, and complementary crops of diverse aquatic foods, fruit and vegetables that contribute to food and nutrition security locally and domestically (Bunting et al., 2017; Mamun, 2016). Development of integrated farming systems in appropriate locations across coastal zones, mega deltas and inland watersheds could contribute to social–ecological resilience and potentially facilitate community-based climate change adaptation (Ahmed et al., 2014; Bunting et al., 2015, 2017; IPCC, 2019). Appropriate redeployment of sediments from ponds and sludge from settlement basins and filters to rehabilitate habitats and enhance soil organic matter in degraded agricultural land could lead to significant increases in carbon storage.

Summary

Poorly planned and managed aquaculture can result in tension and conflict over aquatic resources and have negative social consequences. However, opportunities can be identified concerning approaches to reconcile the multiple demands associated with aquaculture development. Adoption of formal strategies such as EMS could lead to continual improvements in environmental performance, but this may not be appropriate for smaller or poorer producers, especially in developing countries. Consequently, enhanced aquaculture sector planning is needed in many situations to ensure that policies, regulation, monitoring regimes and support to the sector promote responsible production and avoid negative environmental impacts. International organisations, governments and non-governmental organisations have promoted responsible aquaculture development, but interactive stakeholder participation could be critical to implementing appropriate production systems and practices. Measures are necessary to ensure that such processes are equitable, transparent and trustworthy. Accepting that aquaculture production is dependent on supporting ecosystem services and that the extent of development should be governed by the carrying capacity of the environment, where provision is made to maintain environmental stocks and flows is crucial. Aquaculture development can contribute to poverty alleviation, improved nutrition amongst local communities and enhanced food security, but there is a danger that benefits will accrue to, and be captured by, more wealthy and powerful people and groups. The potential of aquaculture to contribute to ecological restoration and facilitate climate change adaptation and mitigation has been described but demands further assessment.

Notes

1 Section adapted from Bunting (2001b) with contemporary references and edits to conform to publishing house style and conventions and ensure coherence with previous text.
2 Section adapted from Bunting (2001b) with contemporary references and edits to conform to publishing house style and conventions and ensure coherence with previous text.
3 Section adapted from Bunting (2001b) with contemporary references and edits to conform to publishing house style and conventions and ensure coherence with previous text.
4 Folke et al. (1994) noted that one U.S. dollar (US$) was equivalent to around six Swedish krona (SEK).

References

1,000 Days (2023) Why 1,000 Days, https://thousanddays.org/why-1000-days/

AccountAbility (2008) AA1000 Accountability Principles Standard 2008, Account-Ability, London.

Ahmed, N., Bunting, S.W., Glaser, M., Flaherty, M.S. and Diana, J.S. (2017) 'Can greening of aquaculture sequester blue carbon?', Ambio, vol 46, no 4, pp. 468–477.

Ahmed, N., Bunting, S.W., Rahman, S. and Garforth, C.J. (2014) 'Community-based climate change adaptation strategies for integrated prawn–fish–rice farming in Bangladesh to promote social–ecological resilience', Reviews in Aquaculture, vol 6, pp. 20–35.

Alexander, K.A., Gatward, I., Parker, A., Black, K., Boardman, A., Potts, T. and Thomson, E. (2014) An Assessment of the Benefits to Scotland of Aquaculture: Research Summary, Marine Scotland, Edinburgh, Scotland.

Bateman, I.J. (2007) 'Valuing preferences regarding environmental change' in J. Pretty, A.S. Ball, T. Benton, J.S. Guivant, D.R. Lee, D. Orr, M.J. Pfeffer and H. Ward (eds) The SAGE Handbook of Environment and Society, SAGE Publications, London.

Belton, B. and Azad, A. (2012) 'The characteristics and status of pond aquaculture in Bangladesh', Aquaculture, vol 358–359, pp. 196–204.

Belton, B., Bush, S. and Little, D. (2016) 'Aquaculture: are farmed fish just for the wealthy?', Nature, vol 538, no 7624, p. 171.

Belton, B., Padiyar, A., Ravibabu, G. and Gopal Rao, K. (2017) 'Boom and bust in Andhra Pradesh: development and transformation in India's domestic aquaculture value chain', Aquaculture, vol 470, pp. 196–206.

Berg, H., Michelsen, P., Troell, M., Folke, C. and Kautsky, N. (1996) 'Managing aquaculture for sustainability in tropical Lake Kariba, Zimbabwe', Ecological Economics, vol 18, pp. 141–159.

Beveridge, M.C.M. (2004) Cage Aquaculture, third edition, Blackwell Publishing, Oxford, UK.

Beveridge, M.C.M., Phillips, M.J. and Macintosh, D.J. (1997) 'Aquaculture and the environment: the supply and demand for environmental goods and services by Asian aquaculture and the implications for sustainability', Aquaculture Research, vol 28, pp. 797–807.

Black, K.D., Ezzi, I.A., Kiemer, M.C.B. and Wallace, A.J. (1994) 'Preliminary evaluation of the effects of long-term periodic sublethal exposure to hydrogen sulphide on the health of Atlantic salmon (*Salmo salar* L)', Journal of Applied Ichthyology, vol 10, pp. 362–367.

Bosma, R., Sidik, A.S., Sugiharto, E., Fitriyana Budiarsa, A.A., Sumoharjo, et al. (2007) Situation of the Mangrove Ecosystem and the Related Community Livelihoods in Muara Badak, Mahakam Delta, East Kalimantan, Indonesia, Mulawarman University, Samarinda, East Kalimantan, Indonesia.

Bosma, R.H., Tendencia, E.A. and Bunting, S.W. (2012) 'Financial feasibility of green-water shrimp farming associated with mangrove compared to extensive shrimp culture in the Mahakam Delta, Indonesia', Asian Fisheries Science, vol 25, no 3, pp. 258–269.

Bostock, J., McAndrew, B., Richards, R., Jauncey, K., Telfer, T., Lorenzen, K., et al. (2010) 'Aquaculture: global status and trends', Philosophical Transactions of The Royal Society, vol B365, pp. 2897–2912.

Bunting, M. (2009) The Plot: A Biography of My Father's English Acre, Granta Publications, London.

Bunting, S.W. (2001a) 'Appropriation of environmental goods and services by aquaculture: a re-assessment employing the ecological footprint methodology and implications for horizontal integration', Aquaculture Research, vol 32, pp. 605–609.

Bunting, S.W. (2001b) A Design and Management Approach for Horizontally Integrated Aquaculture Systems, PhD thesis, Institute of Aquaculture, University of Stirling, UK.

Bunting, S.W. (2008) Ecosystem Processes, Functions and Ecosystem Services as Potential Indicators for Participatory Monitoring of Mangroves and Associated Coastal Areas, University of Essex, Centre for Environment and Society Working Paper 2008-1, Colchester, UK.

Bunting, S.W. (2010) 'Assessing the stakeholder Delphi for facilitating interactive participation and consensus building for sustainable aquaculture development', Society and Natural Resources, vol 23, pp. 758–775.

Bunting, S.W. (2014a) Stakeholder Delphi Assessment of UK Marine Aquaculture Sector Carbon Footprint Management, Final Report for The Crown Estate, University of Essex, Colchester, UK.

Bunting, S.W. (2014b) UK Marine Aquaculture Sector Carbon Budgeting, Final Report for The Crown Estate, University of Essex, Colchester, UK.

Bunting, S.W., Bosma, R.H., van Zwieten, P.A.M. and Sidik, A.S. (2013a) 'Bioeconomic modeling of shrimp aquaculture strategies for the Mahakam Delta, Indonesia', Aquaculture Economics and Management, vol 17, pp. 51–70.

Bunting, S.W., Bostock, J., Leschen, W. and Little, D.C. (2023) 'Evaluating the potential of innovations across aquaculture product value chains for poverty alleviation in Bangladesh and India', Frontiers in Aquaculture, vol 2, no 3, pp. 1–23.

Bunting, S.W. and Edwards, P. (2018) 'Global prospects for safe wastewater reuse through aquaculture' in B.B. Jana, R.N. Mandal and P. Jayasankar (eds) Wastewater Management through Aquaculture, Springer, New York.

Bunting, S.W., Kundu, N. and Ahmed, N. (2017) 'Evaluating the contribution of diversified shrimp–rice agroecosystems in Bangladesh and West Bengal, India to social–ecological resilience', Ocean and Coastal Management, vol 148, pp. 63–74.

Bunting, S.W., Luo, S., Cai, K., Kundu, N., Lund, S., Mishra, R., et al. (2016) 'Integrated action planning for biodiversity conservation and sustainable use of highland aquatic resources: evaluating outcomes for the Beijiang River, China', Journal of Environmental Planning and Management, vol 59, pp. 1580–1609.

Bunting, S.W., Mishra, R., Smith, K.G. and Ray, D. (2015) 'Evaluating sustainable intensification and diversification options for agriculture-based livelihoods within an aquatic biodiversity conservation context in Buxa, West Bengal, India', International Journal of Agricultural Sustainability, vol 13, pp. 275–294.

Bunting, S., Pounds, A., Immink, A., Zacarias, S., Bulcock, P., Murray, F. and Auchterlonie, N. (2022) The Road to Sustainable Aquaculture: On Current Knowledge and Priorities for Responsible Growth, World Economic Forum, Cologny, Switzerland.

Bunting, S.W. and Pretty, J. (2007) Global Carbon Budgets and Aquaculture: Emissions, Sequestration and Management Options, University of Essex, Centre for Environment and Society Occasional Paper 2007-1, Colchester, UK.

Bunting, S.W., Smith, K.G., Lund, S., Bimbao, M.A.P., Punch, S.V., Saucelo, M.D. and Santos, D.A. (2013b) Wetland Resources Action Planning (WRAP) Toolkit: An Integrated Action Planning Toolkit to Conserve Aquatic Resources and Biodiversity by Promoting Sustainable Use, FishBase Information and Research Group, Philippines.

Burbridge, P.R. (1994) 'Integrated planning and management of freshwater habitats, including wetlands', Hydrobiologia, vol 285, pp. 311–322.

Campbell, J. and Townsley, P. (1996) Participatory and Integrated Policy: A Framework for Small-Scale Fisheries, IMM, Exeter, UK.

Carney, D. (1998) Sustainable Rural Livelihoods: What Contribution Can We Make?, DFID, London.

Chambers, R. (2005) Ideas for Development, Earthscan, London.

Chambers, R., Pace, A. and Thrupp, L.A. (1989) Farmer First: Farmer Innovation and Agricultural Research, IT Publications, London.

Corea, A., Johnstone, R., Jayasinghe, J., Ekaratne, S. and Jayawardene, K. (1998) 'Self-pollution: a major threat to the prawn farming industry in Sri Lanka', Ambio, vol 27, pp. 662–668.

Costa-Pierce, B.A. (1996) 'Environmental impacts of nutrients from aquaculture: towards the evolution of sustainable aquaculture systems' in D.J. Baird, M.C.M. Beveridge, L.A. Kelly and J.F. Muir (eds) Aquaculture and Water Resource Management, Blackwell Science, Oxford, UK.

Crona, B.I., Wassénius, E., Jonell, M., Koehn, J.Z., Short, R., Tigchelaar, M., et al. (2023) 'Four ways blue foods can help achieve food system ambitions across nations', Nature, vol 616, pp. 104–112.

de Groot, R.S., Wilson, M.A. and Boumans, R.M.J. (2002) 'A typology for the classification, description and valuation of ecosystem functions, goods and services', Ecological Economics, vol 41, pp. 393–408.

de la Caba, K., Guerrero, P., Trang, T., Cruz-Romero, M.C., Kerry, J., Fluhr, J., et al. (2019) 'From seafood waste to active seafood packaging: an emerging opportunity of the circular economy', Journal of Cleaner Production, vol 208, pp. 86–98.

Dulyapurk, V., Taparhudee, W., Yoonpundh, R. and Jumnongsong, S. (2007) Multidisciplinary Situation Appraisal of Mangrove Ecosystems in Thailand, Faculty of Fisheries, Kasetsart University, Bangkok, Thailand.

Edwards, P. (2000) Aquaculture, Poverty Impacts and Livelihoods, Overseas Development Institute, Natural Resources Perspectives No. 56, London.

Ezban, M. (2020) Aquaculture Landscapes: Fish Farms and the Public Realm, Routledge, Abingdon, UK.

FAO (1995) Code of Conduct for Responsible Fisheries, FAO, Rome.

FAO (2007) The State of World Fisheries and Aquaculture 2006, FAO, Rome.

FAO (2008) Report of the Expert Consultation on Improving Planning and Policy Development in Aquaculture, FAO Fisheries Report No. 858, FAO, Rome.

Ferguson, P., Stone, T. and Young, J.A. (2005) Consumer Perceptions of Aquatic Products to Be Produced from GENESIS Integrated Systems: The UK Perspective, Department of Marketing and Stirling Aquaculture, University of Stirling, UK.

Folke, C. (1988) 'Energy economy of salmon aquaculture in the Baltic Sea', Environmental Management, vol 12, pp. 525–537.

Folke, C., Kautsky, N., Berg, H., Jansson, A. and Troell, M. (1998) 'The ecological footprint concept for sustainable seafood production: a review', Ecological Applications, vol 8, pp. 63–71.

Folke, C., Kautsky, N. and Troell, M. (1994) 'The costs of eutrophication from salmon farming: implications for policy', Journal of Environmental Management, vol 40, pp. 173–182.

Folke, C., Kautsky, N. and Troell, M. (1997) 'Salmon farming in context: response to Black et al.', Journal of Environmental Management, vol 50, pp. 95–103.

Garaway, C., Arthur, R. and Lorenzen, K. (2002) Adaptive Learning Approaches to Fisheries Enhancements, Final Technical Report, MRAG Ltd, DFID Project R7335, London.

Gavine, F.M., Rennis, D.S. and Windmill, D. (1996) 'Implementing environmental management systems in the finfish aquaculture industry', Water and Environment Journal, vol 10, pp. 341–347.

Gephart, J.A., Henriksson, P.J.G., Parker, R.W.R., Shepon, A., Gorospe, K.D., Bergman, K., et al. (2021) 'Environmental performance of blue foods', Nature, vol 597, no 7876, pp. 360–365.

Gilbert, A.J. and Janssen, R. (1998) 'Use of environmental functions to communicate the values of a mangrove ecosystem under different management regimes', Ecological Economics, vol 25, pp. 323–346.

Gillibrand, P.A., Turrell, W.R., Moore, D.C. and Adams, R.D. (1996) 'Bottom water stagnation and oxygen depletion in a Scottish sea loch', Estuaries, Coastal and Shelf Science, vol 43, pp. 217–235.

Guterstam, B. (1997) 'Ecological engineering for wastewater treatment: theoretical foundations and practical realities' in C. Etnier and B. Guterstam (eds) Ecological Engineering for Wastewater Treatment, CRC Press, Boca Raton, FL.

Henriksson, P.J.G., Troell, M., Banks, L.K., Belton, B., Beveridge, M.C.M., Klinger, D.H., et al. (2021) 'Interventions for improving the productivity and environmental performance of global aquaculture for future food security', One Earth, vol 4, pp. 1220–1232.

Hong, P.N. (1999) 'The role of mangroves to sea dykes and in the control of natural disasters' in Proceedings of the National Workshop on Sustainable and Economically

Efficient Utilization of Natural Resources in Mangrove Ecosystem, Nha Trang City, 1–3 November 1998, CRES, VNU, Hanoi and ACTMANG, Japan.

Hong, P.N. and Dao, Q.T.Q. (2005) 'Mangroves in Vietnam', Presentation at the 1st MANGROVE Project Inception Meeting, 9–11 November 2005, Bangkok, Thailand.

Hong, P.N. and Tuan, M.S. (1997) 'Interaction between mangrove ecosystem and coastal aquaculture' in Proceedings of the National Workshop on the Relationship between Mangrove Reforestation and Coastal Aquaculture in Vietnam, Hue City, 31st October to 2nd November 1996, CRES, Hanoi.

Hong, P.N., Tri, N.H., Tuan, L.X., Dao, P.T.A., Dao, Q.T.Q., Tho, N.H., et al. (2007) Situation of the Mangrove Ecosystem and Related Community Livelihoods in Tien Hai, Thai Binh, Vietnam, Vietnam National University, Hanoi.

IPCC (2019) The Ocean and Cryosphere in a Changing Climate, a Special Report of the Intergovernmental Panel on Climate Change, Cambridge University Press, Cambridge, UK, and New York.

Islam, M.M., Mohammed, E.Y. and Ali, L. (2016) 'Economic incentives for sustainable hilsa fishing in Bangladesh: an analysis of the legal and institutional framework', Marine Policy, vol 68, pp. 8–22.

ISO (2011) ISO 14000 Essentials, International Organization for Standardization, Geneva, Switzerland, www.iso.org/iso/iso_14000_essentials

Janssen, R., Gilbert, A. and Padilla, J. (2000) 'Reply to Ronnbach and Primavera', Ecological Economics, vol 35, pp. 141–143.

Johansson, T., Hakanson, L., Borum, K. and Persson, J. (1998) 'Direct flows of phosphorus and suspended matter from a fish farm to wild fish in Lake Southern Bullaren, Sweden', Aquacultural Engineering, vol 17, pp. 111–137.

Kaminski, A. Kruijssen, F., Cole, S.M., Beveridge, M.C.M., Dawson, C., Mohan, C.V., et al. (2020) 'A review of inclusive business models and their application in aquaculture development', Reviews in Aquaculture, vol 12, no 3, pp. 1881–1902.

Kautsky, N., Berg, H., Folke, C., Larsson, J. and Troell, M. (1997) 'Ecological footprint for assessment of resource use and development limitations in shrimp and tilapia aquaculture', Aquaculture Research, vol 28, pp. 753–766.

Klapper, H. (1991) Control of Eutrophication in Inland Waters, Ellis Horwood, New York.

Kruijssen, F., McDougall, C.L. and van Asseldonk, I.J. (2018) 'Gender and aquaculture value chains: a review of key issues and implications for research', Aquaculture, vol 493, pp. 328–337.

Larsson, J., Folke, C. and Kautsky, N. (1994) 'Ecological limitations and appropriation of ecosystem support by shrimp farming in Colombia', Environmental Management, vol 18, pp. 663–676.

Lumb, C.M. (1989) 'Self-pollution by Scottish fish-farms?', Marine Pollution Bulletin, vol 20, pp. 375–379.

Lund, S., Banta, G.T. and Bunting, S.W. (2014) 'Applying stakeholder Delphi techniques for planning sustainable use of aquatic resources: experiences from upland China, India and Vietnam', Sustainability of Water Quality and Ecology, vol 3, pp. 14–24.

Lynch, A.J., Baumgartner, L.J., Boys, C.A., Conallin, J., Cowx, I.G., Finlayson, C.M., et al. (2019) 'Speaking the same language: can the sustainable development goals translate the needs of inland fisheries into irrigation decisions?', Marine and Freshwater Research, vol 70, pp. 1211–1228.

Macintosh, D.J., Aston, E.C. and Havanon, S. (2002) 'Mangrove rehabilitation and intertidal biodiversity: a study in the Ranong mangrove ecosystem, Thailand', Estuarine, Coastal and Shelf Science, vol 55, pp. 331–345.

Mamun, A.A. (2016) Shrimp–Prawn Farming in Bangladesh: Impacts on Livelihoods, Food and Nutritional Security, PhD thesis, University of Stirling, UK.

MEA (2005) Ecosystems and Human Well-Being: Wetlands and Water Synthesis, Millennium Ecosystem Assessment, World Resources Institute, Washington, DC.

Mohammed, E.Y., Steinbach, D. and Steele, P. (2018) 'Fiscal reforms for sustainable marine fisheries governance: delivering the SDGs and ensuring no one is left behind', Marine Policy, vol 93, pp. 262–270.

Moser, C.O.N. (1989) 'Gender planning in the third world: meeting practical and strategic gender needs', World Development, vol 17, pp. 1799–1825.

Moss, B. (1992) 'The scope for biomanipulation for improving water quality' in D.W. Sutcliffe and J.G. Jones (eds) Eutrophication: Research and Application to Water Supply, Freshwater Biological Association, Ambleside, UK.

Muir, J.F. (2005) 'Managing to harvest? Perspectives on the potential of aquaculture', Philosophical Transactions of the Royal Society B, vol 360, pp. 191–218.

Muir, J.F., Brugere, C., Young, J.A. and Stewart, J.A. (1999) 'The solution to pollution? The value and limitations of environmental economics in guiding aquaculture development', Aquaculture Economics and Management, vol 3, pp. 43–57.

Muluk, C. and Bailey, C. (1996) 'Social and environmental impacts of coastal aquaculture in Indonesia' in C. Bailey, S. Jentoft and P. Sinclair (eds) Aquaculture Development: Social Dimensions of an Emerging Industry, Westview Press, Boulder, CO.

Murray, F. and Little, D.C. (2000) Seasonal Tank Management in NW Sri Lanka, DFID AFGRP Working Paper, Institute of Aquaculture, University of Stirling, UK.

NCC (1990) Fish Farming and the Scottish Freshwater Environment, Nature Conservancy Council, Edinburgh, UK.

Newton, R., Zhang, W., Xian, Z., McAdam, B. and Little, D.C. (2021) 'Intensification, regulation and diversification: the changing face of inland aquaculture in China', Ambio, 50, no 9, pp. 1739–1756.

Padiyar, A., Rao, G., Ravibabu, G. and Belton, B. (2014) The Status of Freshwater Aquaculture Development in Andhra Pradesh, WorldFish, Penang, Malaysia.

Panapitukkul, N., Duarte, C.M., Thampanya, U., Kheowvongsri, P., Srichai, N., Geertz-Hansen, O., et al. (1998) 'Mangrove colonization: mangrove progression over the growing Pak Phanang (SE Thailand) mud flat', Estuarine, Coastal and Shelf Science, vol 47, pp. 51–61.

Paphavasit, N., Termvicharkorn, A., Piumsomboon, A., Sivaipram, I., Somkleeb, N., Tongnunui, P. and Pannarak, P. (2004) 'Fish communities in mangrove plantation in Pak Phanang Bay, Nakhon Si Thammarat Province' in S. Aksornkoae, N. Papawasit, S. Angsupanich, K. Wattayakorn, S. Suwannodom and A. Siwaiphram (eds) Integrated Management of Mangrove Plantations for Development of Coastal Resources and Environment of Thailand, Thailand Research Fund. (in Thai)

Phillips, M.J., Lin, C.K. and Beveridge, M.C.M. (1993) 'Shrimp culture and the environment: lessons from the world's most rapidly expanding warm water aquaculture sector' in R.S.V. Pullin, H. Rosenthal, H.and J.L. Maclean (eds) Environment and Aquaculture in Developing Countries, International Centre for Living Aquatic Resources Management, Makati City, Philippines.

Phuong, T.M., Tuan, L.X. and Hong, P.N. (2007) 'Some experience from education activities on mangrove protection in coastal areas, Vietnam', Presentation at MANGROVE Project Review Meeting, 3–10 February 2007, MERD, CRES, Hanoi.

Porter, C.B., Krost, P., Gordin, H. and Angel, D.L. (1996) 'Preliminary assessment of grey mullet (*Mugil cephalus*) as a forager of organically enriched sediments below marine fish farms', Bamidgeh, vol 48, pp. 47–55.

Pretty, J.N. (1995) 'Participatory learning for sustainable agriculture', World Development, vol 23, pp. 1247–1263.

Pretty, J.N., Guijt, I., Thompson, J. and Scoones, I. (1995) A Trainers Guide for Participatory Learning and Action, IIED Participatory Methodology Series, IIED, London.

Primavera, J.H. (1997) 'Socio-economic impacts of shrimp culture', Aquaculture Research, vol 28, pp. 815–827.

Robertson, A.I. and Phillips, M.J. (1995) 'Mangroves as filters of shrimp pond effluent: predictions and biogeochemical research needs', Hydrobiologia, vol 295, pp. 311–321.

Rockstrom, J., Gupta, J., Qin, D., Lade, S.J., Abrams, J.F., Andersen, L.S., et al. (2023) 'Safe and just Earth system boundaries', Nature, vol 619, pp. 102–111.

Ronnback, P. and Primavera, J.H. (2000) 'Illuminating the need for ecological knowledge in economic valuation of mangroves under different management regimes: a critique', Ecological Economics, vol 35, pp. 135–141.

Roth, E., Rosenthal, H. and Burbridge, P. (2000) 'A discussion of the use of the sustainability index: "ecological footprint" for aquaculture production', Aquatic Living Resources, vol 13, pp. 461–469.

SEI and KU (2008) Stakeholder Workshop Report, Thailand. Prepared by Stockholm Environment Institute and Kasetsart University.

Sen, A. (2000) Development as Freedom, Oxford University Press, Oxford, UK.

SFP (2023a) AIP Directory, Sustainable Fisheries Partnership, https://aipdirectory.org/

SFP (2023b) Aquaculture Improvement Projects, Sustainable Fisheries Partnership, https://sustainablefish.org/how-we-work/aquaculture-improvement-projects/

SFP (2023c) Framework for Sustainably Managed Aquaculture, Sustainable Fisheries Partnership, https://sustainablefish.org/how-we-work/aquaculture-improvement-projects/framework-for-sustainably-managed-aquaculture/

Shinji, S. and Haruyoshi, T. (1999) 'Eco-tourism, rural development and environment conservation' in Proceedings of the National Workshop on Sustainable and Economically Efficient Utilization of Natural Resources in Mangrove Ecosystem, Nha Trang City, 1–3 November 1998, CRES, VNU, Hanoi and ACTMANG, Japan.

Shirajee, S.S., Salehin, M.M. and Ahmed, N. (2013) 'The changing face of women for small-scale aquaculture development in rural Bangladesh', Aquaculture Asia, vol 15, no 2, pp. 9–16.

Short, R.E., Gelcich, S., Little, D.C., Micheli, F., Allison, E.H., Basurto, X., et al. (2021) 'Harnessing the diversity of small-scale actors is key to the future of aquatic food systems', Nature Food, vol 2, no 9, pp. 733–741.

Tang, V.T. and Hong, P.N. (1999) 'The role of mangroves to biodiversity and marine resources' in Proceedings of the National Workshop on Sustainable and Economically Efficient Utilization of Natural Resources in Mangrove Ecosystem, Nha Trang City, 1–3 November 1998, CRES, VNU, Hanoi and ACTMANG, Japan.

Taskov, D.A., Telfer, T.C., Bengtson, D.A., Rice, M.A., Little, D.C. and Murray, F.J. (2021) 'Managing aquaculture in multi-use freshwater bodies: the case of Jatiluhur reservoir', Environmental Research Letters, vol 16, no 4, pp. 1–13.

Telfor, T. and Robinson, K. (2003) Environmental Quality and Carrying Capacity for Aquaculture in Mulroy Bay Co. Donegal, University of Stirling, UK and Marine Institute, Galway, Ireland.

Thiao, D. and Bunting, S.W. (2022) Socio-Economic and Biological Impacts of the Fish-Based Feed Industry for Sub-Saharan Africa, FAO Fisheries and Aquaculture Circular 1236, FAO, Rome.

Thilsted, S.H., Thorne-Lyman, A., Webb, P., Bogard, J.R., Subasinghe, R., Phillips, M.J., et al. (2016) 'Sustaining healthy diets: the role of capture fisheries and aquaculture for improving nutrition in the post-2015 era', Food Policy, vol 61, pp. 126–131.

Tho, N.H. and Tri, N.H. (1999) 'Assessment of socio-economic effects of mangrove rehabilitation in Thuy Hai Commune, Thai Thuy District, Thai Binh Province' in Proceedings of the National Workshop on Sustainable and Economically Efficient Utilization of Natural Resources in Mangrove Ecosystem, Nha Trang City, 1–3 November 1998, CRES, VNU, Hanoi and ACTMANG, Japan.

Thompson, A.G. (1990) 'The danger of exotic species', World Aquaculture, vol 21, pp. 25–32.

Toufique, K.A. and Belton, B. (2014) 'Is aquaculture pro-poor? Empirical evidence of impacts on fish consumption in Bangladesh', World Development, vol 64, pp. 609–620.

Toufique, K.A. and Gregory, R. (2008) 'Common waters and private lands: distributional impacts of floodplain aquaculture in Bangladesh', Food Policy, vol 33, no 6, pp. 587–594.

Townsley, P. (1996) Rapid Rural Appraisal, Participatory Rural Appraisal and Aquaculture, FAO, FAO Fisheries Technical Paper No. 358, Rome.

Tran, T.B., Le, C.D. and Brennan, D. (1999) 'Environmental costs of shrimp culture in the rice-growing regions of the Mekong Delta', Aquaculture Economics and Management, vol 3, pp. 31–42.

Tri, N.H. (2005) 'Stakeholder analysis and integrated management of coastal biosphere reserves' Presentation at the 1st MANGROVE Project Inception Meeting, 9–11 November 2005, Bangkok, Thailand.

Troell, M., Joyce, A., Chopin, T., Neori, A., Buschmann, A.H. and Fang, J.G. (2009) 'Ecological engineering in aquaculture: potential for integrated multi-trophic aquaculture (IMTA) in marine offshore systems', Aquaculture, vol 297, pp. 1–9.

Turner, K. (1991) 'Economics and wetland management', Ambio, vol 20, pp. 59–63.

United Nations (2015) Transforming Our World: The 2030 Agenda for Sustainable Development, United Nations, New York.

van Zwieten, P.A.M., Sidik, S.A., Noryadi, I. and Suyatna, I. (2006) 'Aquatic food production in the coastal zone: data-based perceptions on the trade-off between mariculture and fisheries production of the Mahakam Delta and estuary, East Kalimantan, Indonesia' in C.T. Hoanh, T.P. Tuong, J.W. Gowing and B. Hardy (eds) Environment and Livelihoods in Tropical Coastal Zones: Managing Agriculture–Fishery–Aquaculture Conflicts, CABI Publishing, Wallingford, UK, in association with the International Rice Research Institute (IRRI), Philippines and the International Water Management Institute (IWMI), Sri Lanka.

Wackernagel, M. and Rees, W.E. (1997) 'Perceptual and structural barriers to investing in natural capital: economics from an ecological footprint perspective', Ecological Economics, vol 20, pp. 3–24.

Wattayakorn, G. (2004) 'Nutrient status in Pak Phanang Bay, Nakhon Si Thammarat, Thailand' in S. Aksornkoae, N. Papawasit, S. Angsupanich, K. Wattayakorn, S. Suwannodom and A. Siwaiphram (eds) Integrated Management of Mangrove Plantations for Development of Coastal Resources and Environment of Thailand, Thailand Research Fund. (in Thai)

Wattayakorn, G. and Saramul, S. (2004) 'Exchange of nutrients between Klong Paknakorn and Pak Phanang Bay, Nakhon Si Thammarat Province' in S. Aksornkoae, N. Papawasit, S. Angsupanich, K. Wattayakorn, S. Suwannodom and A. Siwaiphram (eds) Integrated Management of Mangrove Plantations for Development of Coastal Resources and Environment of Thailand. Thailand Research Fund. (in Thai)

Willett, W., Rockström, J., Loken, B., Springmann, M., Lang, T., Vermeulen, S., et al. (2019) 'Food in the Anthropocene: the EAT–Lancet Commission on healthy diets from sustainable food systems', Lancet, vol 393, no 10170, pp. 447–492.

Wosten, J.H.M., de Willigen, P., Tri, N.H., Lien, T.V. and Smith, S.V. (2003) 'Nutrient dynamics in mangrove areas of the Red River Estuary in Vietnam', Estuarine, Coastal and Shelf Science, vol 57, pp. 65–72.

WWF (2010) Pangasius Aquaculture Dialogue Standard, World Wildlife Fund, Washington, DC.

Xan, L. and Khuong, D.V. (1999) 'Aquaculture situation in mangrove areas in Hai Phong and measures for improvement' in Proceedings of the National Workshop on Sustainable and Economically Efficient Utilization of Natural Resources in Mangrove Ecosystem, Nha Trang City, 1–3 November 1998, CRES, VNU, Hanoi and ACTMANG, Japan.

Young, J.A., Brugere, C. and Muir, J.F. (1999) 'Green grow the fishes – Oh? Environmental attributes in marketing aquaculture products', Aquaculture Economics and Management, vol 3, pp. 7–17.

4 Sustainable coastal and marine aquaculture

Key points

The aims of this chapter are to:

- Describe the origins of enduring traditional coastal aquaculture practices and identify factors contributing to resilience.
- Review success with rehabilitating and improving traditional coastal aquaculture systems and discuss aspects of social–ecological resilience demonstrated by diversified shrimp–rice agroecosystems in coastal areas of Bangladesh and West Bengal, India (see Bunting et al., 2017), and discuss limitations and possible issues of moving brackish and marine culture inland.
- Note better management practices developed to enhance shrimp production as part of post-tsunami reconstruction efforts in Indonesia.
- Evaluate progress with integrated mangrove aquaculture and alternative strategies to promote more efficient and effective operation.
- Highlight the importance of adopting a One Health framework approach and recent progress on animal welfare, notably avoiding eyestalk ablation in shrimp whilst still producing robust offspring (see Zacarias et al., 2019, 2021; Zacarias, 2020), and discuss implications for the sector globally.
- Consider prospects for enhanced biosecurity (e.g. vaccinations, specific pathogen–free juveniles, better management practices, biorefinery strategies to manage fallen stock and closed and semi-closed containment systems) and note some strategies that could create win–wins of increased production efficiency and greater social licence to operate (see Bunting et al., 2022).
- Introduce innovations to enhance production efficiency and benefit the bioeconomy (e.g. automation and sensors, better data use, nutrition-sensitive aquatic food systems, complete utilisation of aquaculture products, nutrient and waste recovery and valorisation; see Bunting et al., 2022, 2023).
- Review progress in refining horizontally integrated, land-based marine aquaculture, drawing on pertinent examples from Europe and Israel.
- Discuss prospects for open-water integration of salmon with shellfish and seaweed.

DOI: 10.4324/9781003342823-4

- Critically review benefits derived from small-scale marine aquaculture development, notably grouper culture in Indonesia and Vietnam and organic and ethical salmon and shrimp aquaculture.
- Evaluate the potential of artificial reefs for sequestering nutrients from aquaculture and contributing to stock enhancement and ranching initiatives.
- Discuss progress with stock enhancement, culture-based fisheries and ranching and critical issues demanding responsible management practices.

Traditional coastal aquaculture

Coastal aquaculture has a long and illustrious history, with many examples of traditional practices from around the globe being referred to as exemplars of responsible and environmentally sustainable food production. Traditional practices from Hawaii were comprehensively reviewed by Costa-Pierce (2002: 30), who noted that 'ancient mariculture systems of Hawaii are unique in that they connect an isolated island society with sophisticated ocean harvesting and integrated sea farming activities to an entire watershed management/food production system (the *ahupua'a*)'. Massive seawalls were constructed of coral and larval rock, demanding huge labour inputs that are considered indicative of strong social organisation, instituted in this case under hierarchical chiefdoms. Resulting seawater ponds or *loko kuapa* were connected to the sea by canals regulated by timber gates to permit timely water exchange for efficient cleaning, stocking and harvesting. These ponds were predominantly used to culture milkfish and mullet, but twenty-two species were harvested from such systems which have their origins 1500–1800 years BP (Costa-Pierce, 2002). Evidence from Europe indicates that culturing and relaying oysters were established practices in the Adriatic 2100–2200 years BP (Beveridge and Little, 2002). Contemporary with this, the Romans were constructing saltwater ponds throughout their Mediterranean domain to hold live food fish such as eels, mullets, seabass and turbot, a form of proto-aquaculture viewed as a demonstration of status and wealth. Archaeological explorations of Roman settlements in southeast England indicate that oysters were being widely eaten there, but published accounts suggest that culture practice was not instigated until medieval times. A fascinating, yet sobering, account concerning the ebbing fortunes of oyster fisheries in waters surrounding the USA and early attempts to bolster stocks through re-stocking and culture in the early 1800s was presented by Roberts (2007). The author noted:

> The story of oyster fisheries is an oft-repeated one in the history of human exploitation of natural resources. Where a resource is common property, shared by all, there is a tendency for individuals to take more of that resource than is sustainable. Individuals can obtain a private gain but at a cost to the rest of society by acting selfishly.

(p. 226)

Construction of embankments to form coastal lagoons to trap juvenile fish to be on-grown until required in the productive shallows was undertaken by the Etruscans to enable more productive management of coastal waters in the Adriatic and Tyrrhenian seas (Beveridge and Little, 2002). Such practices, developed 2400–2500 years BP, were the forerunners of *vallicoltura* that are still used to culture fish in the Mediterranean around the Tyrrhenian and northern Adriatic as well as in Sardinia and Sicily. Developmental pathways for several of these traditional marine aquaculture systems can be traced to present-day operations. Whilst this may be taken as indicative of resilient and sustainable practices, it is notable that declining seed supplies from wild stocks, climate change impacts, alternative uses of coastal areas, introduced invasive species and diseases and competition from more intensively managed production facilities constitute unprecedented threats to continued production. Faced with such problems, substantial efforts are being made to enhance such traditional practise, bolster their resilience to external pressure, formulate appropriate adaption strategies and promote awareness of their broader value amongst the public and, critically, prospective consumers. Constraints and associated responses being investigated for systems in southern Europe are summarised in Table 4.1.

Traditional marine aquaculture practices in Asia are thought to have their origins around 600–800 years BP with the culture of milkfish (*Chanos chanos*) in coastal lagoons in Java, Indonesia (Beveridge and Little, 2002). Subsequent development of coastal aquaculture in Asia resulted in more widespread and diverse culture practices than occurred in Europe. Traditional practices have not been immune from increasingly globalised economic and market forces, however, putting pressure on producers to opt for more intensive, monoculture production of high-value species. Coastal ponds formed along the coast of the upper part of the Bay of Bengal in Bangladesh and India, locally known as *bheries*, are widely regarded as traditional, but there is a danger that such generic categorisation masks complex and often coincidental patterns of development and loss. Euphemistically termed 'reclamation works' were undertaken to convert large tracts fringing the coast to intensive agricultural production through Green Revolution programmes sponsored by the World Bank, and these also opened up areas for marine aquaculture development (see Box 4.1).

Aquaculture practices perceived as traditional or indigenous often become imbued with cultural or heritage values, which in turn can lead to calls for their maintenance and preservation. As with many agricultural systems, however, reasons why such practices have persisted are probably multifaceted and not based on short-term financial assessments that may hold sway over the viability of new ventures. Costs for establishing traditional culture practices would be recouped over an extended time period, financial returns are probably sufficient to generate modest returns compared with ongoing costs, risks are likely to be minimal and well defined, opportunity costs for labour and other inputs do not warrant change, and sentiment dictates that assets, notably land holdings and property and access rights, would appreciate in value.

Table 4.1 Operation and challenges facing traditional coastal aquaculture practices in southern Europe

System	Operation	Challenges
Semi-intensive polyculture in Portugal and southern Spain	• seabass and seabream cultured in a semi-intensively managed polyculture in earthen ponds; production enhanced owing to recruitment of wild fish larvae that on-grow and are harvested	• production costs higher than intensive cage culture; continued operation depends on product differentiation and enhancing production; protocols needed for successful integration of stocked Senegalese sole culture to complement wild recruitment
Semi-intensive juvenile production in France, Greece, Italy and Portugal	• fingerlings raised under semi-intensive conditions in lagoons and ponds have higher growth potential and enhanced viability, with greater success in avoiding predators, resilience to fluctuating water conditions and ability to freely consume wild feed	• refinement of rearing systems for juvenile marine fish under extensive and semi-intensive conditions needed to ensure supplies of high-quality fingerlings to stock on-growing systems; promising strategies, including deployment of mesocosms and nursery structures, optimisation of larval stocking densities and low-cost strategies to enhance productivity require further assessment
Extensive eel culture in south-western France	• salt works and wetlands were converted to extensive eel culture, leading to enhanced protection of these modified agroecosystems	• rising labour costs and declines in the availability of eels for stocking threaten the economic viability of the system; fishpond–wetlands sustain other fish populations and have 'patrimonial' value (32); policies supporting restoration are needed
Valliculture, Italy	• eel, mullet, seabass and seabream are cultured in coastal wetlands fringing the upper Adriatic in a system valued for its 'unique ecological, landscape and cultural heritage' (32)	• opportunities needed to add value to products in recognition of environmental and cultural attributes; optimised farming protocols for seabream and diversified stocking and harvesting strategies required to enhance productivity and spread predator risks

Source: Developed from Anonymous (2007).

Box 4.1   Genesis of shrimp farming in the Sundarbans

Citing a paper by Paul (1995) written in Bengali, Azad et al. (2009: 801) recounted that 'traditional shrimp farming started in 1829 in the Sundarban (mangrove) area of Bangladesh'. The authors noted that hydrological engineering works associated with Green Revolution[1]– inspired policies adopted by the government in the 1960s to polder and reclaim coastal land for cereal cultivation opened up significant new areas for permanent settlement by migrants. A large part of this land was used for rice cultivation for a few years before being converted to shrimp culture. Burgeoning populations in the newly colonised coastal areas undoubtedly placed great pressure on the remaining coastal ecosystems. Mangroves were cut for timber and fuel wood and terrestrial and aquatic animals and birds were hunted for food. Mangrove clearance outside newly constructed dikes paved the way for rapid shrimp farming development.

Similar development phases are apparent in the adjoining Sundarbans of West Bengal, India. According to Naskar (1985), large areas of mangrove were destroyed when villages were established in the Kulti–Minakhan Sundarbans area by people who had migrated from Midnapore District in West Bengal and the neighbouring State of Orissa (now Odisha). These villages predominantly engaged in agriculture and fishing but were subsequently evicted to permit port development during the early nineteenth century. Irrigation channels and embankments constructed in this area by the Department of Irrigation in 1967 helped to protect some areas; prior to this, the farmers in Minakhan Block were entirely dependent on *kharif* (autumn or monsoon season) paddy cultivation. As Naskar (1985: 117) observed, 'During the summer months those fields were lying fallow for want of any irrigation facilities. Those rural people also suffered without any employment or further engagement for their livelihood; they were forced to spend their nights with hungry stomach'. According to Naskar, farmers with land marooned outside the embankments constructed a secondary series of bunds or dikes to enclose areas where they practiced brackish water aquaculture. As fish and shrimp grew well, other farmers then devised their own systems for channelling brackish water to their paddy fields to permit fish and shrimp culture together with their rice crop and after the rice was harvested.

Source: Bunting et al. (2011).

Playing devil's advocate, contemporary proposals for such systems would probably be viewed with disdain and be criticised for adverse environmental impacts, whilst business cases and anticipated financial returns would be viewed as unattractive. Poor management and infusion of unsustainable practices could, however, also undermine longstanding and apparently resilient culture practice, as could limited capacity to respond appropriately to numerous pervasive and emerging pressures such as urbanisation, migration, population growth, environmental degradation, hydrological regime shifts and climate change. Consequently, a need for mechanisms to enhance and strengthen traditional practices has been identified, for example, through the formulation of better management practices (BMPs) guidelines for *tambak* farming in Aceh, Indonesia, which are discussed further below.

Enhanced traditional practices

Attention has been given to enhancing production and promoting awareness of benefits derived from traditional culture practices, notably from coastal lagoons throughout Southern Europe as well as from extensively managed tropical shrimp ponds in Asia (World Bank et al., 2005; Anonymous, 2007). Challenges concerning the continued operation of extensive and semi-intensive aquaculture in coastal wetlands in southern Europe are summarised in Table 4.1, and trials were undertaken by the SEACASE project to optimise stocking regimes and species combinations, enhance productivity and combine intensive and extensive culture practices. Attention focused on adding value to cultured species in Italy to better differentiate products originating from semi-intensive polyculture systems in Portugal and Spain and to highlighting the potential role of extensive culture systems in enhancing wild fish-stocks, notably European eel (*Anguilla anguilla*), in France (Anonymous, 2007).

Earlier innovations intended to enhance production of valliculture along the shores of the northern Adriatic, Italy, were described by Melotti et al. (1991), who reported that wastewater from intensively managed systems producing seabass and seabream was discharged to valliculture areas where extensive polyculture of eel (*A. anguilla*), mullet (*Mugil* spp.), seabass and seabream was being carried out. Mean average harvest weights for mullet reportedly increased significantly from 302 g in 1987 to 375 g in 1990 when wastewater was being used to fertilise the system; average weights for seabream were 322 g in 1987 and 386 g in 1990, but seabass harvest weights declined from 525 g to 422 g over the same period. Conclusions drawn from this were that integrated intensive–extensive culture in the northern Adriatic valliculture area could increase overall farm production and reduce coastal pollution.

Demonstration systems were developed at the Centre Regional d'Experimentation et d'Application Aquacole (CREAA), Poitou–Charentes Région, France, to assist farmers operating traditional earthen fishponds and oyster-fattening ponds to evaluate the performance of alternative husbandry practices and production strategies. Amongst the principal strategies investigated were integrated fish–phytoplankton–shellfish culture and modified fishponds planted with *Salicornia* spp. to produce a complementary crop whilst reducing nutrient discharges. Financial returns generated from such a strategy were assessed with bioeconomic modelling (Table 4.10); findings are presented comprehensively elsewhere (see Bunting and Shpigel, 2009), and critical operational aspects, financial considerations and broader social–economic implications are discussed further below.

Integrated rice–shrimp culture practices are widespread in the transboundary Sundarbans mangrove forest along the border between Bangladesh and West Bengal, India (Bunting et al., 2017). Reviewing the extent of such practices, Islam et al. (2005: 489) noted that 'coastal shrimp (*Penaeus monodon*) aquaculture in Bangladesh is mostly practiced in a special type of fish/pond situated by the side of a river – called a *gher* – that is used to cultivate rice in winter and shrimp in summer'. Regarding the specific situation in southwest coastal Bangladesh, Azad et al. (2009: 805) observed that where 'brackish water is available in nearby canals and rivers, *P. monodon* and winter rice (January–July) are farmed simultaneously in the *ghers*' and that this is followed by river prawn, *M. rosenbergii* culture from July to December when rainwater is accessible.

Integrated shrimp–rice culture in the Sundarbans, the origins of which are alluded to in Box 4.1, persists and appears to demonstrate a degree of social–ecological resilience (Bunting et al., 2017). Survey results presented by Islam et al. (2005) demonstrated that over 90 per cent of shrimp farmers employed low-input, semi-intensive practices, with water exchange based largely on tidal exchange, stocking based on natural recruitment, limited manure inputs to stimulate production and an absence of inorganic fertiliser and synthetic pesticide applications. As with similar production strategies in the Mahakam Delta, Indonesia (Bunting et al., 2013), farmers appeared content to secure modest financial returns whilst not exposing themselves to excessive risks from disease outbreaks, environmental perturbations, natural disasters and market price fluctuations. Transition from traditional to low-input semi-intensive culture over the preceding two decades was reported by Hoq (2008) and a tendency toward improved culture practices was identified.

Assessment of the production strategies and economic–social–environmental interactions associated with 'diversified shrimp–rice agroecosystems' in coastal areas of Bangladesh and West Bengal, India, demonstrated that they could contribute to enhanced social–ecological resilience (Bunting et al., 2017: 63). Co-culture of shrimp and rice can create synergistic effects, notably enhanced nutrient cycling, reduced requirements for feed and fertiliser, decreased pest and weed problems and enhanced production. Staple crop and vegetable cultivation contributes to food and nutrition security in farming households, local communities and nationally. Export crop production can stimulate economic and social development, by creating income-generating and employment opportunities across value chains. The ability of diversified shrimp–rice and prawn–fish–rice agroecosystems to produce staple and cash crops in intermediate salinity and intermittently flooded conditions may be critical to enabling communities to adapt to climate change induced hydrological fluctuations in tropical coastal zones globally (Ahmed et al., 2014, 2017; Bunting et al., 2017; IPCC, 2019).

Elsewhere, transition to high-input, semi-intensive and intensive shrimp farming resulted in short-term financial gains but was often short-lived owing to disease and water quality problems; producers either abandoned their ponds or reverted to extensive and traditional modes. Devastated by the 2004 tsunami, plans to rehabilitate traditional *tambak*-based shrimp culture in Aceh, Indonesia, included actions to enhance management of these systems. Whether authentic or modified, such intervention appeared to dispel the myth that contemporary production practices labelled as traditional constituted benign activities, highly attuned to local conditions and livelihoods asset portfolios, whilst achieving the most efficient use of natural resources. Appropriate measures to address apparent problems, increase production and mitigate identifiable risks were presented and promoted as BMPs. Specific areas addressed included pond location and design, preparation, water conditioning, seed selection and stocking, feed management, water and health management, harvest and postharvest handling, crop planning and farmer group formation. Prospects for BMPs guiding responsible aquaculture development given different systems, environments and socio-economic contexts are discussed further in chapter 8.

Silvo-aquaculture

Traditional practices to enhance the productivity of mangroves included deploying rock mounds called *ampong* in the Philippines and damming creeks to trap fish and crustaceans in Java, Indonesia (Primavera, 1997; Beveridge and Little, 2002). As aquaculture developed in coastal areas, mangroves were strategically planted for protection, wood and associated non-timber forestry products. Two illustrative historical examples were described by Primavera (1997). Pond embankments in the Philippines were planted with mangrove species (*Rhizophora* spp. and *Sonneratia* spp.) and nipa palm (*Nypa fruticans*) in the early 1900s primarily to protect against soil erosion owing to tidal action and storms. Mangrove species such as *Rhizophora* spp. and *Avicennia* spp. were planted along embankments and on tidal flats surrounding coastal ponds or *tambak* in Indonesia as a source of detritus and firewood and to guard against erosion. Further to productive and protective ecosystem services supported by mangroves, the capacity of these coastal wetlands to process nutrients and organic matter in wastewater has been recognised since the 1970s (Box 4.2).

What was not appreciated perhaps was that material, nutrient and energy budgets within mangrove ecosystems are often finely balanced and ultimate consequences of waste discharges uncertain.

Box 4.2   Capitalising on mangrove ecosystem services to manage aquaculture wastewater

Wastewater treatment processes facilitated by mangroves have the potential to improve the quality of run-off flowing into coastal lagoons, estuaries and near-shore areas. Mangroves have been used to treat sewage and wastewater discharged from societal systems, with the majority demonstrating a high capacity for retaining nutrients and degrading organic matter (Clough et al., 1983; Dwivedi and Padmakumar, 1983; Tam and Wong, 1995; Wong et al., 1995; Machiwa, 1998). Improving water quality in coastal areas would increase the availability of good quality water for shrimp culture and contribute to the maintenance of the trophic status of receiving habitats, ensuring that ecosystem support functions provided by the coastal environment persist and the utility of the resource is preserved. The potential to exploit mangroves to condition water for use in shrimp aquaculture and to treat wastewater from shrimp farms has been considered by a number of authors (Robertson and Phillips, 1995; Rajendran and Kathiresan, 1996; Frederiksen et al., 1998; Wolanski et al., 2000).

Source: Bunting (2001).

Assessing the capacity of mangrove stands to treat aquaculture waste-water, Robertson and Phillips (1995) highlighted several areas of uncertainty, including effects of modified hydrological regimes on ecological processes, long-term assimilative capacity of mangroves for phosphorus, influence of sediment composition on nitrogen retention and transformations, and impact of different silvofishery practices on mangrove species growth and nutrient uptake. Subsequently, more systematic and controllable strategies to combine beneficial aspects of mangrove ecosystem functioning with or within coastal aquaculture development have been conceived, piloted and, to a limited extent, adopted commercially.

Organic standards for shrimp farming published by Naturland (1995: 15) confirm that 'farms (here: independent, coherent production units), which in parts occupy former mangrove area, can be converted to Organic Aquaculture according to Naturland standards if the former mangrove area does not exceed 50 per cent of total farm area', although the standards stipulate further that 'the former mangrove area in property of the farm shall be reforested to at least 50 per cent during a period of maximum five years' (15) and specify that relevant legal requirements for land use and reforestation must be respected.

Regulations have been established in several countries to ensure that mangroves are protected from further encroachment and are re-planted along exposed coasts and inlets for shoreline protection. Authorities in Vietnam have enacted a system of zoning in the lower Mekong River Delta that demarcates Full Protection Zones, Buffer Zones where certain economic activities are permitted whilst retaining 60 per cent mangrove cover, and Economic Zones where no restrictions concerning forest conservation apply (World Bank et al., 2005). Forestry rules in the Philippines stipulate a 20 m-wide mangrove buffer strip along shorelines, increasing to 50 m on coasts vulnerable to typhoons. *Principles for a Code of Conduct for the Management and Sustainable Use of Mangrove Ecosystems* prepared by the World Bank et al. (2005: 67) specify under Principle 11.5 that 'states should promote the development and adoption of integrated mangrove aquaculture systems, which are both environmentally sustainable and suitable (socio-economically viable) to support the livelihoods of poor fishers and farmer communities'. Integrated mangrove planting in shrimp ponds in Mahakam Delta, East Kalimantan, Indonesia, is feasible (Figure 4.1), but this can make pond management more challenging.

Integrated mangrove–shrimp culture in the Mahakam Delata, Indonesia, could potentially generate worthwhile financial returns for operators, whilst rehabilitating coastal ecosystems that sustain diverse ecosystem services, notably provisioning an array of aquatic foods and sequestering carbon

Figure 4.1 Integrated mangrove–shrimp culture ponds in the Mahakam Delta, East Kalimantan, Indonesia.

Photo credit: Stuart Bunting.

(Bunting et al., 2013; Ahmed et al., 2017). Those practicing traditional and extensive production modes may, however, be deterred from adopting mangrove–shrimp culture owing to greater investment costs and financial risks. Prevailing sub-leasing and *de facto* landholding arrangements are likely to complicate the situation, and appropriate interventions (e.g. policies, regulations and guarantees) may be required to create an enabling institutional environment for desirable change. If additional payments for ecosystem services could be guaranteed for producers over the medium to long term, this might encourage them to switch (Bunting et al., 2013). Allocation of sizable resources from inter-governmental and private investment funds (Bunting et al., 2022; World Bank, 2022) could potentially help achieve large-scale transformation of the sector. Similar interventions may be needed elsewhere across the tropics and sub-tropics where integrated mangrove–aquaculture systems could potentially be developed.

Integrated aquaculture–mangrove strategies were comprehensively reviewed by Fitzgerald (2002), who invoked the term silvofisheries to encompass the broad range of contemporary practices, notably in Hong Kong, Indonesia, the Philippines, Thailand and Vietnam. The review concluded that practices must be adapted to local mangrove ecosystem characteristics and site-specific conditions and conform to prevailing polices and regulations. Critical considerations regarding system design for ponds with a platform planted with mangrove are summarised in Table 4.2. Given that the dike

Table 4.2 Critical system design considerations for integrated aquaculture–mangrove culture

System design aspect	Consideration
Mangrove-to–pond surface ratio	• platform planted with mangrove should be 40–80 per cent of the total pond area
Pond surface area–to–dike length ratio	• ratio of pond surface to dike length is a key determinant of production potential versus capital costs
Gate width ratio	• sluice gate should have design capacity of 50 cm ha^{-1} of pond surface, with appropriately designed aprons to minimise erosion and promote efficient flows
Tidal regime	• tidal range should permit pond flushing as required and can be regulated through appropriate sluice gate operation, with a range of 1.5–2.5 m being desirable
Flow regime	• flow rates within ponds should be sufficient to maintain water quality and avoid low oxygen concentrations (<4 ppm)
Inundation of central platform	• depth of inundation should be 20–50 cm, with duration appropriate to mangrove species planted, generally <7 days
Perimeter channel	• channel depth should be 0.8–1.2 m whilst channel width should be 5–8 m

Source: Developed from Fitzgerald (2002).

length and number of sluice gates relative to the productive pond surface area are often central determinants of construction costs and the perimeter canal is employed for culture, it was concluded that square ponds with a size approaching 10 ha would be most cost effective.

Irresponsible aquaculture development was in a large measure responsible for widespread mangrove loss globally. Strategies including zoning to exclude aquaculture or to permit it in restricted areas with the promotion of integrated aquaculture–mangrove practices may help redress the balance of aquaculture with its vital ecosystem support area. Conflicts persist, however; for example, clam culture activities abutting the seaward extent of mangroves may impede coastal processes and prevent natural colonisation by mangrove propagates, thus restricting mangrove development and exacerbating pressures from land-based activities and coastal engineering works. Awareness has grown of the fact that mangroves sustain stocks and flows of ecosystem services that support societal systems, notably their contributions to storm abatement and erosion prevention. Divergent views persist, however, concerning how best to reconcile conservation with wise use and to balance mangrove ecosystem services appropriation with the carrying capacity of environment.

Mangrove establishment and rehabilitation

Mangrove loss and degradation results in denuded and abandoned areas that require rehabilitation; establishment of mangrove areas might also be required for mitigation purposes. Natural regeneration may be deemed desirable from ecological and economic perspectives, resulting in a mix and distribution of mangrove species reflecting the indigenous or local mangrove assemblage and ecological processes governing colonisation and establishment. Several guides provide a schematic view of idealised mangrove transects through the mangrove, from sea to inland, to illustrate the zonation or natural distribution that characterises many mangrove stands (Lewis et al., 2006). Such an approach highlights the fact that mangrove stands are not homogenous; the general pattern of zonation varies between and within sites and is influenced by several factors. Where possible, such transects should be developed with respect to local reference sites composed of pristine or relatively undisturbed mangroves or reconstructed from records and historical accounts to avoid a shifting baseline. However, major barriers and constraints prevent natural mangrove establishment, including:

- inadequate propagule supplies;
- currents dispersing propagules away from site;
- coastal engineering works preventing propagule migration;
- altered topographic and hydrologic regimes rendering site unsuited for re-colonisation;
- ongoing disturbance and deliberate clearance preventing establishment.

Mangrove re-planning has been widely advocated and is happening on a large scale, whilst extensive programmes of restoration are planned in various countries. Rehabilitation programmes have frequently resulted in variable outcomes, however, often demonstrating poor success rates for propagule establishment and thus being criticised as inefficient. Mangrove planting could conceivably result in negative ecological and environmental impacts and possibly cause disruption to livelihoods. Restricting access to mudflats planted with mangroves may prohibit shellfish gathering and impinge on navigation routes and access to fishing grounds. Mudflats, sandflats and seagrass beds, sometimes disregarded as wastelands and unproductive, might be considered for mangrove planting. However, they can be ecologically important, sustain stocks and flows of ecosystem services and be bestowed with non-use values and may therefore need to be preserved.

Reviewing mangrove replanting schemes in the Philippines, Samson and Rollon (2008) noted that extensive areas had failed to regenerate and those plantations that were established were often inappropriately sited and consisted of mono-species stands of limited ecological value, and similar concerns surround replanting programmes in Northern Vietnam (Figure 4.2).

Ecological and economic benefits of well-established mangrove plantations in the Philippines have been reported. It was noted, however, that selective

Figure 4.2 Mangrove replanting in Northern Vietnam with the principal aim of coastal protection.

Photo credit: Stuart Bunting.

Table 4.3 Guidelines for five steps to successful mangrove rehabilitation

Critical step	Guideline
1. Knowledge acquisition	• understand the ecology of individual mangrove species at a particular site or management area, notably patterns of reproduction, propagule distribution and successful seedling establishment with local mangrove forest communities
2. Site survey and ecological processes assessment	• understand normal hydrological patterns and other stress factors that control the distribution and successful establishment and growth of targeted mangrove species
3. Pressures and barriers to regeneration identification	• assess previous modifications to the mangrove environment that currently prevent natural secondary succession, including hydrological change and other stresses (timber cutting, grazing, fires, disease)
4. Phased programme design	• design the restoration programme to initially restore the appropriate hydrology and/or remove any additional stressors that might prevent natural colonisation; then attempt to utilise natural volunteer mangrove propagule recruitment for plant establishment
5. Resort to replanting only when natural recruitment proves impracticable	• resort to utilising planting of propagules, collected seedlings or cultivated seedlings only after determining through steps 1–4 that natural recruitment will not provide the quantity of successfully established seedlings, rate of stabilisation or rate of growth of saplings established as goals for the restoration project

Source: Developed from Lewis (1999).

planting of one or two species does not result in diverse forest regeneration in the short to medium term, and 'if you want to restore diverse mangrove forests, you have to plant diverse mangrove forests' (Walters, 2000: 237). Bodies and institutions responsible for planning and implementing mangrove rehabilitation would benefit, apparently, from better guidance and information concerning site selection and efficient and effective replanting strategies. Guiding principles for mangrove rehabilitation proposed by Lewis (1999) and summarised in Table 4.3 were reiterated in the World Bank–sponsored *Principles for a Code of Conduct for the Management and Sustainable Use of Mangrove Ecosystems* (World Bank et al., 2005) and provided the framework for the manual 'Five Steps to the Successful Ecological Restoration of Mangroves' (Lewis et al., 2006).

Failed mangrove rehabilitation can result in local communities and society generally foregoing substantial benefits in terms of the total economic value (TEV) or ecosystem services associated with the appropriately restored mangrove (Macintosh and Ashton, 2002). Participatory assessment of coastal livelihoods of communities in Thailand indicated that women considered mangrove replanting a more important activity than did men (Dulyapurk et al., 2007). Subsequent enquiry revealed that women were routinely employed to carry out replanting and hence probably valued the activity for short-term financial gain, compared with potential long-term benefits. Data on planned and completed planting are generally available, but concern has been expressed that certain areas are being replanted annually to meet targets and secure continued funding.

Importance of a One Health framework approach

Adoption of a One Health framework approach, linking the health of animals, ecosystems and humans, is essential to ensure that assessments are comprehensive and that outcomes are optimised across the life-cycle of cultured organisms and across food systems (Stentiford et al., 2020; Bunting et al., 2022). Trials focused on the elimination of eyestalk ablation of crustaceans demonstrated that productivity was not affected and that the progeny can be more stress resistant and robust to disease challenges (Zacarias et al., 2019, 2021; Zacarias, 2020). As scrutiny of the life-cycle of sentient animals in captivity intensifies, it is likely that other practices that could have negative animal welfare impacts will need to be curtailed or acceptable mitigation measures established.

Trade-offs have been noted in utilising small pelagic fish species for aquatic and terrestrial animal feed production and potential negative impacts on food and nutrition security of people in selected communities, notably those on the west coast of Africa (Thiao and Bunting, 2022). Specific measures for decision-makers and future research to tackle this situation were identified by these authors and are detailed elsewhere (chapter 7, Table 7.5). Adopting a food systems approach (see chapter 7) that encompasses input suppliers, food producers, value chain actors and consumers is crucial in this regard to ensure that unacceptable practices – for example, human trafficking, modern slavery and illegal labour relations (Marschke and Vandergeest, 2016) – are eliminated from seafood supply chains irrespective of where and when they occur. Coordinated action to address shared issues, such as minimising food loss and waste, could also contribute to achieving the Sustainable Development Goals of the United Nations (2015; see chapter 1).

Opportunities for enhanced biosecurity

Biosecurity is considered here as a comprehensive set of actions, strategies and policies adopted to prevent the spread of diseases, pathogens and pests that could pose serious threats to animal welfare and cause ecological, financial, economic or social harms. Opportunities to enhance biosecurity can be identified on farms and across associated value chains (both upstream and downstream; see Table 4.4). Enhancements that address concerns amongst local communities that aquaculture may be negatively impacting native

Table 4.4 Opportunities for enhanced biosecurity measures in coastal and marine aquaculture

Biosecurity measure	Opportunity	Challenges
Vaccinations	• vaccinations have been developed for several important diseases of temperate and warmwater fish (+++); efficacy of vaccines in preventing problems caused by specific diseases can be high (+++); multiple vaccines can be delivered with a single injection (++)	• vaccination procedures constitute stress-inducing events, and care is needed to minimise handling and injuries (+++); perceived benefit–cost ratio may deter private sector investment in vaccine development for certain diseases (++)
Specific pathogen–free (SPF) stock	• when several key pathogens of local concern are covered by an initiative, this could be more attractive for producers (++); certain diseases of aquatic animals have resulted in major financial losses for individual producers and sectors, and knowledge that juveniles are SPF could encourage reinvestment (+)	• SPF animals may harbour other diseases and pests (++); effectiveness and reliability of SPF producers may be difficult to verify (++); juveniles constitute a source of pathogens, but there are many others in commercial production systems (+++); SPF seed may suffer from inbreeding and juveniles may perform less well in routine culture conditions (++)
BMPs	• BMPs can address several management practices and production issues (++); can yield benefits when developed in conjunction producers (++); may assist clusters of smaller producers begin entry to formal certification schemes (++); can help instigate area management plans to collectively counter disease problems (++)	• generic guidelines may not be appropriate for diverse culture systems (e.g. monoculture, polyculture, diversified, integrated multi-trophic aquaculture or integrated aquaculture–agriculture) and operators (e.g. disaggregated by age, education, gender or wealth) that characterise many coastal zones (++)
Biorefinery strategies to manage fallen stock	• timely and responsible management of fallen stock can help prevent the spread of pests and diseases to other culture animals and to other production facilities (+++); biorefinery processes can add value to fallen stock and recover important nutrients (++)	• appropriate facilities to process fallen stock biomass may be distant, and transport costs, greenhouse gas emissions and risks of disease transfers to new locations may need to be included in decision-making processes (++)
Closed and semi-closed containment systems (e.g. Canada and Norway)	• enclosed floating systems provide a physical barrier to potential pests and diseases entering or leaving systems (++); preventing nutrients from entering surrounding waters could help reduce cumulative environmental impacts and enable recovery and reuse (++); trials are underway to test viability (++)	• diseases and pests may still be introduced inadvertently with seed, feed, intake water, equipment, machinery and personnel (++); need to circulate and treat water adds significant financial costs and brings added risks, demanding backups and safeguards (+++)

Note: an indication is given of the perceived significance thusly: (+++) very important, (++) important, (+) less important, but this may vary, for example, over time and in different contexts and geographies, for example,.

stock and environments could help establish a greater social licence to operate (SLO). Achieving SLO is regarded as imperative when aquaculture operations occur within or depend on shared ecosystems and seascapes (Bunting et al., 2022).

Innovations to enhance production efficiency and benefit the bioeconomy

Innovations are considered here to encompass technological advancements, novel business arrangements and collaborations, adoption of new paradigms to guide collective action and supportive regulation and policies that help achieve sustainable aquaculture development. Technological innovation, combined with appropriate institutional reconfigurations, is often required to achieve radical or transformative change (Bunting et al., 2023). These authors noted that innovations across aquaculture value chains in Bangladesh and India have potential to achieve poverty reduction. Innovations that could enhance feed and nutrient use efficiency on farms, help minimise costs, ensure more complete utilisation of aquaculture products, optimise health outcomes for consumers and prevent nutrients being lost to the biosphere whilst adding value are summarised in Table 4.5.

Origins and evolution of land-based marine IMTA

Trials to establish the feasibility and evaluate the performance of different species combinations, as well as operating parameters and system configurations, for land-based integrated multi-trophic aquaculture (IMTA) prototypes have been conducted in a wide array of geographical settings and environmental conditions. Early experiments conducted by Ryther and co-workers at the Woods Hole Oceanographic Institution, Massachusetts, USA, focused on the culture of marine phytoplankton, shellfish, polychaete worms, amphipods, flounder, lobster and commercial red seaweed species to treat domestic wastewater diluted with seawater (Ryther et al., 1975). Research in South America initially focused on optimising the integration of tank-based salmon (*Oncorhynchus kisutch*) and trout (*Oncorhynchus mykiss*) culture with on-shore seaweed (*Gracilaria chilensis*) cultivation in Marti Bay, Chile (Buschmann et al., 1994). Motivations included reducing the environmental impact of the burgeoning salmon culture industry, product diversification to guard against market uncertainty, cost sharing between production activities and increasing the range of sites suitable for aquaculture development. Constraints to commercial development included high capital costs and operating costs, notably pumping seawater ashore, and the need to supplement carbon dioxide, nutrient and dissolved oxygen concentrations.

Tank- and raceway-based trials were conducted in Israel with the objective of designing productive and efficient production systems for the integrated

Table 4.5 Innovations to enhance production efficiency and benefit the bioeconomy

Innovation	Opportunity	Challenges
Automated feeding and sensors	• formulated feeds constitute valuable production-enhancing inputs, and automated feeding, governed by sensors, can help optimise conversion to harvestable biomass; automated feeding can reduce labour costs	• automatic systems can suffer electrical and mechanical failure, so backups and safeguards may be needed; constant surveillance by experienced staff may still be required to monitor other aspects of animal welfare
Data collection, analysis and sharing	• as sensors become more accessible and reliable and internet connectivity expands, this can enable automated data collection, analysis and sharing, potentially with preferred input suppliers to enhance logistics and minimise costs	• appropriate safeguards are needed to ensure that data collection and storage is ethical and secure and that personal or sensitive data are stored anonymously and used only for the agreed purposes and time
Nutrition-sensitive aquatic food systems	• optimal feed formulation could reduce feed conversion ratios, enhance production efficiency and help minimise costs; tailoring feed composition to ensure aquatic foods have optimal nutrient profiles could maximise beneficial nutritional and health outcomes for consumers	• sourcing feed ingredients for aquaculture must not contribute to food and nutrition insecurity in vulnerable communities elsewhere globally; awareness-raising and behaviour change initiatives may be needed to inform consumers about the health and sustainability gains of an appropriate intake of aquatic foods
100% fish use movement (Sigfússon, 2022) and 'shell biorefinery' (Yan and Chen, 2015)	• recognition is building that utilising all parts of cultured animals and plants has potential to contribute to economic and social development and enhance environmental protection; significant value can be added to by-products and compounds through appropriate biorefinery techniques	• existing uses of less valuable items by poor and marginal groups may be overlooked, with negative social and food and nutrition security outcomes; utilisation of all parts may be limited unless full costs (both monetary and non-monetary) of non-use are included in decision- and policymaking
Nutrient recovery and valorisation	• nutrients that accumulate in sediments and sludge can be used to fertilise terrestrial crops; nutrients entrained in process water can be used to cultivate integrated crops of aquatic animals and plants, adding value and avoiding losses to the biosphere	• crops best suited to reusing nutrients may not match with local customs or consumer demand; intermediate phases (e.g. phytoplankton culture) may be needed to produce organisms to match the requirements of filter-feeding animals (e.g. bivalves)

culture of fish and seaweed (Neori et al., 1996); abalone and seaweed (Shpigel and Neori, 1996; Neori et al., 1998; Langdon et al., 2004); abalone, fish and seaweed (Neori et al., 2000); fish and shellfish (Shpigel et al., 1997); fish, shellfish and seaweed (Shpigel et al., 1993); fish, microalgae and shellfish (Shpigel and Blaylock, 1991); and fish, shellfish, abalone and seaweed (Shpigel and Neori, 1996). Efforts to develop commercially viable production units are discussed further below.

Research at CREAA on the Atlantic coast of France centred on the integration of traditional marine fishponds with phytoplankton bioreactors and ponds for growing and fattening shellfish (Lefebvre et al., 2004). Process water derived from a tidally filled reservoir flowed to ponds stocked with seabass (*Dicentrarchus labrax*) partially fed with supplementary feed and stocked at low densities for overwintering in anticipation of premium early tourist season sales. Fishpond process water flowed to settlement chambers to reduce suspended solids concentrations and then passed to lined phytoplankton bioreactors to fertilise a microalgal (mainly diatoms) culture. Water enriched with microalgae flowed first to a pond holding Pacific oysters (*Crassostrea gigas*) on racks for fattening and then to one stocked with Pacific oysters and clam (*Ruditapes philippinarum*) with on-bottom culture. Prior to discharge, water was directed to a modified fishpond planted with *Salicornia* sp. to improve water quality and sequester nutrients whilst producing a crop valued locally for cooking.

Production of shellfish and seaweed with shrimp farming has been investigated in several tropical countries with a range of integration strategies. Trials conducted by Lin et al. (1993) assessed prospects for culturing green mussels (*Perna viridis*) fastened to bamboo sticks in a drainage canal conveying wastewater from a commercial shrimp farm in Thailand. More contemporary accounts come from Kota Bharu, Malaysia, where Enander and Hasselstrom (1994) cultured hairy cockle (*Scapharca inaequivalvis*) and seaweed (*Gracilaria* spp.) in commercial shrimp farm wastewater. Further experience with integrating seaweeds and bivalve mollusc culture with conventional aquaculture is reviewed below; opportunities for integrating other species groups and prospects for widespread adoption of such strategies by commercial producers are also considered.

Seaweeds[2]

Seaweeds have been proposed as a suitable taxonomic group for horizontal integration because they sequester dissolved nutrients, particularly total ammonia nitrogen (Vandermeulen and Gordin, 1990; Subandar et al., 1993; Neori and Shpigel, 1999). Jiménez del Río et al. (1996) cultured *Ulva rigida* in wastewater from land-based seabream (*S. aurata*) culture units, and results indicated that 153m^2 of seaweed cultivation tanks would be required to assimilate nitrogen discharged from the production of 1 tonne of fish. However, seaweed production and concomitant treatment efficiency decreased during winter, whilst susceptibility to infections increased.

Buschmann et al. (1994) investigated the use of wastewater from an on-shore facility stocked with Pacific salmon (*Oncorhynchus kisutch*) and rainbow trout to cultivate the agarophytic red seaweed (*G. chilensis*) in raceways. The authors considered using wastewater from intensively managed salmonids to cultivate seaweed as an effective strategy in southern Chile. Seaweed production rates (48.9 kg m^{-2} y^{-1}) were reportedly higher than those recorded in traditional open systems or tank cultures in central and northern Chile, and epiphytism did not constitute a major constraint; however, the agar content of the *G. chilensis* decreased during periods of high growth. Martinez and Buschmann (1996) found that the agar content of *G. chilensis* cultured in aquaculture wastewater was lower than that of plants cultured in fresh seawater. These findings may have serious consequences for the financial viability of aquaculture operations that integrate the culture of fish and seaweed as the value of the latter is largely dependent on its agar content (Anonymous, 1996). However, if seaweed produced in integrated systems were to be used in the formulation of feed for herbivorous species (e.g. abalone), the nutritional content, as opposed to the agar content, would also influence its market value.

There are both financial and practical constraints that could limit the application of seaweed to remove nutrients from wastewater discharged from commercial aquaculture. Fouling with particulate matter and biofouling with colonising epiphytes have been widely reported as restricting seaweed production in horizontally integrated units. Phang et al. (1996) found that production of *Gracilaria changii* cultivated in shrimp ponds and an irrigation canal in Morib, Malaysia, was severely affected by a heavy infestation of epiphytes (*Hypnea* sp.); furthermore, siltation, wave action and predation by rabbitfish (*Siganus* spp.) also adversely affected the production of *G. changii*.

The turbid nature of wastewater restricts light penetration into the water column, thereby limiting primary production. Phang et al. (1996) suggested that turbidity in shrimp ponds and the irrigation canal restricted the production of *G. changii*. Faced with this problem, Briggs and Funge-Smith (1996) recommended pre-treatment of shrimp farm effluent using either settlement or bio-filtration using bivalves as a primary stage for integrated seaweed cultivation. Light entering the water column is also rapidly attenuated in productive cultures by plant growth causing self-shading. Problems of self-shading and biofouling can be limited through intensive management; for example, by frequent harvesting and/or agitation by aeration (Briggs and Funge-Smith, 1996). However, the cost of labour required to manage horizontally integrated seaweed culture may represent a substantial burden to the operator, although recent developments such as culturing seaweed in net tubes, as opposed to attaching individual plants to a substrate, may reduce labour demand (Zertuche-Gonzalez et al., 1999). Despite additional costs associated with the integrated cultivation of seaweed, a financial analysis undertaken by Buschmann et al. (1996) suggested that income generated through sale of *G. chilensis* may be significant, making such a venture profitable.

Filter-feeding bivalve molluscs[3]

The negative effect of turbidity on the growth of seaweed may be prevented by using a settlement pond or lagoon to remove particulate matter before it passes into the seaweed culture unit. An alternative approach that develops the concept of horizontal integration a stage further is to use filter-feeding organisms such as bivalves to remove suspended solids, whilst at the same time assimilating nutrients contained within the particulate waste (Helfrich et al., 1995; Soto and Mena, 1999). However, Briggs and Funge-Smith (1996) recommended that the wastewater be pre-settled prior to entering the integrated bivalve culture to reduce fouling with particulate matter.

Pilot-scale experiments conducted by Lin et al. (1993) investigated the culture of the green mussel (*Perna viridis*) stocked on bamboo sticks at a rate of 200–250 and positioned 40–50 cm below the water surface in a drainage canal carrying wastewater from a commercial shrimp farm in Thailand. Survival to harvest was 85 per cent, mean individual mussel weight increased from 12 g to 42 g and mean meat weight increased from 7 g to 28 g during the 113-day grow-out period. The potential of another bivalve, the Sydney rock oyster (*Saccostrea commercialis*), to remove total nitrogen (TN), total phosphorus (TP) and suspended solids (SS) from shrimp (*Penaeus japonicus*) culture wastewater was tested in Moreton Bay, Australia (Jones and Preston, 1999). Pumping wastewater through tanks containing oysters (mean individual wet weight 55 g) at a density of 0.7 per litre resulted in mean inflow concentrations of SS (130 mg l^{-1}), TN (1.4 mg l^{-1}) and TP (0.15 mg l^{-1}) being reduced by 51, 20 and 33 per cent, respectively. Oysters stocked at a density of 0.5 l^{-1} removed only 36, 12 and 17 per cent of SS, TN and TP, respectively. However, the flow rate employed was not reported, so potentially useful indicators such as removal rates per unit time may not be calculated.

Guerrero and González (1991) investigated the culture of the palourde clam (*Ruditapes decussatus*) in wastewater from a turbot farm in Ria de Arosa, Spain. During a 6-month trial, two groups of fifty clams with a total biomass of 1.9 g and 6.4 g suspended in wastewater discharged from the farm increased in weight to 512 g and 1200 g, respectively. The authors suggested that as only 20–30 per cent of N and P added as feed was assimilated in intensively managed farms, unassimilated nutrients and fish metabolites may stimulate phytoplankton blooms, contributing to good growth and survival; mortality rates were below 10 per cent.

Shpigel et al. (1997) studied the particulate removal efficiency of equivalently sized (mean wet weight 5 g) oysters (*Crassostrea gigas*) and clams (*Ruditapes philippinarium*) stocked in 14.4 l troughs receiving 40 l h^{-1} of settled wastewater from ponds stocked with seabream (*S. aurata*). At densities of 7 l^{-1}, the treatment efficiencies of oysters measured against influent turbidity levels of 47 NTU and chlorophyll *a* concentrations of 42 μ g l^{-1} were 87 and 93 per cent, respectively; clams reduced turbidity and chlorophyll *a* by 68 and 83 per cent, respectively. With the same total biomass and mean individual weights, particulate removal by oysters was consistently more efficient than that by clams, and this corresponded with findings from other studies.

Clam seed (*R. decussatus*; mean live weight 0.16 g) stocked at a density of $2.5 \, kg \, m^{-3}$ in 560 l tanks receiving $1120 \, l \, h^{-1}$ of wastewater from a turbot farm in Galicia, Spain, displayed instantaneous growth rates of 0.036 over 60 days, compared with 0.001 in a control tank receiving the same volume of fresh seawater (Jara-Jara et al., 1997). During the study, the Condition Index (CI) of seed in the control tank was lower, with a range of 9–12, whereas that of wastewater-grown clams ranged from 13–19. The levels of all biochemical components measured (e.g. total carbohydrates, total lipids and protein) were higher in wastewater-grown clams, as was the energy content. The CI, biochemical component levels, energy content and survival of seed cultured in wastewater were also higher than those of seed reared under natural conditions. The authors attributed better performance using wastewater to continuous submergence, permitting uninterrupted feeding, compared with the diurnal tidal rhythm and fluctuating food supply due to periodic exposure to air in natural systems.

Integrating bivalve culture into a continuous stream of aquaculture waste-water may, however, cause biochemical changes within the organisms, such as the accumulation of lipids (Jara-Jara et al., 1997), which may affect processing or consumer acceptance. Continuous submergence may also be accompanied by increased fouling with solids and colonising species. Therefore, integration of filter-feeding molluscs may be most appropriate for systems with a variable discharge of wastewater, allowing periodic exposure to the atmosphere, thereby limiting fouling and predation. However, in culture systems where water exchange is restricted, a growing trend with the move to closed water systems for grow-out, it may be necessary to circulate and aerate the water separately in the integrated culture unit to provide dissolved oxygen and limit fouling (Briggs and Funge-Smith, 1996).

Shellfish may harbour bacteria and toxic algae or accumulate chemicals used to treat diseases and remedy water quality problems and therefore should be regarded as a potential human health hazard. This may require operators of integrated systems to avoid chemical treatments in favour of alternative management options and ensure that a period of depuration is observed. Shpigel and Blaylock (1991) raised concerns that horizontally integrated bivalve culture could increase dissolved nutrient loadings due to the excretion of waste products, although in terms of overall mass balance a net reduction in nutrient loadings would be expected. Therefore, where further reductions in nutrient loadings are desirable, seaweed cultivation provides a logical step in developing the integrated system, examples of which are given in the following section.

Combined bivalve mollusc–seaweed culture

Based on a series of pilot studies, Shpigel et al. (1993) modelled the performance of a settlement pond and culture units containing bivalves

(*C. gigas*, *Ruditapes semidecussatus*) and seaweed (*U. lactuca*) to remove particulate matter and dissolved nutrients from wastewater from a land-based seabream (*S. aurata*) culture unit. Fish, seaweed and bivalves were found to assimilate 63 per cent of nitrogen introduced to the system as feed, deposition in faeces accounted for a further 33 per cent and 4 per cent was discharged to the marine environment. Based on nitrogen modelling alone, 3.0 kg of bivalves and 7.8 kg of seaweed could be produced through the assimilation of nitrogen released while producing 1 kg of fish. Although bivalve and seaweed sales potentially could generate income, viability may be limited by the cost of developing the area required as well as the occurrence of markets for bivalves and seaweeds. Prior to commercially applying these research findings, it would be desirable to evaluate the model further, considering the dynamics of other critical nutrients.

Cultivating hairy cockle (*Scapharca inaequivalvis*) and seaweed (*Gracilaria* spp.) in wastewater from a commercial shrimp farm in Kota Bharu, Malaysia, reduced the concentration of ammonium, TN and TP by 61, 72 and 61 per cent, respectively (Enander and Hasselstrom, 1994). Wastewater was passed through two ponds in series, one (12 m long, 2.5 m wide, 0.3 m deep) stocked with 60 kg of cockles and one (12 m long, 1.5 m wide, 0.7 m deep) containing 5 kg of seaweed. Two sets of ponds were operated in parallel and each received 5.5 m^3 d^{-1} of wastewater. Mean TN and TP concentrations in bivalve ponds were reduced by 55 and 67 per cent, respectively, whilst seaweed ponds reduced mean TN and TP concentrations by 36 and 10 per cent, respectively. A mass balance model predicted that during 1 month, bivalve production (30 kg) would sequester 260 g of nitrogen and 9 g of phosphorus, whilst seaweed production (60 kg) would assimilate 5 g of nitrogen and 7 g of phosphorus. The authors noted that bivalves and seaweed are both potentially useful products for shrimp farmers, as they may be either recycled as shrimp feed or sold for supplementary income. Another suggested advantage is that diversifying species may disperse financial risks associated with farming shrimp, although the latter probably would still comprise the most valuable component of the integrated system.

Based on previous studies, Shpigel and Neori (1996) proposed integrating the production of abalone (*Haliotis tuberculata*), macroalgae (*U. lactuca*, *Gracilaria* spp.), Manila clam (*T. tuberculata*) and seabream (*S. aurata*). A modular design was proposed to allow operators to focus on particular components depending on the availability of labour and seed, environmental conditions, market prices and the status of individual components. Commenting on possible operational demands, the authors noted that greater flexibility may be accompanied by greater complexity, requiring more skilled labour and sophisticated management. Neori et al. (1998) conducted a pilot-scale study of one such modular culture unit. Seaweed (*U. lactuca* or *Gracilaria conferta*) was cultivated in 1500 litre tanks receiving wastewater from one of two 600 litre abalone (*H. tuberculata*) culture units. Seaweed biomass produced was used to formulate a mixed feed for the abalone. Nitrogen input to the abalone tanks, each stocked with 235 abalone with a total biomass of 2.2 kg, averaged 494 mg per month, of which 62 mg (14 per cent) was assimilated in biomass; at an areal rate of 105 mg m^{-2} d^{-1}, 284 mg (59 per cent) was discharged in the effluent, while the remainder was unaccounted for. Production of *U. lactuca* (initial standing crop 1.5 kg m^{-2}) averaged 230 g fresh weight d^{-1}, with maximum and minimum values of 412 and 52 g d^{-1} occurring in summer and winter, respectively. Seaweed harvested from the system removed 1621 mg N m^{-2} d^{-1}, accounting for 34 per cent of that entering the system.

Mean annual specific growth rates for abalone ranged from 0.25 to 0.26 per cent d^{-1}, compared with 1.16 per cent d^{-1} reported by Mai et al. (1996) for *H. tuberculata* fed with *U. lactuca*. Poor growth periods were attributed to competition and high feed conversion ratios associated with high summer temperatures, perhaps suggesting that the abalone species cultured was not suitable for the environment. Cumulative annual mortality rates for abalone of 32–39 per cent were attributed partly to smothering and cannibalism, while handling stress following stocking and rapidly declining water temperatures during autumn were considered responsible for periods of high mortality. Growth, and therefore treatment capacity, of *G. conferta* was also affected by variable water temperatures, again indicating a lack of suitability for integration. Although practical management constraints were identified, the authors noted that were integration not practiced, double the water volume would be required to supply the separate culture units, nitrogen from abalone culture units would be discharged to the sea and seaweed cultivation would require nitrogen fertiliser.

Despite the apparent benefits of culturing several species in horizontally integrated systems, conclusions drawn from the pilot studies presented above would require additional validation and testing under a wider range of environmental and operating conditions prior to commercialisation. Horizontally integrated units producing inputs for primary aquaculture activities – for example, feed or seed – have considerable appeal as they potentially increase resilience to external factors. However, management would be subject to similar constraints as those identified for integrating single secondary aquaculture activities, and a higher degree of dependence on 'feedback loop' inputs would increase the need for control. Furthermore, the risk of disease transmission between culture units, both to the primary culture system from feed cultivated downstream and to seed from the on-growing unit would need to be carefully assessed and, where necessary, suitable precautions such as monitoring, depuration or processing should be introduced to avoid such problems.

Complementary species for integration

Assessments conducted under the auspices of the European Commission–sponsored GENESIS (Development of a Generic Approach to Sustainable

Integrated Marine Aquaculture for European Environments and Markets) project identified a range of species, other than seaweeds and bivalve molluscs, that partners deemed appropriate for inclusion in commercial land-based IMTA systems in different climates throughout Europe (Table 4.6).

Species requiring formulated feed inputs included seabass and seabream for temperate conditions, whereas purple sea urchin (*Paracentrotus lividus*) and green tiger shrimp (*Paeneus semisulcatus*) were deemed most promising for warmwater production. Cool temperate production was assumed to focus on intensive, indoor culture of high-value turbot (*Scopthalmus maximus*) and Dover or common sole (*Solea solea*). Shellfish production was not included in the cold-water system, but trials were undertaken in which organisms including *Artemia* spp. and rotifers were harvested from shallow ponds receiving small volumes of discharge water and assessed for their potential for inclusion in vertical integration. No intermediary stages were foreseen for the warmwater system as it was anticipated that constructed wetlands planted with *Salicornia* spp. would perform a dual function, assimilating nutrients and producing biomass with culinary appeal and monetary value. Excepting the warmwater system, phytoplankton production was regarded as an integral component, converting nutrients to biomass to enhance integrated shellfish production.

Outdoor trials with the batch culture of diatoms demonstrated that 90 per cent of inorganic matter was removed from marine fish farm wastewater within 3 to 5 days of inoculation (Lefebvre et al., 1996). Cultures were dominated either by *Skeletonema costatum* or *Chaetoceros* spp. depending on initial nutrient levels, whilst low silicate concentrations were found to limit diatom growth. Silicate supplementation with sodium silicate (Na_2SiO_3) significantly increased production and nutrient removal rates. Issues associated with dilution rates and nutrient supplementation in microalgae cultures are discussed further in Box 4.3. Trials employing completely mixed photobioreactors to culture the non-toxic cyanobacterium (*Phormidium bohneri*) in water from small tanks stocked with rainbow trout resulted in removal rates of 82 and 85 per cent for TAN and soluble orthophosphate, respectively (Dumas et al., 1998). Instantaneous growth rates for *P. bohneri* of 0.03–0.06 d^{-1} were, however, only one-tenth of those for cultures grown on cheese factory waste, attributed to higher TAN and phosphorus concentrations in cheese processing water.

Congruent with findings from other studies and insights from commercial activities, several common constraints can be identified with regards to maintaining microalgae cultures. Dumas et al. (1998) reported that below 15°C, treatment efficiency was significantly reduced. Hence, application of outdoor systems would be restricted to regions with suitable climates or suited only to seasonal operation in temperate environments. Efforts might be made to increase temperatures through coverings, but irrespective of substantial added costs, this would probably impact negatively on incident light penetration and spectral composition. Latitude and day length would

Table 4.6 Species cultured in horizontally integrated units by partners in the GENESIS project

System	Primary (fed)	Biofilter I	Biofilter II (extractive)	Nutrient polishing unit
Warmwater	purple sea urchin (*Paracentrotus lividus*) green tiger shrimp (*Penaeus semisulcatus*)			constructed wetlands planted with *Salicornia* spp.
Warmwater-temperate	gilthead seabream (*Sparus aurata*)	phytoplankton	Pacific cupped oyster (*Crassostrea gigas*) Japanese carpet shell (*Ruditapes philippinarum*)	seaweed (*Ulva lactuca*)
Temperate	European seabass (*Dicentrarchus labrax*)	phytoplankton (*Skeletonema costatum*)	European flat oyster (*Ostrea edulis*) Pacific cupped oyster (*Crassostrea gigas*) Japanese carpet shell (*Ruditapes philippinarum*) hard clam (*Mercenaria mercenaria*)	
Cool temperate	turbot (*Scophthalmus maximus*) common sole (*Solea solea*)	phytoplankton	*Artemia* spp. rotifers	

Source: Bunting (2005).

Box 4.3   Combined phytoplankton–bivalve culture

Hussenot et al. (1998) investigated culturing bivalves (*C. gigas*) on microalgal cultures grown in a 1.2m-deep reactor receiving wastewater from an intensive seabass (*Dicentrarchus labrax*) farm. A mean dilution rate of 70 per cent was maintained in the $48m^3$ reactor. Concentrations of TAN (1.85 mg l^{-1}) and phosphorus (1.1 mg l^{-1}) in wastewater passing through the microalgal reactor were reduced by 67 and 47 per cent, respectively, although total pigment concentrations (0.02 µg l^{-1}) increased by 1139 per cent. Adding sodium silicate to achieve an optimal Si:P ratio of 4:1 ensured that a native diatom (*Skeletonema costatum*) that had shown good growth in a previous study (Lefebvre et al., 1996) proliferated, dominating the algal species leaving the reactor to the bivalve culture unit. Bivalves cultured in $1m^3$ tanks receiving 1.2 m^3 d^{-1} of wastewater and stocked at a biomass of 424 mg l^{-1} removed 50 per cent of total pigments; however, during gametogenesis, the treatment efficiency was lower as food availability was reduced to avoid mortalities. Although promising, the cost of sodium silicate and a system to regulate and monitor its levels in the microalgae reactor may prohibit commercial development, and reduced treatment capacity during gametogenesis may represent a further practical constraint.

Source: Bunting (2001).

---

also influence photosynthesis and govern microalgae production. Poor mixing and stratification can result in sub-optimal growth rates for algae cultures. Competition between nitrifying bacteria and *P. bohneri* was noted as a potential limit to effective operation and treatment of aquaculture wastewater. Grazing by herbivorous zooplankton may also impinge on production as would pests and diseases affecting algae.

Collapses in algal cultures require complete draining, cleaning and reinoculation, resulting in a cessation in food production for shellfish and other filter-feeders downstream; furthermore, associated water conditioning and treatment would be disrupted and mean statutory discharge standards violated. Commissioning multiple photo-bioreactors arranged in parallel may help spread such risks but would introduce new engineering and management challenges as well as cost more. Retention times suited to achieving good algal growth and water quality improvement suggest that large areas of relatively shallow, lined ponds or lagoons would be needed to process water from primary culture systems. Transition to commercial scales may help to introduce more stability and lessen the influence of external environmental conditions, whilst knowledge derived from experiences with commercial microalgae culture facilities in Australia, North America and southern Israel (see Borowitzka, 1993) would assist in designing and operating reliable and efficient integrated systems.

Problems specific to including photo-bioreactors within integrated aquaculture systems can be envisaged. During dark periods, respiration in the culture may significantly reduce dissolved oxygen concentrations, which may have a negative effect on integrated production downstream. Retaining water in relatively shallow and low-volume photo-bioreactors would cause water temperatures to change, and this may negatively impact shellfish physiology and growth. Turbidity originating from fed culture units, a notable problem with earthen fishponds in the CREAA case, necessitates inclusion of settlement lagoons, adding to costs and potentially removing nutrients from water flowing to photo-bioreactors. Water or sediment conditioning inputs and chemotherapeutant applications may have undesirable and unpredictable impacts on other species in integrated systems, whilst integration may result in treatments sanctioned for selected species being prohibited. As the biomass of different species groups within IMTA systems changes owing to growth and cropping and metabolic processes vary in response to environmental change and external stresses, balancing nutrient and energy fluxes between compartments would present major problems. Combined with a need to simultaneously optimise production and mitigation of environmental impacts, managing such systems would constitute an immensely challenging proposition.

Prospects for economically sustainable land-based marine IMTA

Reviewing prospects for IMTA as part of *The State of World Fisheries and Aquaculture 2008*, it was noted that 'by converting solid and soluble waste nutrients from fed organisms and their feed into harvestable crops and/or extractive organisms (thereby reducing the potential for eutrophication) and by increasing economic diversification, integrated multitrophic aquaculture promotes economic and environmental sustainability' (FAO, 2009: 22). Considering integrated tilapia and shrimp culture in America and integrated aquaculture system at Mikhmoret, Israel, an article in *The Economist* ('The Promise of a Blue Revolution') (Anonymous, 2003: 21) concluded that 'such land-based, integrated farming techniques offer great promise but with minimal environmental cost'. Outcomes of a stakeholder Delphi focused on discerning better prospects for horizontal integration and differing stakeholder perspectives regarding such strategies demonstrated that a range of opportunities are apparent across environmental, physical, social, managerial, economic and institutional domains (Table 4.7).

Joint and interdisciplinary assessment by stakeholders concerning potential benefits of IMTA and horizontally integrated aquaculture appears to demonstrate that such strategies should convey a range of benefits at multiple scales. Widely cited attributes include enhanced environmental protection,

Table 4.7 Opportunities participants associate with horizontally integrated aquaculture, frequency of occurrence in Round 1 (n) and both mean score (x) and mean ordinal rank following Round 3

Systems feature	n	x	Rank
Environmental			
Reduced impact on the environment and downstream users; e.g. other farms	8	8.5	1.5
Reduced sediment and nutrient concentrations in wastewater	8	8.3	3
Less energy consumed and waste generated in food production	1	8.3	4
Increased habitat diversity, providing shelter to endemic species	3	6.5	17
Physical			
Improved efficiency of resource use; e.g. nutrients and water	7	8.5	1.5
More attractive in the landscape than conventional treatment systems	2	6.4	16
Better transport around remote areas	1	3.7	27
Social			
Improved public perception; reconciling environmental and economic goals of different groups	6	7.6	5
Increased employment opportunities	5	6.5	15
Enhanced appeal of the primary aquaculture product to consumers	5	6.4	19
Represents a good educational resource	1	6.3	20
Better places to live	1	5.2	26
Managerial			
Reduce the potential for farms to self-pollute	2	7.2	6
Meet expectations of managers regarding environmental protection and waste management	2	7.0	8
Increase opportunities for recirculation	1	7.0	11
Facilitate the development of a new paradigm for aquaculture and other agricultural sectors	2	6.6	13
Contribute to the skill base of managers	1	5.6	24
Economic			
Increased income generated from additional crops based on same inputs	12	7.2	7
Reduce wastewater treatment costs	1	7.0	9
Reduce economic risks through diversification	5	6.5	14
Reduced the level of any potential pollution tax	6	6.1	22
Contributes to the regional economy, adding to tax base	2	5.2	25
Institutional			
Helps meet the standards for quality assurance or organic certification	1	7.2	10
Enabling more stringent discharge standards to be satisfied in the future	6	6.8	12
Potential site for research and development	2	6.5	18
New opportunities to develop commercial partnerships	2	6.2	21
Improved discharge standards that reduce penalties; e.g. court action or closure	2	6.1	23

Source: Bunting (2008).

product diversification and economies of integration (Whitmarsh et al., 2006; FAO, 2009). Integration that facilitates the productive use of nutrients released from fed culture units should reduce the ecological footprint or ecosystem support area required to process and assimilate waste. Combining production of complementary aquatic species could potentially make more efficient use of resources (by-products, nutrients and water), farm infrastructure (land, buildings, machinery and services), husbandry skills and marketing opportunities (Bunting, 2001; Muir, 2005). Culturing several species may contribute to more regular cash flows and, in certain cases, earlier returns on investments when species with shorter grow-out periods (e.g. seaweed and fish) are combined with slow-growing species (e.g. abalone and sea urchin). Overall returns on capital and operating costs are generally expected to be higher, especially where integration results in enhanced public perceptions that translate into price premiums (Ferguson et al., 2005; Bunting and Shpigel, 2009).

Although IMTA has gained a great deal of attention as a prospective strategy to achieve more sustainable aquaculture production, limited development in terms of geographical spread and production volume signifies enduring constraints to commercial development. Joint stakeholder assessment identified financial costs associated with development as the principal constraint to adoption of horizontally integrated production (Table 4.8). This highlights the need to restrict costs to a minimum, optimise financial returns and provide appropriate incentives to producers where proposed integration strategies are founded on principles of responsible and sustainable production.

Financial considerations are likely to be foremost in the minds of potential producers and investors and govern decision-making when contemplating alternative production strategies. Acknowledging this, several assessments of IMTA strategies have adopted a bioeconomic modelling approach (discussed further in chapter 7) combining biological stock models and production estimates with financial assessments that employ standard accounting approaches and economic indicators. Outcomes from bioeconomic modelling assessments dealing with open-water integration are discussed further below.

Bioeconomic modelling of integrating mangrove constructed wetlands to assimilate nutrients from shrimp culture wastewater predicted that water quality would be improved and stocks and flows of ecosystem services enhanced but that financial returns from selectively harvesting mangrove biomass would probably not attract investment (Bunting, 2001). Bioeconomic modelling was employed in the European Commission–funded GENESIS project to evaluate warmwater and temperate, land-based IMTA strategies in Israel and France, respectively (Bunting and Shpigel, 2009). Based on prototypes developed by the Israel Oceanographic and Limnological Research (IOLR) Centre, the warmwater baseline case combined sea urchin (*Paracentrotus lividus*), shrimp (*Paeneus semisulcatus*) and samphire (*Salicornia* spp.) cultivation. Stocking densities, survival rates, product values and land and labour costs assumed for the warmwater baseline are summarised in Table 4.9.

Table 4.8 Constraints stakeholder Delphi participants associated with horizontally integrated aquaculture, frequency of occurrence in Round 1 (*n*) and mean score (*x*) and mean ordinal rank following Round 3

Systems feature	*n*	*x*	*Rank*
Economic			
Financial costs associated with development	8	7.3	1
Lack of funding, access to venture capital	2	6.9	4
Market analysis and stimulation required	3	6.5	8
Costs associated with management and operation	3	5.9	10
Holders of economic power not interested	1	5.3	21
Limited revenue generated by the integrated system	4	5.0	25
Insufficient techniques for analysis of operating costs and accounting for broader issues such as opportunity costs	2	4.9	27
Physical			
Availability of suitable land or water	11	7.4	2
Wastewater supply; e.g. nutrient flows not optimal or presence of chemicals, antibiotics or pathogens	4	7.1	6
Biological treatment cannot remove nutrients effectively throughout the growing season	1	5.8	16
Engineering of wastewater flows difficult; e.g. high volumes and distance	2	5.4	22
Facility designed primarily for original aquaculture species	1	4.7	28
Managerial			
Decisions based on short-term financial appraisal	1	7.2	3
Limited knowledge base regarding design and management; e.g. optimal loading rates, harvesting strategies and disease management	6	7.0	5
Managers lack leadership, resist change and adopt a non-systematic approach	5	6.4	9
Lack of skills and training for existing managers and workers	4	6.0	14
Inadequate supplies of seed and inputs for integrated system	1	4.9	26
Environmental			
Discharge requirements; by-products from integrated systems may not comply with regulations	2	6.5	7
Site-specific design required; e.g. climate vulnerability	4	5.7	14.5
External pollution could affect the integrated system	1	5.4	20
Translocation of organisms suitable for integrated production restricted	1	4.3	29
Institutional			
Inertia; i.e. a lack of research and development	1	6.0	11
Environmental laws and planning restrictions constrain integration	3	5.8	14.5
Constrained by existing paradigms of food production; i.e. no structures exist to facilitate systems thinking	2	5.8	17
Absence of instruction and training support	1	5.9	18
Lack of structures for monitoring and auditing integrated systems	1	5.6	19

(Continued)

Table 4.8 (Continued)

Systems feature	n	x	Rank
Social			
Public acceptance of produce because of a perception as 'dirty' food	4	5.8	13
Conflicts with stakeholder groups; e.g. user groups and environmentalists	4	5.3	23
Limited systems thinking education and information exchange	1	5.2	24

Source: Bunting (2008).

Table 4.9 Baseline assumptions and financial parameters for the warmwater prototype

Parameter	Value	Units
Sea urchin stocking density	350	no. m^{-2}
Sea urchin mortality	15	$\% \ y^{-1}$
Sea urchin value	21	€ kg^{-1}
Shrimp stocking density	281.25	no. m^{-2}
Shrimp mortality	40	%
Shrimp value	6.9	€ kg^{-1}
Salicornia spp. stocking density	7.5	no. m^{-2}
Salicornia spp. value	10	€ kg^{-1}
Land cost	2759	€ ha^{-1}
Labour cost	5	€ h^{-1}

Source: Bunting and Shpigel (2009).

Estimated financial returns for the baseline warmwater system producing 29 t y^{-1} of shrimp, 9.5 t y^{-1} of samphire and 50 t y^{-1} of sea urchin from year 3 onwards generated a net profit of €432,420, whilst the net present value (NPV) of discounted cash flows employing a discount rate of 10 per cent was €459,090 for the first 10 years of operation (Table 4.10). Future scenarios assumed for the purposes of the assessment demonstrated that a fall in value of sea urchin from €21 to €18 kg^{-1} resulted in a negative 10-year NPV of −€83,060, indicating that the venture would not be viable. Action to reduce the mortality rate for sea urchin from 15 to 9 per cent annually would increase the NPV of the investment to €1,305,780 over 10 years. Assuming extrapolated samphire yields of 33 kg m^{-2} from trials at IOLR were attained from the commercial systems considered here resulted in a significant increase in the NPV to €6,147,690 over 10 years. Further work is required to assess whether such samphire yields are achievable under commercial conditions and to better define the likely costs of production.

Bioeconomic modelling enabled simultaneous assessment of key physical, managerial and financial aspects of IMTA systems. Previous assessments of

Table 4.10 Financial parameters and key indicators of performance for warm water prototypes

Parameter		Baseline	Falling sea urchin price	Reduced sea urchin mortality	Salicornia spp. production increase
Capital costs (€)					
Land and site development		27,380	27,380	27,380	27,380
Farm services and infrastructure		176,420	176,420	176,420	176,420
Culture unit materials		1,746,490	1,746,490	1,746,490	1,746,490
Total (€)		1,950,290	1,950,286	1,950,290	1,950,290
Operating costs (€)					
Stocking and marketing		391,820	391,820	395,060	638,800
Feed		243,500	243,500	243,500	243,500
Electricity		163,400	163,400	163,400	163,400
Labour		82,500	82,500	82,500	82,500
Maintenance		19,500	19,500	19,500	19,500
Total (€)		900,720	900,720	903,970	1,147,710
Income (€)	Sea urchins	1,037,550	889,330	1,273,140	1,037,550
	Shrimp	200,090	200,090	200,090	200,090
	Salicornia spp.	95,500	95,500	95,500	1,400,680
Profit excluding depreciation (€)		432,420	284,200	664,760	1,490,610
Rate of return on capital costs (%)		22.2	14.6	34.1	76.4
Rate of return on operating costs (%)		48	32	74	130
Payback period (y)		4.5	6.9	2.9	1.3
NPV at:	5%	949,870	208,990	2,107,680	8,278,500
	10%	459,090	−83,060	1,305,780	6,147,690
	15%	137,680	−267,780	770,450	4,662,320
	20%	−75,440	−384,620	406,700	3,600,670
IRR (%) over:	10 years	18.0	8.3	29.4	133.4

Source: Bunting and Shpigel (2009).

IMTA systems focused on technical constraints or optimising treatment performance or productivity of selected integrated components; few considered practical and financial implications for operators or broader economic implications. Moreover, joint assessment and model formulation with commercial operators produced a tool permitting them to independently assess design modification, alternative production scenarios and the sensitivity of financial returns to changing costs, varying production levels and evolving market conditions.

Modelling outcomes demand consideration from a systems-based perspective. Although bioeconomic modelling permits important considerations such as treatment performance, productivity, financial returns and economic indicators to be assessed, these constitute only parts of the systems context for aquaculture development; environmental, legislative, social and ethical aspects must be considered also. Consumer perceptions and preferences would be critical and targeted market research would be advisable to sample consumer opinion and gauge likely demand and prospects for price premiums. Broader assessments of production trends, returns from substitute capture fisheries, import and export statistics and prevailing market arrangements for products amenable to culture in temperate and warmwater IMTA systems in Europe shed light on the most promising candidate species and areas for further investigation (Table 4.11). Similar assessments should be contemplated when considering IMTA development in other geographical settings to identify promising native species combinations. Renewed analysis would be warranted also in the European context, as opportunities and constraints vary and new candidate species for culture may emerge.

Open-water IMTA

Prospects for sequestering nutrients through integrated rope-based red seaweed (*G. chilensis*) cultivation adjacent to cages producing 227 t y^{-1} of rainbow trout (*O. mykiss*) and Coho salmon (*O. kisutch*) in Chile were evaluated by Troell et al. (1997). Extrapolated findings from field trials conducted in Metri Bay, Puerto Montt, indicated that 1 ha of seaweed culture would sequester amounts of nitrogen and phosphorus equivalent to 5 and 27 per cent of that released from the cages annually. Production efficiency gains and feed formulation enhancements since this study probably mean a greater proportion of nutrients released from the farm would be retained were this area of seaweed to be cultured.

Bioeconomic modelling of integrated kelp (*Laminaria saccharina*) culture adjacent to a salmon farm in Canada producing 250 t y^{-1} predicted that arrays of ten 60 m ropes positioned at each end of the cage farm would yield 32t of wet biomass. Production at this level would equate to a yield of 1.6t of dried seaweed annually with a farm-gate value of CA$35 kg^{-1} (equivalent to US$41.65) being assumed (Petrell et al., 1993). With initial capital costs of CA$61,000 (equivalent to US$72,590),[4] the pay-back period was calculated at 6 years, whilst the NPV of cash flows was 25 per cent over the same period and comprising 1 year for setup and 5 years of sales. Options identified for increasing financial returns, thus making kelp production in association with salmon farming a more economically viable proposition, included increasing the number of 60 m kelp ropes to thirty at each end of the farm whilst reducing spacing to 1 m apart, adopting a vertical rope-based kelp culture system and spreading investment costs across several sites.

Modelling estimated that with current velocities of 0.1 m s^{-1} and kelp oxygen consumption rates of 0.026 mg O$_2$ g^{-1} h^{-1}, less than 1 per cent of the available oxygen would be consumed. Even if higher temperatures doubled

Table 4.11 Marketing opportunities and constraints for products from integrated units in Europe

Culture species	Opportunities	Constraints
Fed species seabass (*D. labrax*)	• growth in aquaculture sector good • established market • opportunity to capitalise on green credentials with labelling	• capture fishery stable • competition from wild catches and other producers
seabream (*S. aurata*)	• growth in aquaculture sector good • opportunity to capitalise on green credentials with labelling	• capture fishery stable • modest product value • competition from wild catches and other producers • modest growth in aquaculture sector
turbot (*S. maximus*)	• product value high • wild fishery in decline • natural distribution extends to Northern European countries	• newly established aquaculture sector with potentially limited capacity for expansion
common sole (*S. solea*)	• wild fishery in decline	• aquaculture sector not established
green tiger prawn (*P. semisulcatus*)	• significant import trade of shrimp and prawns to Europe	• aquaculture sector not established • absence of traditional market based on wild catch
purple sea urchin (*P. lividus*)	• marked decline in wild harvest of urchins • established export market, although in decline	• aquaculture sector not established
Biofilter I phytoplankton	• may represent an important input to other integrated units	• aquaculture sector not established • external market channels highly specialised • primary production seasonal in Northern Europe
Biofilter II European flat oyster (*O. edulis*)	• marked decline in wild harvest • established market • plan to produce 'spéciale' oysters that command a premium	• modest growth in existing aquaculture sector • modest product value • possible sanitary problems • uncertainty over consumer acceptance

(Continued)

Table 4.11 (Continued)

Culture species	Opportunities	Constraints
Japanese carpet shell (*R. philippinarum*)	• product value high • growth in aquaculture sector good	• capture fishery for other carpet shell species stable
Pacific cupped oyster (*C. gigas*)	• rapid decline in wild harvest • established market	• no recent growth in aquaculture sector • product value low • possible sanitary problems • uncertainty over consumer acceptance
hard clam (*M. mercenaria*)	• established market	• aquaculture sector not established • capture fishery for other clam species stable • uncertainty over consumer acceptance
brine shrimp (*Artemia* spp.)	• may represent an important input to other integrated units	• dedicated aquaculture sector not established in Europe • external market channels highly specialised
Polishing unit *Salicornia* spp.	• small local market established	• aquaculture sector not established • limited and seasonal market for wild production • possible sanitary problems and uncertainty over consumer acceptance
Seaweed (*U. lactuca*)	• recent increase in wild harvest of green seaweed • may represent an important input to other integrated units	• aquaculture sector shows signs of decline • primary production seasonal in Northern Europe

Source: Adapted from Bunting (2005).

kelp respiration rates, it was concluded that this would not pose a threat to salmon production. Whilst predicted oxygen consumption would not preclude deployment of thirty ropes 60 m in length at each end of the cages, it was noted that diminishing dissolved inorganic nitrogen levels further from the cages may limit kelp production. Potential benefits to salmon farmers regarding enhanced public perception and potential price premiums associated with integrated production were not considered in the assessment and would have enhanced predicted financial returns.

Monitoring mussels (*Mytilus edulis*) positioned adjacent to salmon cages on the west coast of Scotland, Stirling and Okumus (1995) observed that temperature and food availability governed growth. Consequently, mussel growth was seasonal – higher from May to October but significantly lower at other times. Growth was better in proximity to the salmon cages compared with a local shellfish farm; this was attributed to higher suspended organic matter concentrations at the cage site. Thus, seasonal variation in growth and concomitant nutrient assimilation may limit the effectiveness of mussels as biofilters for open-water aquaculture operations. Seasonal changes in daylight duration in temperate areas and generally lower light levels at higher latitudes would restrict the use of plants and seaweeds to sequester nutrients and assimilate them in biomass suitable for feeding to other integrated species groups.

Economic prospects for integrated open-water culture of 600 t y^{-1} of salmon in cages and 16,000 m of mussel ropes were assessed using bioeconomic modelling by Whitmarsh et al. (2006). They concluded that this arrangement had commercial potential with net cash flows having an NPV of £1.425 million over 20 years, but assessing the sensitivity of returns to a downward trend in salmon prices of 2 per cent annually failed to generate a positive NPV, indicating that the venture would not be financially viable. Enhanced mussel productivity of 20 per cent owing to the proximity of salmon cages for the integration baseline was predicted to result in an increase of £150,243 in the NPV of the investment compared with farming mussels and salmon separately.

Although experimental results and financial modelling appeared to demonstrate that integration strategies for seaweed and shellfish culture with open-water aquaculture are feasible and would generate attractive returns for investors, empirical evidence to substantiate claims concerning production rates, management demands, financial returns and economic performance is limited. Trials with commercial partners were conducted on the west coast of Scotland to evaluate integrated Atlantic salmon (*Salmo salar*) culture with sea urchin (*Psammechinus miliaris* and *Paracentrotus lividus*) and seaweed (*Palmaria palmate, Laminaria digitata, L. hyperborea, Saccharina latissima* and *S. polyschides*), organic salmon with oyster (*Crassostrea gigas*) and king scallops (*Pecten maximus*) and sea urchin (*P. lividus*) and mussel (*Mytilus edulis*; Barrington et al., 2009). Collaboration between researchers and producers is essential to adequately test and evaluate whether proposed integration strategies would be commercially viable. Commercial-scale deployment of IMTA approaches incorporating seaweed and mussel culture has been undertaken by salmon farmers in the Bay of Fundy, Canada (Barrington et al., 2009). Important indicators of success would be continuity of operation and adoption of such strategies by other producers in Canada or elsewhere based on the experiences of these innovative producers. A comprehensive review of IMTA conducted by Sickander and Filgueira (2022) found that the practice 'is currently not applied as a mitigation measure in Atlantic salmon farms' (1). These authors noted that capital and maintenance costs constituted barriers to adoption, as did a lack of government support.

Small-scale, low environmental impact innovations

Calls for appropriate, low-impact marine aquaculture development to promote economic development in poor coastal communities, diversify livelihoods away from damaging fishing activity and bolster food security came from several highly respected institutions, notably FAO (1995), the United Nations Environment Programme (UNEP, 2006) and the World Bank (2007). Initiatives promoted small-scale cage culture in Indonesia and Vietnam, but issues with seed and feed supplies constituted a constraint, environmental impacts were of concern and prospects for poverty alleviation were unclear (Bulcock et al., 2000; Briggs, 2003). A notable limitation with advocating open-water marine aquaculture development for social–economic development is that coastal conditions demand robust construction and anchorage, even for small-scale developments, costs of which raise immediate concern regarding financial viability and associated risks to producers. Ventures to cultivate sedentary species and plants, thus avoiding containment costs, appear to be more suitable for marginal coastal communities. Despite lower capital costs, technical challenges remain to be overcome, as well as ongoing costs and demands on time and resources and risks demanding assessment and appropriate mitigation and management.

Several species have been proposed for small-scale marine aquaculture to diversify local livelihoods dependent on fishing include: open-water seaweed (*Eucheuma spinosum*) farming in Zanzibar and Tanzania (Pettersson-Lofquist, 1995); intermediate culture of topshell (*Trochus niloticus*) in reef-based cages in Vanuatu (Amos and Purcell, 2003); small-scale slipper oyster (*Crassostrea iredalei*) and green mussel (*Perna viridis*) in Western Visayas, Philippines (Siar et al., 1995); village-based giant clam (*Tridacna* spp.) culture in the Solomon Islands (Hart et al., 1999); and pen culture of mud crab (*Scylla serrata*) in Mtwapa mangrove, Kenya (Mwaluma, 2002). Supplying inputs and processing products to reduce transport costs and add value could potentially contribute further to enhanced livelihood opportunities (Short et al., 2021).

Continued success of such activities, however, often becomes dependent on multi-actor marketing arrangements, frequently beset by patron–client issues to regional and export markets. Concern is warranted, consequently, over the proportion of revenue generated that reaches the producers and increased vulnerability of aquaculture-adopting coastal communities to fluctuating prices on global markets and competition from emerging producer groups. Locally inappropriate innovations can result in undesirable environmental and social impacts. Apparently successful establishment of backyard hatcheries for milkfish in Gondol region, Bali, was accompanied by an influx of migrant workers who overwhelmed local infrastructure and services and disrupted prevailing social arrangements. Considering the legacy of seaweed farming in Zanzibar, Pettersson-Lofquist (1995: 491) speculated that 'local economies will increasingly become tied to the future price of the algae and, consequently, the means for power and control over local livelihoods will be handed over to market forces'. Widespread negative health impacts were found amongst female seaweed cultivators in Zanzibar, East Africa (Frocklin et al., 2012), thus highlighting the need for appropriate risk assessments and corrective actions to avoid such problems, especially when small-scale female producers are involved, as such groups may be particularly vulnerable.

Organic and ethical marine aquaculture

Organic salmon farming has been practiced for 2 decades and may be regarded as established. Despite supplying premium products to market, however, organic producers must still compete with a growing array of salmon products that are certified as responsibly or sustainably produced. Certification schemes often address different aspects of production, and associated standards may demonstrate notable variations (Bunting et al., 2022). Publication of standards for organic Atlantic salmon farming by the Soil Association, the UK certification body, proved controversial as some felt this occurrence threatened to undermine the founding principles of the organic movement. Opponents objected to cages restricting the natural migratory behaviour of wild salmon and permitted releases of untreated cage waste to receiving water bodies and raised concern that pathogens from farmed fish would still infect wild fish (BBC, 2006). Organic salmon produced in the Shetland Isles, Scotland, has received numerous awards and accolades, but a venture to produce organic farmed cod in the Shetland Isles failed owing to financial troubles, demonstrating that attaining organic certification does not necessarily guarantee financial security or peace of mind.

Progress with organic aquaculture certification and standards for shrimp feed, for example, that stipulated that 'fishmeal content as well as the total protein content of compound feed shall be reduced as far as possible' and maximum levels set at '25 per cent for fishmeal/oil content and 25 per cent for total protein' (Naturland, 2005: 17) have highlighted the major problem of unsustainable fisheries supplying fish meal and oil to the global aquaculture industry. Criticisms levelled at fisheries producing fish meal and oil are that fish-stocks are being depleted to dangerously low levels and are unable to replenish, food security is affected in poor and marginal coastal communities where fish are landed but are not accessible for local consumption and feeding fish meal and oil from capture fisheries to farmed carnivorous species can be inefficient, making some aquaculture practices net consumers of energy, protein and fatty acids deemed essential for human development and nutrition (Naylor et al., 2000; Roberts, 2007). Given demand for responsibly sourced fisheries products for feed formulation, several capture fisheries targeting species destined for meal and oil production are undergoing or have completed assessments and have been certified. Standards governing certification include the independently audited 'Global Standard for Responsible Supply of Fishmeal and Fish Oil' from the International Fishmeal and Fish Oil Organisation (IFFO, 2012) and the Marine Stewardship Council's 'MSC Environmental Standard for Sustainable Fishing' (MSC, 2010).

Mounting pressure to adopt responsible aquaculture practices, spreading coverage of codes of conduct and growing awareness concerning good, best and better management practices mean that increasing numbers of producers are moving closer to conforming with certification scheme standards. Joining a formal scheme that acknowledges good practice might mean that producers could benefit from a price premium and guaranteed market access. Smaller producers and those with limited assets may require assistance and support, however, to meet transaction costs, modify facilities and husbandry practices and fulfil administrative obligations. Strategies developed to overcome barriers to organic certification and benefits experienced by one shrimp producer in Ecuador are described in Box 4.4. Apparent non-compliance with organic standards by certain shrimp producers in Indonesia and Ecuador was reported by Ronnback (2003) and the Swedish Society for Nature Conservation (2006); these included marginalisation of local inhabitants and failures to replant 50 per cent of the former mangrove area within 5 years of certification. Thus, this brings into sharp focus major issues that, if left un-checked, could bring organic certification of shrimp into disrepute and highlights the need for more rigorous compliance checks.

Artificial reefs

Deployment of artificial reefs in conjunction with aquaculture development has been proposed to sequester nutrients released from conventional open-water aquaculture and provide extra habitat for aquatic animals as part of stock enhancement and ranching initiatives. When aquatic animals, notably shellfish seeded on artificial reefs, are harvested, this could be considered a form of IMTA or horizontally integrated aquaculture. Results of artificial reef deployment in the Baltic Sea are presented below, and considerations concerning artificial reef configuration and positioning are discussed.

Artificial reefs for stock enhancement

Assessing retention of tagged lobsters on an artificial reef constructed from gypsum, pulverised fuel ash, gravel and cement in Poole Bay, UK, it was

---

Box 4.4   Organic shrimp farming, Bahía de Caráquez, Ecuador

Shrimp farming in Ecuador, once the world's largest producer, has been affected by a number of problems: self-pollution, shrimp disease outbreaks, contamination with chemicals and social injustice as traditional resource users were excluded and land ownership was transferred to a few powerful families. In response to these problems and protests and boycotts in Europe, the owners of EcoCamaronera Bahía sought advice to address the situation. They began by planting mangrove trees on the embankments and in ponds; other vegetation in the ponds was also allowed to develop as a wildlife habitat. In the late 1990s, the initiative was taken to convert the farm to organic production, multi-cropping of embankments was established, organic feeds were formulated and effluent monitoring was initiated. A notable problem in converting to organic production was identifying a suitable replacement source of protein for feed formulation. The solution found was to use the protein-rich seed pods of leguminous trees planted on embankments. Following these developments, the farm was inspected and certified by Naturland, Germany, as organic; subsequently, two neighbouring farms adopted similar management approaches and were inspected and certified. Adopting organic practices also led to wider benefits; modified management approaches such as collecting seed pods resulted in increased employment for local community members, and the diverse bird fauna resident at the farm attracted birdwatchers and tourists who contributed to the local economy.

Source: Bunting (2007).

---

noted that there was a 'varying degree of loyalty to the reef', with some animals rapidly dispersing, whilst others, including berried females, remained for several months (Lockwood et al., 1991: 39). Site selection involved negotiation with representatives for local fishermen, and successful monitoring was attributed to professional fishermen respecting an agreement not to target the area with lobster pots. Evaluating release strategies for hatchery-reared lobsters, including release by divers, pumping through a tube to the sea bed or directly from the shore at spring low tides and by stakeholders (fishermen) deploying creels containing juvenile lobsters, it was observed that stakeholder involvement enhanced the durability, flexibility, reliability, suitability and range of releases (Ellis and Boothroyd, 2008). Reviewing practices at the lobster hatchery operated by Orkney Sustainable Fisheries, Scotland, these authors noted that 'increasing the fishery stakeholders' contributions to a stock enhancement scheme has been shown to lead to the fishing community beginning to take ownership of the stock enhancement programme' and contribute a levy on mature lobsters caught to cover hatchery running costs (Ellis and Boothroyd, 2008: 5).

Early trials with enhancing lobster (*Homarus gammarus*) stocks in the UK were conducted in Bridlington Bay, England and Cardigan Bay, Wales. During the period from 1983 to 1988, some 49,000 juveniles were released in Bridlington Bay, whilst in Cardigan Bay just over 19,000 were released between 1984 and 1988 (Burton et al., 2001). Over 600 lobsters ranging from 3 to 9 years in age were recaptured in Bridlington Bay within 5 km of the release site by 1993, although this accounted for just over 1 per cent of those released. Promisingly, from the perspective of local fishing interests, a significant proportion of the lobsters landed were large enough to be kept, and ten were gravid with eggs and were hence returned. This demonstrated that released juveniles could reach maturity and reproduce successfully. Despite a measure of success, it was concluded from an economic perspective that lobster ranching was 'non-commercial at present and the decision as to whether to proceed with stocking initiatives is largely political' (Burton et al., 2001: 4).

Artificial reefs for water quality enhancement and nutrient sequestration[5]

Employing artificial reefs to remove nutrients from aquatic ecosystems is closely related to ranching as it is a strategy that facilitates horizontal integration in association with open culture facilities. Laihonen et al. (1997) outlined the theory of removing nutrients from aquatic ecosystems using artificial reefs. These represent an increased area capable of supporting colonial organisms and other assemblages (e.g. grazers and predators). The growth of these assimilates nutrients in biomass that may be harvested, resulting in a net removal of nutrients and output of value from the ecosystem. Laihonen et al. (1997) reported the effectiveness of using a variety of artificial reefs to remove nutrients from the Baltic Sea, proposing that sessile filter-feeding organisms and seaweeds that are capable of efficient and rapid assimilation of dissolved nutrients are likely to be most promising.

Concrete pipes (0.6 m in diameter and 1 m long) were used by Laihonen et al. (1997) to construct star-, tube- and pyramid-shaped artificial reefs at depths of 9–12 m in both inshore and offshore locations in Pomerian Bay, Poland. Offshore, colonisation of the star-shaped reef was highest at 216,030 individuals m^{-2}; inshore the highest settlement level (112,500 individuals m^{-2}) was observed on the pyramid reef. Mussel (*M. edulis*) and barnacle (*Balanus improvisus*) dominated both reefs. The filtration rate of the organisms colonising the reef was estimated at 5500 m^3 m^{-2} y^{-1}. The authors linked colonisation of the artificial reefs in Pomerian Bay with increased visibility, rising from 2–5 m in June 1991 to 5.5–7 m in October 1993, and decreased nitrate concentrations, falling from 0.5–2 mg l^{-1} to <0.5 mg l^{-1} during the same period. However, data from control sites and monitoring prior to establishing the reefs were not presented, making it impossible to assess the influence of the reefs with respect to regional water quality trends.

Several factors need to be addressed when considering artificial reefs (Laihonen et al., 1997). The physical, chemical and biological conditions at the site must be assessed, temperatures should be high enough to allow a reasonable level of biological activity, light penetration into the water column should be sufficient to allow photosynthesis and nutrients circulating around the reef should be readily bioavailable. The reef should also be positioned as close as possible to the nutrient source as the dilution of nutrients would reduce the removal efficiency of the reef system. The dispersal of waste discharged from open culture systems – for example, cages and pens – is a dynamic process and depends on a range of factors. The physical properties of the waste, the topography and composition of the substrate underlying the site, tides and currents that affect the hydrology of the site and the interaction of natural biota all influence the ultimate dispersion pattern. Detailed information is therefore required to produce a reliable model of expected dispersion, thereby allowing the horizontally integrated culture unit or artificial reef to be positioned correctly.

Fisheries stock enhancement, culture-based fisheries and ranching

Historically, people living in coastal communities benefitted significantly from the capture and harvest of natural resources along the coast and, as fishing technology developed, further offshore. They were vulnerable, however, to the vagaries and ravages of the sea and endured seasonal resource scarcity and less predictable fluctuations in annual catches. As early as the 1880s, coastal communities and administrations determined to release huge numbers of artificially propagated Atlantic cod (*Gadus morhua*) larvae to counter natural fluctuations in cod stocks. Enlightening and detailed accounts of developments at Flodevigen, Arendal, Norway, and Woods Hole, Massachusetts, USA, were provided by Svasand and Moksness (2004). Captain Gunder Mathisen Dannevig reportedly established a hatchery for Atlantic cod in Flodevigen in 1882 with funds received from the local community and Norwegian government. Mature male and female cod were held in enclosed outdoor basins and viable eggs were obtained from natural spawning and then transferred to the hatchery for incubation. Several million yolk-sac larvae were released annually, but an absence of suitable tagging methods meant that it was impossible to assess the success or otherwise of their endeavours. Interest waned, but cod releases continued until 1971 in Norway.

Trials more recently with genetically marked yolk-sac-stage larvae revealed that benefits are very small. Alongside the development of new tagging techniques, however, interest in stock enhancement has increased. Instead of using larvae, more developed juvenile stages are being released. According to

Svasand and Moksness (2004), survival depends on several factors, including time and size at release, habitat selection, carrying capacity of the release area, prey availability, predator abundance and ability to adapt to the wild. Furthermore, based on comprehensive review of accounts of stocking Atlantic cod, Svasand et al. (2000) identified a number of key findings:

- cultured cod juveniles adapted well and after only a few weeks prey selection was close to that of wild fish;
- released cod suffered higher mortalities, and time spent in captivity influenced migration patterns and anti-predator and feeding behaviour;
- genetic selection and genetic drift were considered to be potentially minor problems, with few differences observed between genotype distribution and gene frequencies for wild and reared cod;
- restocking programmes prompted development of efficient tagging and marking methods;
- most cod stocked remained in the release areas, although there was some variation;
- positive correlation between size at release and survival was noted in most release areas;
- recapture rates varied from 0 to 30 per cent depending on area, time and size of release;
- growth rates were highest in outer coastal areas.

Juveniles released in Norway during the 1980s and 1990s, it was concluded, did not significantly increase cod production and catches, although it was proposed that future cod releases could increase the attractiveness of coastal areas for fishing and tourism. Lobster stock enhancement was considered by Ellis and Boothroyd (2008: 4) to 'complement more traditional fisheries management tools such as minimum landing sizes, V notching, closed season fisheries, closed fisheries for berried hens and permit systems'. Together such measures were regarded as a safety net preventing stock collapses, but opting for stock enhancement is beset with ecological, social and financial risks.

Advocated by Blankenship and Leber (1995), a responsible approach to marine stock enhancement encompassed ten key elements ranging from species selection to genetic resources, disease and health management plans to programme evaluation and continuous improvement (Table 4.12). Each element was regarded as critical to control and optimise the process of marine stock enhancement to restore and increase harvest levels using cultured fish whilst preserving existing wild stocks.

Where wild stocks are declining and unable to sustain commercial fishing pressure, release of cultured juveniles has been advocated to bolster wild stocks or establish new fisheries for introduced species to sustain fishing communities. Such culture-based fisheries are increasingly seen as important against a backdrop of declining returns from wild capture fisheries and limitations to intensive aquaculture development. Culture-based fisheries

Table 4.12 Guidelines for responsible stock enhancement and ranching

Component	Guidelines
Species selection	• establish methods to prioritise and select target species to be enhanced
Species management plans	• formulate species management plans with long- and short-term goals, harvest regimes and genetic conservation objectives
Monitoring and evaluation	• define quantitative measures of success and assess the enhancement project in terms of stated objectives in the management plan
Genetic resources management	• adopt genetic resources management to minimise in-breeding and out-breeding depression and ensure conservation of genetic resources
Disease and health management	• formulate and implement disease and health management plans
Ecological considerations	• incorporate life history considerations and ecological attributes in enhancement strategies and tactics
Stock identification and differentiation	• establish and implement protocols to identify hatchery-reared animals and assess success against quantifiable indicators
Optimise release strategies	• define and use an empirical process to determine optimal release strategies
Economic and policy objectives	• define and integrate socio-economic objectives and evaluate programme outcomes
Adaptive management	• use adaptive management principles to evaluate and improve management strategies and tactics

Source: Developed from Blankenship and Leber (1995).

have been promoted and developed in several situations to sustain the livelihoods of fisherfolk and enhance food security in vulnerable coastal communities. However, those proposing culture-based fisheries should ensure that native populations of species are released, that biodiversity generally is not negatively affected, that social and cultural impacts are assessed fully prior to implementation and that activities are monitored following commencement.

Whereas stock enhancement and culture-based fisheries include both activities and actors associated principally with either aquaculture or fisheries, ranching is a term increasingly used to describe culture, release and capture activities undertaken by a single entity that retains ownership or is granted exclusive rights to exploit organisms being ranched, whilst restrictions are placed on access and exploitation by others. Ranching salmon was described by Thorpe (1980) as 'an aquaculture system in which juvenile fish are released to grow, unprotected, on natural foods in marine waters from which they are harvested at marketable size'. Reviewing shellfish production strategies in the UK, Burton et al. (2001: 1) defined ranching as 'extensively grown species are placed for all or part of their life onto natural habitats, usually the seabed, and grown to market size with little or no further human input or husbandry'.

Reviewing progress with ranching in Japan, Arnason (2001) reported that operations dealing with seventy species were underway, with ten being ranched on a significant scale, including salmon (*Oncorhynchus keta*), Japanese flounder (*Paralichthys olivaceus*), prawn (*Penaeus japonicus*), abalone (*Haliotis discus*) and crab (*Portunus trituberculatus*). Red seabream (*Pagrus major*) are ranched in Japan, and operations based in Kagoshima Bay were cited as economically viable, with benefits exceeding costs. Furthermore, according to Arnason (2001), up to 80 per cent of the red seabream catch in inshore areas is made up of ranched fish. Ranching of homarid lobsters is practised in various countries, including France, Ireland, Norway, the UK and USA (Nicosia and Lavalli, 1999). Arnason (2001) reported that scallop ranching was established in Hervey Bay, Australia, and noted that by the late 1980s over 3 billion juvenile scallops were being released annually in Japan.

Previous experience offers important insights regarding high potential management strategies and species suitable for ranching. Responsible ranching operations should be restricted to using native species, ideally originating from local populations, but exemptions might be permitted when species have become naturalised owing to past introductions. Prospective operators may find that securing exclusive rights to harvest ranched animals is difficult to secure and enforce. Limiting or denying access by other resource users to areas designated for ranching may be problematic, especially where areas further from shore are not easily monitored. Operators proposing to undertake ranching could buy out existing fishing rights or agree to pay compensation to traditional user groups to secure access to suitable areas. Ranching operations proposed in conjunction with more conventional open-water aquaculture might help to avoid undesirable effects of nutrient enrichment, and authorities might be persuaded to take this into account when setting production limits and permissible discharge levels. With stock enhancement through culture-based fisheries and ranching, identification of released animals is critical to enable evaluation and adaptive management, attribute ownership or responsibility, provide a basis for equitably transferring benefits of enhanced catches and harvests and detect impropriety.

Scallop culture cooperatives in Japan

Yesso scallop (*Patinopecten yessoensis*) production accounts for about 20 per cent of global scallop production. Culture is almost exclusively confined to northern Japan where water temperatures ranging from 10°C to 17°C are ideal for this coldwater species (Gosling, 2003). Production is divided equally between bottom culture and suspended culture, with both strategies being dependent on natural spat collection. Collectors are hung from long lines, usually onion bags (0.5 mm mesh size) loosely filled with plastic mesh or netting. Settled scallop spat (shell height 5–10 mm) are removed from collectors from mid-July to mid-August and are transferred to pearl nets that are suspended from long lines 5–10 m below the surface. During this intermediate or nursery stage, the initial stocking density is 100 spat per net, and this is reduced to 20 spat per net by September–October for suspended culture and 50 spat per net if destined for bottom culture. Scallops in suspended culture are grown by ear-hanging or in various nets, but this method is not suited to areas susceptible to excessive wave action.

Bottom culture is mainly concentrated on the north coast of Hokkaido and Mutsu Bay, Honshu. Tokoro Cooperative, operating in Hokkaido, has divided their allotted coastal area into four units averaging 25–30km^2 with water 30–60 m deep; smaller areas are held in reserve in case of under-production. Each unit is harvested every 4 years and then reseeded at 10 m^{-2}, equating to 300 million wild spat per year. Predators, notably starfish and sea urchin, are removed by dredging prior to seeding spat for bottom culture; scallop cultured this way takes 3 to 4 years to reach market size. The ratio of adults to juveniles was 1:4 when the cooperative started, but this has improved to 1:2 owing to better management, notably stocking regimes and predator control. Although the cooperative harvests 30,000t of live weight per year, and despite good organization and a relatively long period of operation suggesting economic viability and resilience, Gosling (2003) noted that continued success would be guaranteed only when:

- reliable supplies of hatchery reared seed are secured, thus removing reliance of uncertain wild spat fall;
- carrying capacity of bays is not exceeded;
- density of bottom culture and suspended culture is regulated, thus increasing food availability and consequently growth and quality;
- additional sites suited to bottom culture are identified and developed;
- routine bed rotation is adopted more widely;
- better methods of predator control on seeding plots are developed.

Summary

Enhancements to traditional or indigenous coastal aquaculture practices have been proposed to ensure their viability despite economic pressure on producers to abandon them. Continued operation of such culture systems is increasingly valued for ecological attributes associated with wetland agroecosystems and broader ecosystem services and cultural values attributed to such practices. Acknowledging the role that ecological processes in ponds and coastal wetlands can play in treating aquaculture process and waste-water, integrated production strategies have been proposed to capitalise on this, whilst also potentially enhancing financial returns and reducing production risks. Despite apparent benefits of such approaches, commercial application has been limited, and widespread adoption may only occur as evidence of effective operation grows and anticipated financial returns become more favourable. Integrated production in open-water settings has been proposed, but technical challenges are apparent, whilst the general dispersion of nutrients in the environment suggests that extensive areas of extractive production or more ambitious strategies such as ranching may be contemplated. There are possible negative environmental impacts associated with extensive, extractive production systems and significant risks associated with interventions involving the stocking of cultured animals and plants in open-water settings.

Notes

1 Green Revolution was a term applied retrospectively to transfers of technological packages and industrial approaches to agriculture from international research centres to developing countries, predominantly in Asia and Latin and South America, from the 1940s to 1970s (Bunting et al., 2011).
2 Section adapted from Bunting (2001) with contemporary references and edits to conform to publishing house style and conventions and ensure coherence with previous text.
3 Section adapted from Bunting (2001) with contemporary references and edits to conform to publishing house style and conventions and ensure coherence with previous text.
4 CA$1 equivalent to US$1.19 in May 1992 (Petrell et al., 1993).
5 Section adapted from Bunting (2001) with contemporary references and edits to conform to publishing house style and conventions and ensure coherence with previous text.

References

Ahmed, N., Bunting, S.W., Glaser, M., Flaherty, M.S. and Diana, J.S. (2017) 'Can greening of aquaculture sequester blue carbon?', Ambio, vol 46, pp. 468–477.

Ahmed, N., Bunting, S.W., Rahman, S. and Garforth, C.J. (2014) 'Community-based climate change adaptation strategies for integrated prawn–fish–rice farming in Bangladesh to promote social–ecological resilience', Reviews in Aquaculture, vol 6, pp. 20–35.

Amos, M.J. and Purcell, S.W. (2003) 'Evaluation of strategies for intermediate culture of *Trochus niloticus* (Gastropoda) in sea cages for restocking', Aquaculture, vol 218, pp. 235–249.

Anonymous (1996) 'Uses and markets for seaweed products: Malaysia and Thailand', Infofish International, vol 4, pp. 22–26.

Anonymous (2003) 'The promise of a blue revolution', The Economist, vol 368, no 8336, pp. 19–21.

Anonymous (2007) 'Sustainable extensive and semi-intensive coastal aquaculture in southern Europe', Aquaculture Europe, vol 32, no 3, pp. 30–33.

Arnason, R. (2001) The Economics of Ocean Ranching: Experiences, Outlook and Theory, FAO Fisheries Technical Paper 413, FAO, Rome.

Azad, A.K., Jensen, K.R. and Lin, C.K. (2009) 'Coastal aquaculture development in Bangladesh: unsustainable and sustainable experiences', Environmental Management, vol 44, pp. 800–809.

Barrington, K., Chopin, T. and Robinson, S. (2009) 'Integrated multi-trophic aquaculture (IMTA) in marine temperate waters', in D. Soto (ed) Integrated Aquaculture: A Global Review, FAO Fisheries and Aquaculture Technical Paper No 529, FAO, Rome.

BBC (2006) Concern over Organic Salmon Farms, British Broadcasting Corporation, http://news.bbc.co.uk/1/hi/sci/tech/5409434.stm

Beveridge, M.C.M. and Little, D.C. (2002) 'The history of aquaculture in traditional societies' in B.A. Costa-Pierce (ed) Ecological Aquaculture, Blackwell Science, Oxford, UK.

Blankenship, H.L. and Leber, K.M. (1995) 'A responsible approach to marine stock enhancement', American Fisheries Society Symposium, vol 15, pp. 167–175.

Borowitzka, M.A. (1993) 'Products from microalgae', Infofish International, vol 5, pp. 21–26.

Briggs, M.R.P. (2003) 'Destructive fishing practices in south Sulawesi Island, East Indonesia and the role of aquaculture as a potential alternative livelihood' in G. Haylor, M.R.P. Briggs, L. Pet-Soede, H. Tung, N.T.H. Yen, B. Adrien, B. O'Callaghan, et al. (eds) Improving Coastal Livelihoods through Sustainable Aquaculture Practices. A report to the Collaborative APEC Grouper Research and Development Network (FWG/01/2001).

Briggs, M.R.P. and Funge-Smith, S.J. (1996) Coastal Aquaculture and Environment: Strategies for Sustainability, Final Technical Report, Institute of Aquaculture, University of Stirling, ODA Project R6011 UK.

Bulcock, P., Beveridge, M. and Hambrey, J. (2000) The Improved Management of Small-Scale Cage Culture in Asia, University of Stirling, Final Technical Report for DFID Project R7100, UK.

Bunting, S.W. (2001) A Design and Management Approach for Horizontally Integrated Aquaculture Systems, PhD thesis, Institute of Aquaculture, University of Stirling, UK.

Bunting, S.W. (2005) Second Market Analysis, European Commission Project GENESIS [EC INNOVATION IPS-2000-102], Institute of Aquaculture, University of Stirling, UK.

Bunting, S.W. (2007) 'Regenerating aquaculture: enhancing aquatic resources management, livelihoods and conservation' in J. Pretty, A. Ball, T. Benton, J. Guivant, D. Lee, D. Orr, M. Pfeffer and H. Ward (eds) The SAGE Handbook of Environment and Society, SAGE Publications, London.

Bunting, S.W. (2008) 'Horizontally integrated aquaculture development: exploring consensus on constraints and opportunities with a stakeholder Delphi', Aquaculture International, vol 16, pp. 153–169.

Bunting, S.W., Bosma, R.H., van Zwieten, P.A.M. and Sidik, A.S. (2013) 'Bioeconomic modelling of shrimp aquaculture strategies for the Mahakam Delta, Indonesia', Aquaculture Economics and Management, vol 17, pp. 51–70.

Bunting, S.W., Bostock, J., Leschen, W. and Little, D.C. (2023) 'Evaluating the potential of innovations across aquaculture product value chains for poverty alleviation in Bangladesh and India', Frontiers in Aquaculture, vol 2, no 3, pp. 1–23.

Bunting, S.W., Kundu, N. and Ahmed, N. (2011) Rice–Shrimp Farming Ecocultures in the Sundarbans of Bangladesh and West Bengal, India, New EcoCultures Case Study, Interdisciplinary Centre for Environment and Society, University of Essex, Colchester, UK.

Bunting, S.W., Kundu, N. and Ahmed, N. (2017) 'Evaluating the contribution of diversified shrimp–rice agroecosystems in Bangladesh and West Bengal, India to social–ecological resilience', Ocean and Coastal Management, vol 148, pp. 63–74.

Bunting, S., Pounds, A., Immink, A., Zacarias, S., Bulcock, P., Murray, F. and Auchterlonie, N. (2022) The Road to Sustainable Aquaculture: On Current Knowledge and Priorities for Responsible Growth, World Economic Forum, Cologny, Switzerland.

Bunting, S.W. and Shpigel, M. (2009) 'Evaluating the economic potential of horizontally integrated land-based marine aquaculture', Aquaculture, vol 294, pp. 43–51.

Burton, C.A., MacMillan, J.T. and Learmouth, M.M. (2001) 'Shellfish ranching in the UK', Hydrobiologia, vol 465, pp. 1–5.

Buschmann, A.H., Mora, O.A., Gómez, P., Böttger, M., Buitano, S., Retamales, C., et al. (1994) '*Gracilaria chilensis* outdoor tank cultivation in Chile: use of land-based salmon culture effluents', Aquacultural Engineering, vol 13, pp. 283–300.

Buschmann, A.H., Troell, M., Kautsky, N. and Kautsky, L. (1996) 'Integrated tank cultivation of salmonids and *Gracilaria chilensis* (Gracilariales, Rhodophyta)', Hydrobiologia, vol 326–327, pp. 75–82.

Clough, B.F., Boto, K.G. and Attiwill, P.M. (1983) 'Mangroves and sewage: a re-evaluation' in H.J. Teas (ed) Tasks in Vegetation Science, Dr. W. Junk Publishers, Hague, The Netherlands.

Costa-Pierce, B.A. (2002) 'The Ahupua'a Aquaculture Ecosystems in Hawaii', in B.A. Costa-Pierce (ed) Ecological Aquaculture, Blackwell Publishing, Oxford, UK.

Dulyapurk, V., Taparhudee, W., Yoonpundh, R. and Jumnongsong, S. (2007) Multidisciplinary Situation Appraisal of Mangrove Ecosystems in Thailand, Faculty of Fisheries, Kasetsart University, Bangkok, Thailand.

Dumas, A., Laliberté, G., Lessard, P. and de la Noüe, J. (1998) 'Biotreatment of fish farm effluents using the cyanobacterium *Phormidium bohneri*', Aquacultural Engineering, vol 17, pp. 57–68.

Dwivedi, S.N. and Padmakumar, K.G. (1983) 'Ecology of a mangrove swamp near Juhn Beach, Bombay with reference to sewage pollution' in H.J. Teas (ed) Tasks in Vegetation Science, Dr. W. Junk Publishers, Hague, The Netherlands.

Ellis, C.D. and Boothroyd, D. (2008) Developing Stakeholder Participation in Lobster Stock Enhancement Projects, Department for Environment, Food and Rural Affairs, London.

Enander, M. and Hasselstrom, M. (1994) 'An experimental wastewater treatment system for a shrimp farm', Infofish International, vol 4, pp. 56–61.

FAO (1995) Code of Conduct for Responsible Fisheries, FAO, Rome.

FAO (2009) The State of World Fisheries and Aquaculture 2008, FAO, Rome.

Ferguson, P., Stone, T. and Young, J.A. (2005) Consumer Perceptions of Aquatic Products to Be Produced from GENESIS Integrated Systems: The UK Perspective, Department of Marketing and Stirling Aquaculture, University of Stirling, UK.

Fitzgerald, W.J. (2002) 'Silvofisheries: integrated mangrove forest aquaculture systems' in B.A. Costa-Pierce (ed) Ecological Aquaculture, Blackwell Science, Oxford, UK.

Frederiksen, T.M., Sorensen, K.B., Finster, K. and Macintosh, D.J. (1998) 'Implications of shrimp pond waste in mangrove environments', Aquaculture Asia, vol 3, no 2, pp. 8–11.

Frocklin, S., de la Torre-Castro, M., Lindstrom, L., Jiddawi, N.S. and Msuya, F.E. (2012) 'Seaweed mariculture as a development project in Zanzibar, East Africa: a price too high to pay?', Aquaculture, vol 356–357, pp. 30–39.

Gosling, E. (2003) Bivalve Molluscs, Blackwell Publishing, Oxford, UK.

Guerrero, S. and González, X.O. (1991) 'Clam nursery (*Tapes decussatus*) in the effluent of a fish farm in Ria de Arosa, Spain' in N. De Pauw and J. Joyce (eds) Aquaculture and the Environment, European Aquaculture Society, Special Publication No. 14, Bredene, Belgium.

Hart, A.M., Bell, J.D. and Foyle, T.P. (1999) 'Improving culture techniques for village-based farming of giant clams (Tridacnidae)', Aquaculture Research, vol 30, pp. 175–190.

Helfrich, L.A., Zimmerman, M. and Weigmann, D.L. (1995) 'Control of suspended solids and phytoplankton fish fishes and a mussel', Water Resources Bulletin, vol 31, pp. 307–316.

Hoq, M.E. (2008) Sundarbans Mangrove: Fish & Fisheries – Ecology, Resources, Productivity and Management Perspectives, Graphic Media, Dhaka, Bangladesh.

Hussenot, J., Lefebvre, S. and Brossard, N. (1998) 'Open-air treatment of wastewater from land-based marine fish farms in extensive and intensive systems: current technology and future perspectives', Aquatic Living Resources, vol 11, pp. 297–304.

IFFO (2012) Global Standard for Responsible Supply: Requirements for Certification, International Fishmeal and Fish Oil Organisation, St Albans, UK.

IPCC (2019) The Ocean and Cryosphere in a Changing Climate, a special report of the Intergovernmental Panel on Climate Change, Cambridge University Press, Cambridge, UK, and New York.

Islam, M.S., Milstein, A., Wahab, M.A., Kamal, A.H.M. and Dewan, S. (2005) 'Production and economic return of shrimp aquaculture in coastal ponds of different sizes and with different management regimes', Aquaculture International, vol 13, pp. 489–500.

Jara-Jara, R., Pazos, A.J., Abad, M., Garcia-Martin, L.O. and Sanchez, J.L. (1997) 'Growth of clam seed (*Ruditapes decussatus*) reared in the wastewater effluent from a fish farm in Galicia (N.W. Spain)', Aquaculture, vol 158, pp. 247–262.

Jiménez del Río, M., Ramazanov, Z. and García-Reina. G. (1996) '*Ulva rigida* (Ulvales, Chlorophyta) tank culture as biofilters for dissolved inorganic nitrogen from fishpond effluents', Hydrobiologia, vol 326/327, pp. 61–66.

Jones, A.B. and Preston, N.P. (1999) 'Sydney rock oyster, *Saccostrea commercialis* (Iredale & Roughley), filtration of shrimp farm effluent: the effects on water quality', Aquaculture Research, vol 30, pp. 51–57.

Laihonen, P., Hanninen, J., Chojnacki, J. and Vuorinen, I. (1997) 'Some prospects of nutrient removal with artificial reefs', in A.C. Jensen (ed) European Artificial Reef Research, Proceedings of the 1st EARRN Conference, Ancona, Italy, March 1996. Southampton Oceanography Centre, Southampton, UK.

Langdon, C., Evans, F. and Demetropoulos, C. (2004) 'An environmentally-sustainable, integrated, co-culture system for dulse and abalone production', Aquacultural Engineering, vol 32, pp. 43–56.

Lefebvre, S., Hussenot, J. and Brossard, N. (1996) 'Water treatment of land-based fish farm effluents by outdoor culture of marine diatoms', Journal of Applied Phycology, vol 8, pp. 193–200.

Lefebvre, S., Probert, I., Lefrancois, C. and Hussenot, J. (2004) 'Outdoor phytoplankton continuous culture in a marine fish–phytoplankton–bivalve integrated system: combined effects of dilution rate and ambient conditions on growth rate, biomass and nutrient cycling', Aquaculture, vol 240, pp. 211–231.

Lewis, R.R. (1999) 'Key concepts in successful ecological restoration of mangrove forests' in Proceedings of the TCE-Workshop No. II, Coastal Environmental Improvement in Mangrove/Wetland Ecosystems, 18–23 August 1998, Danish–SE Asian Collaboration in Tropical Coastal Ecosystems (TCE) Research and Training, NACA, Bangkok, Thailand.

Lewis, R.R., Quarto, A., Enright, J., Corets, E., Primavera, J., Ravishankar, T., Stanley, O.D. and Djamaluddin, R. (2006) Five Steps to the Successful Ecological Restoration of Mangroves, Mangrove Action Project and Yayasan Akar Rumput Laut, Yogyakarta, Indonesia.

Lin, C.K., Ruamthaveesub, P. and Wanuchsoontorn, P. (1993) 'Integrated culture of the green mussel (*Perna viridis*) in wastewater from an intensive shrimp pond: concept and practice', World Aquaculture, vol 24, no 2, pp. 68–73.

Lockwood, P., Jensen, A., Collins, K. and Turnpenny, A. (1991) 'The artificial reef in Poole Bay', Ocean Challenge, vol 2, no 2, pp. 35–39.

Machiwa, J.F. (1998) 'Distribution and remineralization of organic carbon in sediments of a mangrove stand partly contaminated with sewage waste', Ambio, vol 27, pp. 740–744.

Macintosh, D.J. and Ashton, E.C. (2002) A Review of Mangrove Biodiversity Conservation and Management, Centre for Tropical Ecosystems Research, University of Aarhus, Denmark.

Mai, K., Mercer, J.P. and Donlon, J. (1996) 'Comparative studies on the nutrition of two species of abalone, *Haliotis tuberculata* L. and *Haliotis discus hannai* Ino. V. The role of polyunsaturated fatty acids of macroalgae in abalone nutrition', Aquaculture, vol 139, pp. 77–89.

Marschke, M. and Vandergeest, P. (2016) 'Slavery scandals: unpacking labour challenges and policy responses within the off-shore fisheries sector', Marine Policy, vol 68, pp. 39–46.

Martinez, L.A. and Buschmann, A.H. (1996) 'Agar yield and quality of *Gracilaria chilensis* (Gigartinales, Rhodophyta) in tank culture using fish effluents', Hydrobiologia, vol 326–327, pp. 341–345.

Melotti, P., Colombo, L., Roncarati, A. and Garella, E. (1991) 'Use of waste-water from intensive fish farming to increase the productivity in north Adriatic lagoons (valli)' in N. De Pauw and J. Joyce (eds) Aquaculture and the Environment, European Aquaculture Society, Special Publication No. 14, Bredene, Belgium.

MSC (2010) MSC Fishery Standard: Principles and Criteria for Sustainable Fishing, Marine Stewardship Council, London.

Muir, J.F. (2005) 'Managing to harvest? Perspectives on the potential of aquaculture', Philosophical Transactions of the Royal Society B, vol 360, pp. 191–218.

Mwaluma, J. (2002) 'Pen culture of the mud crab *Scylla serrata* in Mtwapa mangrove system, Kenya', Western Indian Ocean Journal of Marine Science, vol 1, no 2, pp. 127–133.

Naskar, K.R. (1985) 'A short history and the present trends of brackishwater fish culture in paddy fields at the Kulti–Minakhan areas of Sundarbans in West Bengal', Journal of the Indian Society of Coastal Agriculture Research, vol 3, no 2, pp. 115–124.

Naturland (2005) Naturland Standards for Organic Aquaculture, Naturland, Grafelfing, Germany.

Naylor, R.L., Goldburg, R.J., Primavera, J.H., Kautsky, N., Beveridge, M.C.M., Clay, J., et al. (2000) 'Effect of aquaculture on world fish supplies', Nature, vol 405, pp. 1017–1024.

Neori, A., Krom, M.D., Ellner, S.P., Boyd, C.E., Popper, D., Rabinovitch, R., et al. (1996) 'Seaweed biofilters as regulators of water quality in integrated fish: seaweed culture units', Aquaculture, vol 141, pp. 183–199.

Neori, A., Ragg, N.L.C. and Shpigel, M. (1998) 'The integrated culture of seaweed, abalone, fish and clams in modular intensive land-based systems: II. Performance and nitrogen partitioning within an abalone (*Haliotis tuberculata*) and macroalgae culture system', Aquacultural Engineering, vol 17, pp. 215–239.

Neori, A. and Shpigel, M. (1999) 'Using algae to treat effluents and feed invertebrates in sustainable integrated mariculture', World Aquaculture, vol 30, no 2, pp. 46–49.

Neori, A., Shpigel, M. and Ben-Ezra, D. (2000) 'A sustainable integrated system for culture of fish, seaweed and abalone', Aquaculture, vol 186, pp. 279–291.

Nicosia, F. and Lavalli, K. (1999) 'Homarid lobster hatcheries: their history and role in research, management, and aquaculture', Marine Fisheries Review, vol 61, no 2, pp. 1–57.

Petrell, R.J., Mazhari Tabrizi, K., Harrison, P.J. and Druehl, L.D. (1993) 'Mathematical model of *Laminaria* production near a British Columbian salmon sea cage farm', Journal of Applied Phycology, vol 5, pp. 1–14.

Pettersson-Lofquist, P. (1995) 'The development of open-water algae farming in Zanzibar: reflections on the socioeconomic impact', Ambio, vol 24, pp. 487–491.

Phang, S.-M., Shaharuddin, S., Noraishah, H. and Sasekumar, A. (1996) 'Studies on *Gracilaria changii* (Gracilariales, Rhodophyta) from Malaysian mangroves', Hydrobiologia, vol 326/327, pp. 347–352.

Primavera, J.H. (1997) 'Socio-economic impacts of shrimp culture', Aquaculture Research, vol 28, pp. 815–827.

Rajendran, N. and Kathiresan, K. (1996) 'Effect of effluent from a shrimp pond on shoot biomass of mangrove seedlings', Aquaculture Research, vol 27, pp. 745–747.

Roberts, C. (2007) The Unnatural History of the Sea: The Past and Future of Humanity and Fishing, Gaia, London.

Robertson, A.I. and Phillips, M.J. (1995) 'Mangroves as filters of shrimp pond effluent: predictions and biogeochemical research needs', Hydrobiologia, vol 295, pp. 311–321.

Ronnback, P. (2003) Critical Analysis of Certified Organic Shrimp Aquaculture in Sidoarjo, Indonesia, Swedish Society for Nature Conservation, Stockholm.

Ryther, J.H., Goldman, J.C., Gifford, C.E., Huguenin, J.E., Wing, S.S., Clarner, J.P., et al. (1975) 'Physical models of integrated waste recycling: marine polyculture systems', Aquaculture, vol 5, pp. 163–177.

Samson, M.S. and Rollon, R.N. (2008) 'Growth performance of planted mangroves in the Philippines: revisiting forest management strategies', Ambio, vol 37, no 4, pp. 234–240.

Short, R.E., Gelcich, S., Little, D.C., Micheli, F., Allison, E.H., Basurto, X., et al. (2021) 'Harnessing the diversity of small-scale actors is key to the future of aquatic food systems', Nature Food, vol 2, no 9, pp. 733–741.

Shpigel, M. and Blaylock, R.A. (1991) 'The Pacific oyster, *Crassostrea gigas*, as a biological filter for a marine fish aquaculture pond', Aquaculture, vol 92, pp. 187–197.

Shpigel, M., Gasith, A. and Kimmel, E. (1997) 'A biomechanical filter for treating fishpond effluents', Aquaculture, vol 152, pp. 103–117.

Shpigel, M. and Neori, A. (1996) 'The integrated culture of seaweed, abalone, fish and clams in modular intensive land-based systems: I. Proportions of size and projected revenues', Aquacultural Engineering, vol 15, pp. 313–326.

Shpigel, M., Neori, A., Popper, D.M. and Gordin, H. (1993) 'A proposed model for "environmentally clean" land-based culture of fish, bivalves and seaweeds', Aquaculture, vol 117, pp. 115–128.

Siar, S.V., Samonte, G.P.B. and Espada, A.T. (1995) 'Participation of women in oyster and mussel farming in Western Visayas, Philippines', Aquaculture Research, vol 26, pp. 459–467.

Sickander, O. and Filgueira, R. (2022) 'Factors affecting IMTA (integrated multi-trophic aquaculture) implementation on Atlantic Salmon (*Salmo salar*) farms', Aquaculture, vol 561, 738716.

Sigfússon, T. (2022) 'Promising opportunities for fish by-products', New Food, https://www.newfoodmagazine.com/article/167579/promising-opportunities-for-fish-by-products/

Soto, D. and Mena, G. (1999) 'Filter feeding by the freshwater mussel, *Diplodon chilensis*, as a biocontrol of salmon farming eutrophication', Aquaculture, vol 171, pp. 65–81.

Stentiford, G.D., Bateman, I.J., Hinchliffe, S.J., Bass, D., Hartnell, R., Santos, E.M., et al. (2020) 'Sustainable aquaculture through the One Health lens', Nature Food, vol 1, no 8, pp. 468–474.

Stirling, H.P. and Okumus, I. (1995) 'Growth and production of mussels (*Mytilus edulis* L.) suspended at salmon cages and shellfish farms in two Scottish sea lochs', Aquaculture, vol 134, pp. 193–210.

Subandar, A., Petrell, R.J. and Harrison, P.J. (1993) 'Laminaria culture for reduction of dissolved inorganic nitrogen in salmon farm effluent', Journal of Applied Phycology, vol 5, pp. 455–463.

Svasand, T., Kristiansen, T.S., Pedersen, T., Salvanes, A.G.V., Engelsen, R., Naevdal, G. and Nodtvedt, M. (2000) 'The enhancement of cod stocks', Fish and Fisheries, vol 1, no 2, pp. 173–205.

Svasand, T. and Moksness, E. (2004) 'Marine stock enhancement and sea-ranching' in E. Moksness, E. Kjorsvik and Y. Olsen (eds) Culture of Cold-Water Marine Fish, Blackwell Publishing, Oxford, UK.

Swedish Society for Nature Conservation (2006) Eco-labelling of Shrimp Farming in Ecuador, Swedish Society for Nature Conservation, Stockholm, Sweden.

Tam, N.F.Y. and Wong, Y.S. (1995) 'Mangrove soils as sinks for wastewater-borne pollutants', Hydrobiologia, vol 295, pp. 231–241.

Thiao, D. and Bunting, S.W. (2022) Socio-Economic and Biological Impacts of the Fish-Based Feed Industry for Sub-Saharan Africa, FAO Fisheries and Aquaculture Circular 1236, FAO, Rome.

Thorpe, J.E. (1980) 'The development of salmon culture towards ranching' in J. Thorpe (ed) Salmon Ranching, Academic Press, London.

Troell, M., Halling, C., Nilsson, A., Buschmann, A.H., Kautsky, N. and Kautsky, L. (1997) 'Integrated marine cultivation of *Gracilaria chilensis* (Gracilariales, Rhodophyta) and salmon cages for reduced environmental impact and increased economic output', Aquaculture, vol 156, pp. 45–61.

UNEP (2006) Annotated Guiding Principles for Post-Tsunami Rehabilitation and Reconstruction, Global Programme of Action for the Protection of the Marine Environment from Land-Based Activities, United Nations Environment Programme, Nairobi, Kenya.

United Nations (2015) Transforming Our World: The 2030 Agenda for Sustainable Development, United Nations, New York.

Vandermeulen, H. and Gordin, H. (1990) 'Ammonium uptake using *Ulva* (Chlorophyta) in intensive fishpond systems: mass culture and treatment of effluent', Journal of Applied Phycology, vol 2, pp. 363–374.

Walters, B.B. (2000) 'Local mangrove planting in the Philippines: are fisherfolk and fishpond owners effective restorationists?', Restoration Ecology, vol 8, no 3, pp. 237–246.

Whitmarsh, D.J., Cook, E.J. and Black, K.D. (2006) 'Searching for sustainability in aquaculture: an investigation into the economic prospects for an integrated salmon-mussel production system', Marine Policy, vol 30, pp. 293–298.

Wolanski, E., Spagnol, S., Thomas, S., Moore, K., Alongi, D.M., Trott, L. and Davidson, A. (2000) 'Modelling and visualising the fate of shrimp pond effluent in a tidally flushed mangrove creek', Estuarine, Coastal and Shelf Science, vol 50, pp. 85–97.

Wong, Y.S., Lan, C.Y., Chen, G.Z., Li, S.H., Chen, X.R., Liu, Z.P. and Tam, N.F.Y. (1995) 'Effect of wastewater discharge on nutrient contamination of mangrove soils and plants', Hydrobiologia, vol 295, pp. 243–254.

World Bank (2007) Changing the Face of the Waters: The Promise and Challenge of Sustainable Aquaculture, The World Bank, Washington, DC.

World Bank (2022) Project Appraisal Document on a Proposed Loan in the Amount of US$400 Million to the Republic of Indonesia for a Mangroves for Coastal Resilience Project, The World Bank, Washington, DC.

World Bank, ISME and cenTER Aarhus (2005) Principles for a Code of Conduct for the Management and Sustainable Use of Mangrove Ecosystems, World Bank, ISME, Japan, cenTER Aarhus, Denmark.

Yan, N. and Chen, X. (2015) 'Don't waste seafood waste', Nature, vol 524, no 7564, pp. 155–157.

Zacarias, S. (2020) Use of non-ablated shrimp (*L. vannamei*) in commercial scale hatcheries, thesis, University of Stirling, UK.

Zacarias, S., Carboni, S., Davie, A. and Little, D.C. (2019) 'Reproductive performance and offspring quality of non-ablated Pacific white shrimp (*Litopenaeus vannamei*) under intensive commercial scale conditions', Aquaculture, vol 503, pp. 460–466.

Zacarias, S., Fegan, D., Wangsoontorn, S., Yamuen, N., Limakom, T., Carboni, S., et al. (2021) 'Increased robustness of postlarvae and juveniles from non-ablated Pacific whiteleg shrimp, *Penaeus vannamei*, broodstock post-challenged with pathogenic isolates of *Vibrio parahaemolyticus* (VpAHPND) and white spot disease (WSD)', Aquaculture, vol 532, 736033.

Zertuche-Gonzalez, J.A., Garcia-Lepe, G., Pacheco-Ruiz, I., Chee-Barragan, A. and Gendrop-Funes, V. (1999) 'A new approach to seaweed cultivation in Mexico', World Aquaculture, vol 30, no 2, pp. 50–51.

5   Sustainable rural aquaculture

Key points

The aims of this chapter are to:

- Define what constitutes rural aquaculture in the context of this review and reflect on recent trends and progress concerning traditional and sustainable rural aquaculture development.
- Describe aquaculture practices occurring in semi-intensive static water ponds and multipurpose household ponds and consider prospects for sustainable development.
- Highlight recent accounts of producers transitioning to semi-intensive production modes in China, notably feed dynamics and the transition to on-farm production of feed ingredients (see Newton et al., 2021).
- Review promising examples from current farmer practice and development projects that constitute an important element of the evidence-base concerning prospects for sustainable rural aquaculture.
- Review the ecological, social and economic benefits of culturing indigenous freshwater species in Latin America and South America.
- Review strategies for integrated aquaculture–agriculture focusing on examples from Argentina and Vietnam and reflect on associated opportunities and threats.
- Describe traditional practices of integrating aquaculture with rice farming and review recent innovations, notably established rice–fish culture systems in Egypt, *gher* (trenched rice fields) farming in Bangladesh and prospects for the System of Rice Intensification (SRI).
- Assess the potential roles of constructed wetlands in managing aquaculture wastewater and the relative merits of surface and sub-surface flow regimes.
- Evaluate aquaculture development in irrigation and hydroelectric schemes.
- Discuss prospects for culture-based fisheries and cage aquaculture development.
- Review establishment of cage-based aquaculture in the African Great Lakes region and consider aspects of carrying capacity and disease prevention and implications of trial results comparing mixed-sex and mono-sex cage culture of tilapia (see Bostock et al., 2022).

DOI: 10.4324/9781003342823-5

- Describe promising coldwater and highland culture systems and threats to production including anticipated climate change impacts and discuss prospects for future development.

Traditional rural aquaculture

Rural is a term commonly invoked by geographers to describe areas with relatively low population densities and economies dominated by agriculture, whilst inaccessible forest, wetland and mountain ecosystems not directly supporting communities may be classified as natural or wilderness. Rural areas may contain provincial towns and cities and be subjected to urban encroachment; furthermore, unequal population distribution across the globe may result in some communities with seemingly high population densities being classified as rural, as agriculture continues to constitute the predominant source of livelihoods.

Rural aquaculture encompasses a vast array of practices from high-altitude trout culture and culture-based fisheries to lowland landscapes dominated by vast expanses of fishpond to aquaculture in transition zones between freshwater and marine systems, including tilapia culture in brackish water ponds and prawn–rice culture in *ghers* in coastal Bangladesh. Case studies of rural aquaculture presented by Edwards et al. (2002) ranged from small-scale pond culture in the Red River Delta, Vietnam, and northwest Bangladesh, to floating cages in Lake Maninjau, Indonesia, to aquaculture in upland forest buffer zones in Quirino Province, Philippines. The comprehensive review compiled by these authors highlighted the variation in developed practices and demonstrated the influence that geophysical settings, environmental conditions, socio-economic circumstances and market opportunities have on the evolution of rural aquaculture systems. Distinctions between rural communities and cultures of practice and those in other settings are often conceptually rather arbitrary; further distinctions are commonly drawn between highland and lowland environments and between coastal and inland areas. Elsewhere, communities not classified as urban and peri-urban may be regarded as rural. Aquaculture systems are often classified on the basis of being freshwater, brackish water or marine; here the focus is on freshwater production systems, as coastal brackish water and marine culture were covered in chapter 4.

Alluding to rural communities often conjures up visions of traditional, low-input, mixed farming systems and commonly invokes sentimental feelings for seemingly idyllic landscapes and fulfilling livelihoods. Such nostalgic recollection is often misplaced, however, coloured by unrepresentative historical accounts and poor knowledge of the facts surrounding the often miserable and short-lived existence of the general populace. Commonly, rural communities remain poorly served by health and education provision, are disconnected from government support and decision-making and are increasingly vulnerable to food and water shortages and limited employment and livelihoods options.

It must be borne in mind that with enhanced connectivity, knowledge and skills entering rural communities with returning economic migrants and graduates, inward investment by entrepreneurs and commercial operators and increasingly pervasive market networks, aquaculture development in predominantly rural settings can still entail intensive production of high-value species destined for urban or export markets. Rural aquaculture is increasingly connected through input streams, modes of services provision, financial and marketing arrangements and infrastructure development to urban centres, influences and pressures.

Dike–pond farming: zenith, decline, renaissance

Land under dike–pond cultivation in the Zhujiang Delta, Guangdong Province, China, declined significantly in the second half of the last century as farmers removed dikes to increase the pond area for intensive monocultures of high-value species, including eels, prawns and terrapins destined for urban and export markets (Wong Chor Yee, 1999). Confounding factors leading to these changes were prospects for greater financial returns to operators and declines in the availability of locally sourced production enhancing inputs owing to the concentration of livestock production in larger commercial units and falling demand for silk, an important co-product of the traditional dike–pond system.

Reclamation of productive land from the sea in the Zhujiang Delta to feed a burgeoning population during the Song Dynasty (960–1279 AD) was a major driving force, giving rise to vast tracts of impounded, low-lying land suited to dike–pond culture (Lo, 1996). According to Lo (1996), around 1350 ha were under integrated mulberry–dike fishpond cultivation by 1581 AD, although the dikes were originally planted with fruit trees. Mulberry–dike fishpond cultivation expanded to around 64,860 ha in the Zhujiang Delta by 1926 and was close to its zenith in 1925, occupying close to 94,000 ha of the delta. Onset of the global depression of the 1930s affected demand for silk, and Lo (1996: 192) noted that when the price of silk 'dropped in 1929, mulberry was replaced by sugar cane on the dike' and that 'mulberry dikes have continued to decline in recent years, following the economic reforms of the 1978, and a greater variety of crops has been grown on the dikes, and new species of fish introduced'.

Variation in the physical environment of the Zhujiang Delta combined with uneven market access prompted some farmers to integrate horticulture and rice farming in paddy fields with sugarcane and fish production (Lo, 1996). Elephant grass cultivation was included in similar systems to provide additional fodder, whilst elsewhere producers grew paddy rice, sugarcane and horticulture crops in conjunction with fishponds but the dikes were not cultivated. Planting trees on dikes can present additional risks in situations where storms and high winds are common; trees may be blown over, threatening the integrity of dikes and embankments. Elsewhere, concern has been expressed that tree planting may hinder watch and ward activities, thus encouraging theft from fishponds.

Sometimes considered self-contained systems, reliant on internal nutrient cycling to sustain yields, dike-based production of 30–80 t ha^{-1} annually of crops and 10–15 t ha^{-1} annually of fish actually requires significant externally derived nutrient inputs (Colman and Edwards, 1987). Estimates suggested that 10 tonnes of fish production could require 550 tonnes of cattle manure, 454 tonnes of pig manure or 75 tonnes of duck manure and, consequently, the dike–pond system can constitute a major sink for manure and other organic waste. Furthermore, Korn (1996: 11) proffered that modified dike–aquaculture principles could be applied more widely and that such a system with 'many sub-units of ditches and stagnant [static] waters, has perhaps not yet been properly appreciated as a waste-management method in communities that cannot afford a sewage system'.

Assorted dike–pond and canal–dike agroecosystems emerged throughout Asia, notably canal–dike systems associated with poldering in Thailand and Vietnam, whilst the historic *chinampas* situated in Mexico offer an important example of such practices that evolved and persist independent of developments in Asia. Conception of dike–aquaculture systems was a common response, it seems, to opportunities presented by work to reclaim land from wetlands, lakes and the sea for agriculture and settlement. Fingerponds, excavated from wetlands as water recedes in the dry season, were constructed in Kenya and Uganda in the Lake Victoria basin and in Tanzania on lakesides in the Rufiji River floodplain to facilitate 'wise use' of wetland resources (Bailey et al., 2005). Fingerponds were not stocked, and production relied on the natural ingress of fish during the wet season; trials with manure applications were undertaken and vegetables were cultivated on dikes between ponds.

Integrated dike–aquaculture systems are often promoted as exemplars for sustainable aquaculture development. However, construction that is not associated with investment in land reclamation may represent a considerable cost for prospective operators. Geographical areas suited physically and hydrologically to development outside of areas prone to unacceptable flooding may be limited. Reiterating constraints to adopting horizontally integrated and multi-trophic aquaculture, operators must contend with different challenges presented by aquatic and terrestrial crops and acquire distinct skills and knowledge to permit efficient and financially viable production of an array of products from integrated dike–aquaculture. Despite potential environmental gains, dike–aquaculture operations are not immune from economic and market forces, and intense competition is likely to come from specialist producers operating monocultures who probably stand to prosper unfairly owing to subsidised inputs and not being called to account or providing recompense for negative externalities.

Considering cage and pen culture in lakes, reservoirs, rivers and streams in China, the authorities have introduced regulations outlawing such activities in an attempt to improve water quality for all stakeholder groups (Newton et al., 2021). These authors noted that pond-based polyculture of carp species (involving bighead carp, crucian carp, grass carp and silver carp) continues to dominate the sector, whilst producers with only one or two ponds may opt to produce crayfish (*Procambarus clarkii*). Producers are also seeking to increase efficiency and reduce production costs and financial risks and lessen their dependence on formulated feeds. Some farmers are reportedly reverting to traditional practices, and this can involve planting crops (e.g. rye grass, soybeans and wheat) on embankments for use as supplementary feeds in semi-intensive systems (Newton et al., 2021). Sediments from ponds were used by some farmers to fertilise crop-growing areas. It was observed that increasing numbers of more affluent consumers are strengthening demand for higher value species (i.e. black carp, crayfish, swamp eel and Wuchang bream), with fresh, often live animals, being preferred. As sales of fresh and live animals are especially time sensitive, and as wet markets in Wuhan, China, have been implicated as the source of the COVID-19 pandemic, this could lead to consumers opting to purchase aquatic foods in new ways, potentially using the 'internet and other high-tech vending options' (Newton et al., 2021: 1750).

Semi-intensive static water fishponds

Freshwater fishponds in many contexts were not conceived as such; frequently, they were formed by natural process and evolved in conjunction with wetland reclamation[1] and large-scale hydraulic engineering, flood defence and waste-water discharge schemes or were developed from borrow pits and excavations left by clay and aggregate extraction. Purpose-built fishponds, in contrast, are generally more uniform and afford operators greater control over water levels, and many can be drained to enable maintenance and sediment conditioning.

Despite a long tradition in many instances, established principles of pond management, encompassing stocking and husbandry practices, are frequently less resource efficient, productive and resilient than what may be termed best, good or better management practices (BMPs; see chapter 8). Reasons for this will be specific to individual producers and difficult to discern, but access to capital, perceived opportunities to optimise the allocation of available resources, assessment of risks and specific motivations for individual farmers and other household members will govern the type of management regime adopted.

Government extension services for agriculture and fisheries were routinely tasked with promoting improved aquaculture practices for semi-intensively managed fishponds. Commonly they called on farmers to adopt a formulaic approach to pond management that dictated pond preparation, involving eradication of possible pest and predator species and standard dosage rates for lime and organic and inorganic fertiliser to stimulate phytoplankton production; ratios of fish species to stock and stocking densities to use; and application rates for manure and fertiliser and feeding regimes to maximise production. Often farmers were unable to drain their ponds and hence could not prepare ponds as prescribed, seed of specified fish species were not available and households did not have access to manure or agricultural by-products or funds to purchase inorganic fertiliser inputs.

Culturing carp species in semi-intensively managed ponds receiving organic and inorganic fertiliser and supplementary feed inputs continues to account for a significant source of freshwater aquaculture production in many Asian countries. Pressure to intensify production is, however, likely to grow owing to driving forces including international trade, population growth and urbanisation. Whilst this may achieve a higher income from a specific pond area, it may well increase financial and other risks for producers and threaten the maintenance of stocks and flows of ecosystem services sustained by aggregations of semi-intensively managed fishponds forming wetland agroecosystem landscapes. Semi-intensively managed fishponds can sustain an array of ecosystem services including provisioning fish and other aquatic animals, hydrological regulation, providing habitat for wildlife, processing waste resources such as manure and assimilation of nutrients and contributing to climate change mitigation (Box 5.1). Considering the top 100 priority research questions to guide sustainable agriculture development globally, question 32 asks, 'How can long-term carbon sinks best be created on farms (e.g. by soil management practices, perennial crops, trees, ponds, biochar)?' (Pretty et al., 2010: 226).

Appropriate BMPs could potentially assist in making semi-intensive pond management and associated husbandry more efficient, thus enhancing economic returns and helping to minimise financial and fish health risks and bolstering the resilience of such systems. Care is required, however, when formulating management practice guidelines; whereas measures to exclude wild fish from entering ponds may be advocated to reduce disease risks and competition, elsewhere interventions to promote retention and recruitment of wild stocks may be desirable for food security and conservation reasons. Culture of small indigenous species (SIS) in Bangladesh and self-recruiting species (SRS) in both Bangladesh and Vietnam has been proposed as an appropriate and sustainable means of aquaculture development posing much reduced risks to the environment, whilst generating enhanced returns for operators. Drainage of ponds followed by drying and liming was routinely recommended but was often difficult to achieve and expensive, whereas regular ploughing of sediments for conditioning without draining and closer attention to seed quality, feed and fertiliser management and water quality and fish health can reduce disease risks and water consumption. A degree of intensification beyond traditional practices, combining supplementary feed and organic and inorganic fertiliser inputs, is often deemed desirable as it can reduce the amounts of land and water needed to produce similar amounts of aquatic products compared to extensive culture practices.

An alternative paradigm to promoting phytoplankton blooms to sustain autotrophic food-webs to enhance aquaculture production is to promote

Box 5.1   Potential of fishponds in climate change mitigation

Freshwater pond-based aquaculture accounts for the largest proportion of finfish production globally, mostly focused on carp and tilapia culture in semi-intensively managed ponds. Although agro-industrial formulated pelleted feed is increasingly being used, traditionally these were fertilised with manure and agricultural by-products, but increasingly farmers use inorganic fertilisers. Kautsky et al. (1997) noted that primary production (2.3 g C m^{-2} d^{-1}) within semi-intensively managed fishponds in Zimbabwe, stocked with tilapia at rates of 0.1–0.5 kg m^{-3} and fed locally available plant and agriculture by-products, was sufficient to sustain fish production at 1.3 g m^{-2} d^{-1} or 4.75 t ha^{-1} y^{-1}. Furthermore, the ecosystem support areas for oxygen production and phosphorus assimilation both equated to 0.9 m^2 per m^2 of fishpond and this could probably be sustained within the pond.

However, in the context of this review, it is appropriate to consider whether the capacity of semi-intensively managed fishponds to absorb atmospheric carbon and sequester this in harvestable biomass or sediments could be enhanced and, indeed, whether it would be reasonable to request the majority of producers culturing fish in ponds semi-intensively not to intensify production, with its attendant higher use of, and dependence on, inorganic fertiliser, in pursuit of greater returns. Considering the first point, comparisons with other semi-intensively managed aquaculture systems are difficult as the methodologies employed in primary production studies vary considerably (Melack, 1976; Liang et al., 1981; Yusoff and McNabb, 1989; Knud-Hansen et al., 1993), resulting in production estimates which are not readily comparable.

Yusoff and McNabb (1989) recorded rates of net primary production of 1.48 and 2.41 g C m^{-2} d^{-1} in fishponds fertilised monthly with either triple superphosphate (5.7 kg P ha^{-1}) or triple superphosphate (1.4 kg P ha^{-1}) and urea (16.6 kg N ha^{-1}), respectively. However, the ease with which such production rates could be replicated remains to be tested, as does whether or not such pond management strategies are conducive to culturing other fish species. As semi-intensively managed pond aquaculture constitutes one of the most productive sectors, enhanced carbon management and potentially capture in such systems could make a significant contribution to the sequestration of carbon in freshwater systems. However, whether operators might be persuaded to adopt more carbon-conscious management strategies, against a trend towards intensive production of more commercial species, sometimes destined for export markets, remains questionable.

Source: Bunting and Pretty (2007).

attached algae growth, through the addition of appropriate substrates, thus making feeding by selected species more efficient. Attached algae, along with other attached and sessile organisms found on submerged surfaces, is technically termed periphyton, but the more descriptive German term *aufwuchs* is perhaps more illustrative concerning the potential of such systems as it encompasses both attached and motile organisms, including bacteria, fungi, invertebrates and protozoans (Azim et al., 2005). Potential benefits of enhanced periphyton production have been noted for a range of culture systems, including prawn (*Macrobrachium rosenbergii*) grow-out in ponds (Tidwell et al., 2000), tilapia (*Oreochromis mortimeri*; *O. niloticus*; *Tilapia rendalli*) cage culture (Norberg, 1999) and African catfish (*Clarias gariepinus*) fry rearing (Nwachukwu, 1999).

Production of native calbaush (*Labeo calbasu*) stocked at a rate of 10,000 fingerlings per hectare in trial ponds in Bangladesh containing 93,000 bamboo poles per hectare was 713 kg ha^{-1} over 120 days, significantly higher than production without substrate (399 kg ha^{-1}) in replicate ponds over the same period (Wahab et al., 1999a). Similarly, production of rohu (*Labeo rohita*) stocked at the same rate in ponds with a bamboo substrate density of 90,000 poles per hectare was 1899 kg ha^{-1} over 4 months, compared with 1089 kg ha^{-1} in replicate ponds (Wahab et al., 1999b). Despite comparatively higher production levels with substrate additions, extra costs for purchasing, deploying and managing the bamboo would probably make a comparative financial assessment less compelling for potential producers. Elsewhere, it was postulated that where fertiliser inputs were low and consequently suspended phytoplankton production was limited, substrate additions might promote more efficient feeding by Nile tilapia (*Oreochromis niloticus*) stocked in ponds (Shrestha and Knud-Hansen, 1994).

Upper sections of bamboo poles employed in the experimental *acadja* system instigated by Hem et al. (1995) in Benin were packed with chicken manure to enhance periphyton growth. Wastewater from aquaculture facilities that contains relatively low nutrient concentrations compared with food processing and agricultural effluents, for example, might similarly promote sufficient attached algae growth to permit integration of secondary culture of suitable grazers. Elevated nutrient concentrations may tend to promote suspended algal blooms to the detriment of attached algae that fish selected for stocking would presumably have been more adept at exploiting. Decision-making over whether or not to adopt periphyton-based aquaculture strategies would depend, therefore, on expected ambient or planned enhanced nutrient levels, access to candidate culture species suited to feeding on periphyton, and their market potential, anticipated financial returns and perceived production risks.

Multipurpose household ponds

Aquaculture-orientated extension services and development projects conceived for poverty alleviation, enhanced food security and livelihood diversification routinely have mainly focused on farming households with ponds as potential beneficiaries. Generally, such ponds were already being used by the household for multiple purposes ranging from bathing and washing cooking utensils to watering vegetable crops and fruit trees and often for low-input aquaculture. Consequently, promoting a predefined package of interventions to establish improved aquaculture practices risked disrupting established access and use patterns and imposing aquaculture production strategies that may not have been appropriate for particular ponds or households or taken into account likely market demand locally for species being cultured.

A more refined approach was to introduce and trial options to increase aquaculture production with farmers as part of broader social development programmes. Implemented by CARE-Bangladesh, the LIFE project focused on Locally Intensified Farming Enterprises that facilitated participatory action research (PAR) amongst farmer groups and encompassed trials with home-stead gardening, seedling nurseries and integrated pest management (IPM) and enhanced horticultural crop production methods; support in producing added-value products; and joint assessment of opportunities for aquaculture develop-ment with interested farmers. Carp-based polycultures, rice–fish culture, community fish culture fields, dike cropping and *in situ* composting of organic residues in ponds to enhance aquaculture production were adopted by farmers.

Periphyton-based aquaculture was included as part of the suite of management options promoted to rural farming households in Kishoreganj and Rajshahi districts, Bangladesh. Meeting farmers and discussing their situation demonstrated that the approach had been modified by farmers to fit their circumstances and often recommended species combinations and substrate densities could not be implemented. Irrespective of this, farmers reported several apparent advantages of introducing substrate to their ponds as well as some drawbacks, notably possible unintended consequences and probable areas of conflict with other users and uses of multipurpose ponds (see Figure 7.9). Several practical opportunities to further refine and enhance the management of substrate in rural ponds to optimise production were evident (Table 5.1).

Culturing indigenous freshwater species in Latin America and South America

Aquaculture in Latin America and South America accounts for a relatively small proportion of the global total; however, four of the top-twenty producing countries (Chile, Ecuador, Brazil and Mexico) are found here (see chapter 1; Table 1.2). With rapidly increasing numbers of more affluent consumers, it can be anticipated that demand for animal source foods (i.e. dairy and meat), including aquatic animal source foods, is set to increase markedly (Godfray et al., 2010).

There is a rich history of proto-aquaculture systems in the region, exemplified by the 'Savanna Weirs of Baures, Bolivia' (Ezban, 2020: 173).

Table 5.1 Opportunities for enhanced substrate management in rural ponds in Bangladesh being managed for periphyton-based aquaculture

Issue	Observation	Management opportunity
Timing of substrate introduction	• substrate was often introduced up to 1 month after fish were stocked	• ideally, substrate should be deployed as soon as possible prior to stocking, to permit colonisation by periphyton, ensuring that newly introduced fish can graze it immediately; in highly seasonal ponds, this may only be practical immediately prior to stocking
Depth of substrate	• once saturated, branches sank to the bottom of deep (2–3 m) seasonal ponds in both project sites	• optimal substrate use in the farming systems evaluated was important; therefore, substrate should be positioned at a depth where periphyton growth will be maximised; some farmers suggested using trellis-like structures
Vertical extent of substrate	• bamboo poles and tops inserted vertically into ponds extend well below the light and well-oxygenated upper reaches where periphyton growth mainly occurs and often protruded 2–3 ft above the surface	• substrate at the bottom of deep ponds and protruding well above the surface will not contribute much to production, therefore, limited substrate resources should be deployed carefully to maximise periphyton growth
Substrate density	• substrate, such as branches, was deployed by farmers randomly in their ponds to give an impression of complete coverage	• substrate density, and specifically the surface area for periphyton growth, should be matched to the potential grazing intensity of fish stocked; more trials may be required in this regard
Relative contribution of periphyton	• periphyton growth may be inhibited in ponds with dense phytoplankton blooms resulting from high supplementary feed and fertiliser inputs	• further trials are required to understand the relative contribution of periphyton, phytoplankton and other sources of nutrition in ponds receiving regular feed and fertiliser inputs, especially where polycultures including exotic species are stocked, an approach favoured by farmers in both project sites

(Continued)

Table 5.1 (Continued)

Issue	Observation	Management opportunity
Substrate type	• substantial hardwood branches and bamboo poles were often employed	• opportunities to use thinner branches and bamboo tops or alternative substrates such as palm leaves should be investigated, reducing pressure on substrate material that may have other uses or be in short supply, and encouraging farmers to exploit unused on-farm resources, as opposed to purchasing additional substrate unnecessarily
Management during harvest	• during harvesting, substrate is usually removed from ponds for several hours, which may adversely affect periphyton growth; reintroducing substrate in a different orientation may inhibit or retard periphyton production	• care could be given to ensuring that substrate removed from the pond is stored in a manner which limits any possible negative impacts on the growth of periphyton and substrate should generally be reintroduced to the pond with the same orientation; further trials may be required to assess the relative benefits of adopting such approaches

Source: Bunting et al. (2005).

Dikes and weirs were constructed across the savanna by the people of Baures in the sixteenth century to intercept and corral fish for collection, and reservoirs (up to 30 m in diameter) were constructed to hold and raise fish until they were required. Indeed, such strategies could inspire a new era of capture-based aquaculture (Lovatelli and Holthus, 2008) and could constitute innovative opportunities as 'ocean food production systems' (Costa-Pierce and Chopin, 2021: 26). As noted in the case of Brazilian reservoirs, it will be crucial to pursue the equitable, optimal and sustainable development of both aquaculture and capture fisheries to enhance food and nutrition security and provide meaningful and dignified employment opportunities (Lopes et al., 2018).

Latin America and South America are endowed with immense aquatic biodiversity that has huge potential in provisioning nutritious aquatic foods and establishing vibrant and successful aquaculture sectors for indigenous and native species (van Beijnen and Yan, 2019). These authors noted that the culture of indigenous and native aquatic animals and plants could help eliminate risks associated with introducing potentially invasive non-native aquatic species. Culture of native fish species (e.g. arapaima, *Arapaima gigas*; tambaqui, *Colossoma macropomum*; and tambacu, a hybrid of *Colossoma macropomum* and *Piaractus mesopotamicus*) has become established in Brazil and accounts for over 40 per cent of production (van Beijnen and Yan, 2019). These authors noted that culture of indigenous species is appreciated as it matches with prevailing customs, meets domestic consumer demand and enables farmers to sell through shorter supply chains with fewer intermediaries, resulting in better financial returns and lower risks.

Integrated aquaculture–agriculture

Recognition of animal manure, human waste and food and beverage processing by-products as promising nutrient sources for pond-based aquaculture led to such locally accessible inputs being employed widely in traditional culture systems. Employing unexploited resources derived from within farming systems or appropriated from local environs to intensify production is often termed 'integrated aquaculture' (Little and Muir, 1987). Recently, increasingly widespread and reliable inorganic fertiliser and formulated feed supplies have encouraged many farmers to intensify production of commonly cultured species as well as to culture more valuable species to generate greater financial returns. Demand will persist, however, for cheaper aquatic products, whilst costs and attendant risks of extensive and low-input production are generally lower and, as such, certain producers have resisted pressure to intensify production or have selectively adopted elements from externalising technology packages. Strategies developed by farmers to integrate aquaculture with agriculture, including culture practices adapted to irrigation water storage and conveyance structures and saline drainage water, are summarised in Table 5.2.

Table 5.2 Integration of aquaculture practices with other activities to optimise water-use efficiency

Integration strategy	Management practices	Constraints and conditions	Potential water-use efficiency outcomes
Livestock–aquaculture	• ducks and geese foraging on ponds • wildfowl and poultry housed over fishponds • waste from pigs and cattle directed to fishponds for treatment and nutrient recycling • plant and fish biomass cultivated using solid and liquid waste fed to livestock	• risks have been identified concerning possible pathogen and disease transfers within integrated systems • chemical treatments and dietary supplements for livestock may affect production and accumulate in aquaculture components • excessive waste loadings or perturbations affecting the ecological balance of the pond can result in low oxygen levels and fish health problems and mortality	• multiple crops from ponds and lakes with products having lower water footprints • enhanced environmental protection of receiving waters owing to better on-farm waste management and nutrient recycling • aquaculture of biomass and fodder crops helps avoid possible problems associated with public health risks and consumer acceptance of aquatic products grown using waste resources
Aquaculture in irrigation and water management schemes	• fish cages in irrigation channels in India and Sri Lanka • culture-based fisheries in domestic supply and irrigation reservoirs • aquaculture in traditional irrigation structures within micro-catchments in Sri Lanka • fish culture in irrigated ricefields and farmer-managed systems in Africa and Asia	• excessive flow rates can impact of animal welfare and make food unavailable • debris can block mesh, reducing flow rates, and cause physical damage to cages • agrochemicals used in irrigation systems and adjacent areas can affect aquaculture and may constitute a public health concern • management must balance irrigation and aquaculture demands, and new structures may be needed to sustain fish populations during low-water periods • agencies responsible for irrigation are opposed to the installation of cages in canals	• nature of aquaculture means little water is consumed and instead is conserved, potentially with a higher nutrient content, enhancing crop production • aquatic species being cultured may predate on disease vectors and crop pests and weeds • integration of aquaculture activities may enhance nutrient cycling and uptake by plants under irrigation

Aquaculture in water storage reservoirs	• fish cages in reservoirs for hydroelectric power generation • culture-based fisheries in water storage and hydro-electric reservoirs • polyculture in urban and peri-urban water bodies primarily for floodwater discharge and amenity	• inappropriate reservoir bed preparation, presence of submerged structures and drowned trees and routine dropdown may reduce the area suited to aquaculture development • rapid dropdown may damage physical cage structures • changes in access and use rights associated with aquaculture development may cause social problems	• multiple use of water in reservoirs could contribute to increased revenue generation and alternative livelihoods for displaced or marginal communities • appropriate species selection for aquaculture could contribute to weed control and enhance water quality in reservoirs
Aquaculture in saline drainage and wastewater	• aquaculture in saline groundwater evaporation basins in Australia • fish culture in saline wastewater from industrial processes and desalinisation	• variation in salinity levels and possible extremes may constrain species selection or culture duration • low production rates as compared with prevailing commercial operations suggest need for further assessment of financial and economic attributes	• exploitation of saline water resources through integration of aquaculture can contribute to overall farm productivity and generate new income streams • economic benefits of integrating aquaculture, salt-tolerant crop production and salt harvesting could help offset costs of controlling saline groundwater problems
Aquaculture in multipurpose household ponds	• fish culture in small ponds used primarily for domestic and agricultural purposes • composite fish culture in rainwater harvesting structures	• introduction of aquaculture can cause conflicts with other agricultural and domestic uses of household ponds • inclusion of aquaculture in rainwater harvesting ponds may constrain the use of water for other crops and exacerbate food insecurity and financial risks faced by operators and their families	• appropriate integration of aquaculture in household ponds can contribute to food security and enhanced livelihood outcomes without reducing water availability for other purposes • aquaculture in ponds can help reduce pressure on provisioning ecosystem services of natural water bodies

With prices for inorganic fertiliser and formulated feeds rising in line with increasing energy costs and expected to continue to do so over the coming years, it is conceivable that high-input and intensive culture of valuable aquatic species will be curtailed, with many producers forced to revert to more extensive production and culture of less demanding omnivorous species. Potential strategies permitting livestock–fish integration were comprehensively reviewed by Little and Edwards (1999) and a classification scheme was proposed based on the location of production, either rural or peri-urban; feeding system employed; production environment; and anticipated value of livestock waste for aquaculture. These authors reported that 'intensification of both traditional livestock and fish production components through the strategic use of imported nutrients is identified as a promising strategy for smallholder livestock–fish production' (Little and Edwards, 1999: 118) and that fishponds could be used to enhance nutrient use efficiency and recycling. Livestock and poultry can potentially be housed directly over fishponds to facilitate more efficient nutrient cycling, but there are inherent health risks and added costs and management demands associated with such strategies. During avian flu outbreaks, farmers are often compelled to confine ducks indoors as a pre-emptive measure to halt the spread of the disease. Implications for integrated poultry–aquaculture systems in urban and peri-urban contexts are discussed in Box 6.3.

Recalling the part geo-political isolationist policies played in the development of wastewater-fed aquaculture in Eastern Europe and the influence that measures adopted to promote economic self-reliance in post-Soviet-era Cuba had on homestead aquaculture development (chapter 6), it seems that adversity can be an important driving force in stimulating aquaculture innovation. Further evidence may be gleaned perhaps from the evolution and proliferation of integrated VAC (*vuon-au-chuong*) agroecosystems in the Red River Delta during the Vietnamese or American War as it is known within the country (Luu et al., 2002). VAC is an acronym of the first letter of the Vietnamese words for garden (*vuon*), pond (*au*) and livestock housing (*chuong*) and succinctly describes the sub-systems that constitute this iconic integrated agroecosystem. Despite having been promoted widely as an effective means to enhance food security and nutrition in small-scale farming communities, these authors noted that there was considerable scope in the Red River Delta to enhance fish production but that 'intensification of aquaculture systems may result in a surplus beyond local needs' (Luu et al., 2002: 74). Indicative of the limited reach of market networks and actors at the time, this is probably much less of an issue now, and today the system is described as 'improved VAC', with increasing intensification through stocking tilapia and increasing use of pelleted feed.

Adoption of integrated agriculture–aquaculture (IAA) has been cited as a response amongst family farmers in Misiones Province, Argentina, to the economic crisis in the country during the 1990s that was precipitated by implementation of neoliberal policies (Zajdband, 2011). Seeking to diversify production away from cash crops such as tea and tobacco, traditionally grown in this part of the country, which is characterised by low-lying hills, many farmers opted for small-scale aquaculture being proposed by various institutions. An estimated 2500 farmers adopted aquaculture, and credit provision by the public sector was deemed the main factor explaining rural aquaculture development in Misiones.

Integrated production of aquatic plants and animals in low-input culture units to produce food for more valuable, intensively cultured species could conceivably become commonplace. One million tonnes of carp were reportedly purposefully cultured in China in 2008 as live-feed to produce 230,000t of more valuable mandarin fish (*Siniperca chuatsi*; FAO, 2010). Elsewhere, assorted freshwater aquatic plants, notably duckweed and animals (e.g. chironomid larva, snails and zooplankton), are cultured as live feed or ingredients for formulated aquaculture feed as well as for feed for the aquarium trade (see Box 5.2).

Box 5.2   Freshwater zooplankton facilitating integrated aquaculture

Zooplankton play an important role in the nutrition of various fish and prawn species (Kibria et al., 1997, 1999). The integrated production of cladocerans, copepods, rotifers and euphausiids in aquaculture wastewater could represent an important source of zooplankton for use as either a live feed for crustaceans and fish larvae or as a component in a formulated diet. Kibria et al. (1997, 1999) described the harvest of zooplankton from wastewater treatment lagoons at Werribee sewage treatment plant, the largest in Australia. Zooplankton (*Daphnia carinata* and *Moina australiensis*) proliferate in the final treatment lagoons and are harvested at a rate of $40–84$ kg h^{-1} by filtering the water through screens mounted on a floating platform. The harvest of 100 kg of fresh zooplankton removes approximately 0.1 kg of phosphorus and 1 kg of nitrogen. Furthermore, these zooplankton were rich in protein, essential amino acids, lipids and phosphorus; the specific growth rate (2.97), feed conversion ratio (1.18) and protein efficiency ratio (1.57) for silver perch (*Bidyanus bidyanus*) fed frozen *D. carinata* were not significantly different to those of fish receiving a commercial diet (Kibria et al., 1999).

Gnudi et al. (1991) cultivated zooplankton (*Daphnia* spp.) in sixteen 1m-deep ponds (1200m^2) that received microalgae from cultures produced in four 0.5m-deep basins (2000m^2). The microalgae culture was fertilised with $1.5–6$ l m^{-2} d^{-1} of pig manure, depending on climatic conditions. The dry weight microalgae standing crop decreased during the four-month study from 62 mg l^{-1} in October to 41 mg l^{-1} in January; this was attributed to lower solar radiation, 940 cal cm^{-2} d^{-1}

in October as compared with 615 cal cm^{-2} d^{-1} in January, and reduced mean daily temperatures: 12°C in October and 3°C in January. The dry weight of microalgae transferred to the zooplankton cultures also declined during the study from 4.4 to 1.6 g m^{-3} d^{-1} and in combination with declining temperatures resulted in mean weekly wet-weight zooplankton harvests (40 per cent of standing biomass) falling from 80 to 28 g m^{-3}. Groeneweg and Schlüter (1981) fed rotifer (*Brachionus rubens*) cultures in 50 l indoor tanks on algae diets produced in shallow 9.7m^2 raceways fertilised with diluted pig manure. Each day 12.5 l of pond water was introduced to the rotifer culture, giving a retention time of 4 days. Cultures were produced containing 200–580 rotifers per millilitre, depending on the concentration of algal–bacterial biomass in the pond water; thus:

$$Y = 1.54X - 226.86, \text{ where: } Y = \text{rotifers (ml}^{-1}\text{); } X = \text{total SS (mgl}^{-1}\text{).}$$

Schlüter and Groeneweg (1981) described the environmental require-ments – for example, nitrite, salinity, dissolved oxygen and pH – for culturing *B. rubens*; dissolved oxygen levels were considered especially important, with minimum concentrations of 1.15 mg l^{-1} being recom-mended; below this, reproduction and survival decreased. Although other food sources have been successfully used for culturing rotifers, those produced with algae diets had a higher nutritive value for juvenile fish.

Source: Bunting (2001b).

Polyculture, introduced in chapter 2, constitutes a principal strategy to optimise the use of space and exploitation of feeding niches in aquaculture systems. Inter-cropping, however, holds potential to facilitate the optimal temporal use of a particular culture area (Wenhua and Qingwen, 1999), especially where prevailing hydrological regimes or access to water has a strong seasonal component. Although fish culture may be suspended during the dry season, staple crops such as rice or cash crops such as maize and vegetables may be cultivated on accessible areas (Little and Muir, 1987; Murray et al., 2002). Other high-potential strategies for sequential aquacul-ture activities include on-growing air-breathing catfish in ponds with low water levels after the main aquaculture crop has been harvested, holding juveniles in ponds to permit decentralised seed distribution and earn extra income before stocking a main crop and maintaining broodstock to enhance recruitment in seasonally interconnected wetland agroecosystems.

Developed in conjunction with traditional *chinampas* farming systems in Xochimilco–Chalco, Mexico, an alternative strategy to integrate terrestrial plant cultivation in wetland agroecosystems was to construct floating rafts

(6–9 m long) of cattail and reed and cover them with nutrient rich mud dredged from the bottom to form a nursery for vegetable plants (Armillas, 1971). In low-lying areas of Bangladesh, floating planting beds made from water hyacinth (*Eichhornia crassipes*) have been used to raise rice and vegetable seedlings and cultivate vegetable crops and could potentially help vulnerable communities cope with extended periods of waterlogging and recover from damaging floods (Irfanullah et al., 2011). Elsewhere, recent assessments have exploited artificial materials such as polystyrene sheets floating on the surface to cultivate cereals, fibre-producing plants, flowers and vegetables, but costs are likely to prohibitive. Furthermore, structures positioned in ponds would likely constrain pond management activities and could give rise to unintended impacts on water quality and associated aquaculture productivity.

Aquaculture integrated with rice farming

Fish culture in flooded rice-fields or paddies has a long history in Asia and more recently has been proposed as an appropriate means to enhance protein production in conjunction with cereal crops for added food security, nutrition and income for poor and marginal farming communities (Box 5.2). Integrated rice–fish production is practiced in several Asian countries in settings where hydrological and environmental conditions permit combined culture. Husbandry practices, production intensity and management arrangements vary depending on household and community assets, market opportunities, perceptions of risk and the relative merits of alternative cropping arrangements. A comprehensive review of rice–fish culture in Thailand by Little et al. (1996) addressed factors prompting development, culture systems facilitating integration and prospects for wider adoption. This was, however, over 25 years ago, and the conclusions may have less relevance today with increased off-farm employment opportunities for small-scale farming households.

Box 5.3 Traditional and contemporary rice–fish culture practices

Culturing fish in rice-fields can help control pests and weeds, promote nutrient availability to rice plants and enhance nutritional benefits and financial returns from what are widely regarded as low-input, environmentally friendly and more sustainable farming systems. Integrating fish culture in irrigated and rain-fed rice-fields also makes more effective use of appropriated water resources. Culturing fish in rice-fields is a traditional practice in China, Japan and Java, Indonesia; more recently, attempts to introduce rice–fish culture have been made by development agencies and extension services to many countries in Asia and a growing number in Africa. Integrated culture of rice and fish requires more refined management approaches, with farmers having to coordinate rice production and fish culture practices. Poor quality and unreliable fish seed supplies and limited financial returns have often constrained widespread and long-lasting adoption.

Rice–fish culture was adopted widely in northeast Thailand and West Java, Indonesia, and it can sometimes make an important contribution to incomes and food security in poor and marginal farming communities. A large proportion of global rice production is not under irrigation, and integrating of fish culture in rain-fed fields unsuited to intensive rice cropping could potentially help in developing more sustainable livelihoods for farming communities. Collection of wild food from rain-fed rice-fields has been shown to be important and is often overlooked; however, perceived declines in the availability of wild fish and well-developed trading networks for fish seed from private hatcheries stimulated the initial adoption of rice–fish culture in northeast Thailand.

As concern grows over the sustainability of high-input, irrigated, monoculture rice production and farmers face increasing bills for fertilisers and pesticides to maintain yields, the viability of low-input rice–fish culture measured in conventional financial terms and based on standard risk assessment criteria could increase. Farmers should be supported in assessing their prospects with regards to adopting rice–fish culture and, where there is demand, action must be taken to ensure that an enabling institutional environment exists (Bunting et al., 2023). Successful development of rice–fish culture has been attributed to adaptation of traditional water management approaches to accommodate fish culture; appropriate extension services, training and capacity-building; access to quality fish seed of appropriate species; and willingness of producers to adapt and innovate (Pounds et al., 2023).

Discussing development trends, FAO (2010: 25) noted that 'rice–fish culture, often operating at family scale with renovated paddy fields, has expanded rapidly among rice farmers in China in recent decades' and reported that 1.47 million hectares were managed for rice–fish culture in 2008, producing 1.2 million tonnes of food fish annually. Reviewing rice–fish culture practices globally, Halwart and Gupta (2004) included examples from Africa, Asia, Australia, Europe, South America, the Caribbean, the former Soviet Union, the Middle East and the USA. Rice–fish production in Egypt was reported at 27,900 tonnes in 2008 but equated to only 4 per cent of national aquaculture output (FAO, 2010). To maximise financial returns from freshwater resources appropriated for farming, fish and rice are cultured simultaneously (Shaalan et al., 2018). These authors noted that production had increased to 34,537 t y^{-1}, with Nile tilapia accounting for around half and the remainder comprising catfish and common carp.

Figure 5.1 *Ghers* have been developed widely in southwest Bangladesh and produce both high value prawns and staple rice crops.

Photo credit: Dr Nesar Ahmed.

Farming freshwater prawns (*Macrobrachium rosenbergii*) in modified rice-fields, referred to locally as *ghers*, is widespread in southwest Bangladesh; dikes surrounding the fields are raised and a peripheral canal is dug around the inside of the dike to retain water during the dry season (Ahmed et al., 2008; Figure 5.1). During the wet season, prawn and fish are cultured together; rice is planted on the central terrace at the onset of the dry season whilst fish culture continues in the peripheral trench. Prawn farming has developed extensively in Bangladesh, and over 100,000 farms covering 50,000 ha have been recorded, whilst an estimated 1.2 million people are engaged in prawn and shrimp farming and allied activities (Muir, 2003; Ahmed et al., 2008).

Gher farming is concentrated in densely populated southwest Bangladesh where, according to Ahmed et al. (2008: 208), 'families tend to be resource-poor, income-poor and vulnerable to environmental, climatic, and economic variability' and prawn culture is widely regarded as a vital opportunity for livelihoods enhancement. Situated principally in floodplain areas, *gher* dikes are commonly planted with fruit trees and vegetables to increase production of these crops prone to water-logging in an otherwise low-lying environment (Ahmed et al. 2014; Bunting et al., 2017; Faruque et al., 2017). Threats facing prospective producers include worsening climate change impacts, poor water quality, pollution, floods, drought, theft and poisoning, whilst accessing finance to meet escalating operating costs constitutes a major constraint. Catching wild prawn seed and collecting snails for feed negatively impact biodiversity and ecological processes and functions.

Elsewhere, crab, crayfish, frog and snail culture have been integrated in rice-fields, and recognition of rice–fish culture as a Globally Important Agricultural Heritage System (GIAHS) has been credited with arresting the decline in cultivation of traditional rice varieties in Wannian County, Zhejiang Province, China. Policy change and increased public awareness were cited as influencing decision-making by farmers, resulting in the area under traditional rice varieties increasing from 7.4 ha in 2003 to 15.8 ha in 2009, but this remains modest given historical coverage of around 100 ha before 1949 (FAO, 2023). Mitten crab (*Eriocheir sinensis*) culture has developed rapidly in China, reaching over 400,000t in 2004, and rice fields are routinely used as nurseries and for on-growing (FAO, 2023). Crabs feed extensively on naturally occurring pests and weeds, releasing nutrients to stimulate rice production; the presence of crabs in paddy fields means that farmers must take considerable care regarding the type and timing of agrochemical applications.

Newly developed rice varieties intended to assist in climate change adaptation by surviving prolonged submersion or periods of drought, respectively, may have unintended consequences where action and investment to maintain stable hydrological conditions with consistent and continuous flooding periods is forgone. Adopting the System of Rice Intensification (SRI) integrated with aquaculture in small-scale water storage infrastructure could increase financial returns for producers, contribute to more diverse and nutritious diets, enhance agrobiodiversity and bolster resilience to adverse shocks and trends (Bunting et al., 2015). Intermittent periods of field flooding that characterise SRI cultivation could potentially help avoid greenhouse gas emissions, notably methane, that is released when paddy fields are continuously flooded (Mboyerwa et al., 2022). Failure to appropriately manage the inter-connectedness of natural wetlands and managed wetland agroecosystems may well restrict further integration of aquaculture with rice production and have adverse consequences for aquatic biodiversity and aquatic food production by communities in farmer-managed systems.

Horizontally integrated constructed wetlands

Constructed wetlands routinely planted with common reed (*Phragmites australis*) were developed initially to treat human wastewater but have been proposed in various settings as a low-maintenance approach to treating aquaculture wastewater that does not depend on external energy supplies to function, whilst achieving fairly reliable treatment under a range of environmental operating conditions. Preliminary trials were conducted in Hale County, Alabama, USA, to condition channel catfish (*Ictalurus punctatus*) pond water in horizontal surface-flow wetlands (Schwartz and Boyd, 1995). Wetlands planted with California bulrush (*Scirpus californicus*), giant cutgrass (*Zizaniopsis miliacea*) and Halifax maidencane (*Panicum hemitomon*) and loaded with 77–91 l m^{-2} d^{-1} of pond water removed 37–67 per cent of biochemical oxygen demand, 75–87 per cent of suspended solids, 45–61 per cent of total Kjeldahl nitrogen and 59–84 per cent of total phosphorus.

Cognisant of such promising findings, bioeconomic modelling was used to assess the potential of comparable wetlands to treat wastewater from an Atlantic salmon (*Salmo salar*) smolt unit in Scotland (Bunting, 2001b). Outcomes demonstrated that wetlands could potentially retain nutrients, enhance treatment performance and routinely comply with regulatory discharge standards. Capital costs, physical area required and prospects for variable treatment effect owing to environmental perturbations and pest and disease attack constituted major constraints. Options to enhance viability included reducing discharge volumes requiring treatment through recycling or supplementary aeration, harvesting biomass from the wetland as a renewable energy source and complementary wetland-based income-generating activities.

Sub-surface-flow wetlands, in which water passes through the rooting substrate, have been shown to offer a more space efficient and practical option for aquaculture wastewater treatment. Hydraulic loading rates of 13.6 m per day were employed by Sindilariu et al. (2007) to treat wastewater from a small trout farm in Bavaria, Germany, and it was concluded that wetlands designed in this manner constituted an effective alternative to pre-existing settlement basins, giving better effluent treatment efficiencies. Dissolved nutrient removal was, however, variable, with limited removal or additional internal loadings of nitrate nitrogen and phosphate phosphorus being generated.

Construction costs per unit area are higher for vertical-flow wetlands compared with surface-flow wetlands, but these may be offset by the smaller area required. Moreover, physical limits to land availability for wetland development at most aquaculture sites would favour more compact treatment options. Compared to mechanical drum filters which may routinely achieve a good level of treatment, wetlands would be much more costly to develop but could require less maintenance, avoid sludge production demanding disposal and eliminate risks associated with mechanical failures or power cuts. Wetland engineering designs could permit routine maintenance and more comprehensive renovation without severe disruption to treatment performance. Where constructed wetlands are proposed for the treatment of aquaculture discharges in lower-order catchments, attention must be paid to potential impacts with regards to heat flux and hydrological regimes.

Aquaculture in conjunction with irrigation and hydroelectric schemes

Construction of minor and micro irrigation systems, classified by Murray et al. (2002) as having water spread areas of 1–50 ha and <0.1–1 ha, respectively, to increase the productivity of marginal agricultural land has a long history in areas such as Karnataka and Tamil Nadu, India, and the North West Province, Sri Lanka. Although it might be proposed that aquaculture development could be integrated into such systems, these authors noted that 'traditional power structures may undermine attempts to utilise irrigation systems for novel uses' (Murray et al., 2002: 38). Difficulty in managing small-scale irrigation schemes because they are less predictable than larger systems was also cited as a potential constraint to aquaculture, although benefits were envisaged concerning seasonality, helping to facilitate predator control, enhancing the effectiveness of fishing activity and permitting crops and pasture to be planted on exposed areas.

Rivers and streams were dammed throughout Europe in the Middle Ages to impound water to supply overshot millwheels, whilst fish such as bream and carp were stocked in the resulting millponds. Given the profusion of watermills across Europe, it was surmised that this 'contributed to the decline of many anadromous and cold-water species by presenting a physical barrier to migration, concentrating fish and making them easier to catch, and producing standing pools of water where the water was warmer and oxygen levels lower' (Bunting and Little, 2005: 121). Contemporary schemes to regulate water flows, permitting irrigated agriculture development and harnessing water for hydro-electric power generation have fundamentally affected the hydrology of globally significant Asian and South American river systems. As with proto-aquaculture development in Europe where fish were stocked and harvested from millponds, opportunities presented by much more massive, modern-day hydrological engineering works for aquaculture have been widely postulated and indeed cited to garner support for development amongst local communities and policymakers alike. Whilst intuitively it might be supposed that impounding large volumes of water would present opportunities for aquaculture development, notable constraints have been identified (Box 5.4).

In view of the complexities surrounding the installation of cages, a framework for research concerning the potential of aquaculture in irrigation systems was proposed by Murray et al. (2002: 29) encompassing 'approaches that use technical and social methodologies both to understand current systems and to develop innovative approaches in participation with stakeholders'. Large-scale irrigation and hydroelectric schemes present particular challenges regarding the integration of aquaculture, notably a tendency for individual households to be excluded from management decisions, with groups and external agencies exerting greater control. It is imperative that irrigation schemes be designed and operated with potential production from both aquaculture and fisheries in mind because optimal yields of both irrigated crops and aquatic foods could make the greatest contribution to achieving Sustainable Development Goal 2 'Zero Hunger' of the United Nations by 2030 (Lynch et al., 2019). Considering medium and major irrigation schemes, classified by Murray et al. (2002) as having water spread areas of 50–200 ha and above 200 ha, respectively, a common strategy employed to enhance fish production is culture-based fisheries, but, again, there are notable challenges and risks associated with such practices (Box 5.5).

Box 5.4   Integration of aquaculture in irrigation schemes

New capture fisheries are often cited as a secondary benefit associated with reservoirs developed for irrigation purposes; however, timely colonisation by species suited to static water reservoir conditions and valued by fishermen is not guaranteed. Furthermore, unrestricted and unregulated fishing could limit the establishment of a substantial, self-reproducing stock of a desirable species (Munro et al., 1990). The latter constraint could be overcome through the establishment of a culture-based fishery (see Box 5.5) or fish culture in pens or cages. Construction of pens and cages could be used to divide the available water resource, potentially enabling displaced or landless people to gain some form of employment and security; however, the costs of constructing and stocking such structures could be prohibitive, often leading to better-off individuals and commercial enterprises dominating the available resources.

Cages in inland waters in Southeast Asia are commonly between $10 \, m^2$ and $100 m^2$ (Beveridge and Muir, 1993), although smaller $1 m^2$ cages were developed in Bangladesh, for example, primarily to permit poor people to engage in cage culture (Brugere et al., 2001). The two major categories of cages utilised are fixed and floating, with fixed cages generally being smaller and used in shallow areas (less than 10m in depth). Management and input requirements for cage culture may be extensive, semi-intensive or intensive, although in Southeast Asia the most commonly practised form of cage culture and can be classified as intensive as it depends on complete feed inputs. Tilapia and common carp are often used for cage aquaculture in freshwater reservoirs. Cages can also be deployed in irrigation channels, but flow rates and regimes must be suitable, and the requirements of cage operators must be considered in the overall planning and management of the irrigation scheme. Furthermore, national irrigation authorities often legislate against the installation of structures in the canal perceived to impede water flow.

Rapid, uncontrollable expansion of aquaculture in larger irrigation reservoirs could potentially result in access to fishing grounds and navigational routes being disrupted, and this could in turn lead to social tension (Beveridge and Phillips, 1993). Drawdown and the presence of submerged trees could restrict the area available for cage-based aquaculture development. Drawdown could also cause physical damage to cages and lead to upwelling of deoxygenated hypolimnetic waters, which could cause mortalities in overlying fish cages. Fluctuating water levels can be a serious problem in reservoirs used for irrigation and hydroelectric power generation. Uncontrolled development of aquaculture and associated waste discharges could lead to serious water quality problems (Beveridge and Muir, 1993), and this could be compounded by reduced water exchange owing to the physical presence of pen and cage structures.

Box 5.5   Culture-based freshwater fisheries

Culture-based fisheries, stocking fish and other aquatic organisms in large water bodies to on-grow for harvest with little further intervention, have been promoted and established in seasonal wetlands, lakes and reservoirs, including water bodies in upland and highland areas. Often developed to sustain livelihoods in fishing communities and enhance food security in poor and vulnerable rural communities, culture-based fisheries have also been proposed to increase employment and income from tourism and angling or enhance food fish production to alleviate fishing pressure on wild stocks. Stocking fish and their subsequent harvest with objectives of reducing invasive macrophyte communities, increasing water clarity or sequestering nutrients have been proposed as potential strategies to facilitate the biomanipulation of water bodies to enhance water quality characteristics.

Culture-based fisheries have been proposed for both marine and freshwater environments but have perhaps been most widely established in South and Southeast Asia, where they have been developed in a diverse array of seasonal and perennial water bodies. Assessment of culture-based fisheries in twenty reservoirs throughout Thai Nguyen and Yen Bai provinces in Northern Vietnam showed that mean yields of stocked fish over two production cycles ranged from 287 to 325 kg ha^{-1} (Nguyen et al., 2005). The catch of wild fish from these water bodies was also assessed, with their contribution to harvests over two production cycles ranging from 51.8 to 95.1 kg ha^{-1} or 2.9–54.4 per cent of the total yields recorded.

Although requiring significantly less inputs than most rural aquaculture strategies, stocking of juvenile fish constitutes a major cost. There are ecological, social and economic risks associated with culture-based fisheries, but no systematic impact assessment of this emerging aquatic resource management strategy has been undertaken. According to Gurung (2003: 15), carp have been stocked in several lakes in upland areas of Nepal to enhance production and reduce 'fishing pressure on thinly populated native species' whilst safeguarding employment and income for traditional fishing communities 'until measures for conservation practices of locally vulnerable species are developed'. When planning and implementing such a strategy, appropriate risk assessments and control measures should be employed to protect native fish populations and ensure that other species are not negatively affected. Potential social and cultural impacts of such interventions also demand careful assessment prior to implementation.

Cage aquaculture

Cage aquaculture has been developed widely for high-value marine finfish, and this is associated with major capital costs and technologically advanced command and control processes. The origins of cage aquaculture are, however, comparatively humble and perhaps more instructive in terms of guiding sustainable aquaculture development. Beveridge (2004) reviewed the origins of cage culture, with accounts of holding fish for extended periods in fine-meshed cloth cages from China between 2200 and 2100 years BP. Floating wooden cages were in operation in Cambodia from the end of the nineteenth century and were often employed for the transhumance of fish between the Great Lakes region and markets in Phnom Penh. Cages were constructed as boats with accommodation included above the cages, and these designs spread widely throughout the lower Mekong River (Figure 5.2). In Indonesia, bamboo cages were used to on-grow small fish caught from Mungdung Lake, Sulawesi, prior to marketing and to culture common carp in organically enriched canals and rivers in West Java (Beveridge, 2004). According to this authoritative source, traditional cage-based aquaculture can be differentiated from modern cage culture owing to continued 'reliance on natural construction materials and on natural or waste feeds' (Beveridge, 2004: 6).

Figure 5.2 Cage-based tilapia culture in Southern Vietnam with accommodation and storage positioned above.

Photo credit: Stuart Bunting.

The CARE CAGES (Cage Aquaculture for Greater Economic Security) project was instigated in Bangladesh based on the knowledge that small-scale, low-input cage culture had previously afforded rural and peri-urban communities with access to streams, rivers and lakes a means to earn a worthwhile income and enhance food security. A range of project sites were selected to test the feasibility of poor households operating small ($1m^3$) cages to on-grow fish and aquatic animals caught locally and to raise fish purchased from traders. Outcomes were disappointing, but one group that it was hoped would benefit from small-scale cage culture were women, as this was not a traditional activity within pre-determined gender roles, and some groups of women were able to set up cages in communal water bodies close to their homes (Brugere et al., 2001). Survey results presented by these authors indicated that benefits attributed by community members to cage culture included income source diversification and increased household fish consumption, although it was noted that the distance between the household and potential culture site could constitute a major constraint to the participation of women. General constrains experienced by operators included fouling of nets; pests and disease; variable growth; loss due to storms and reportedly theft; recurring maintenance costs, exacerbated by restricted access to appropriate net materials; and unrealistic expectations regarding financial returns as initial investment was subsidised.

Widespread concerns surround many forms of intensive cage culture due to pollution of the water body in which the cage is installed and to continued dependence on formulated feeds containing significant levels of fish meal and fish oil derived from sources that have not been certified as sustainably managed. Intensive cage-based culture can, however, constitute a comparatively efficient use of aquatic resources and supporting ecosystem areas providing that the carrying capacity of the water body is not exceeded. Converting ecological footprints associated with intensive tilapia farming in Lake Kariba, Zimbabwe, from an area-based assessment to a production unit of 1 kg indicated that phosphorus and nitrogen assimilation capacity were used more efficiently in cage culture than in semi-intensive tilapia pond culture (Figure 5.3). It was noted, however, that the area required for oxygen production was slightly smaller for semi-intensive pond culture than for intensive cage culture, at 1 and 1.18 m^2 kg^{-1}, respectively.

Overall, the cumulative ecological footprint for feed input production and waste assimilation remained relatively high at 166.7 m^2 kg^{-1} for intensive cage culture compared to 3.8 m^2 kg^{-1} for pond-based production exploiting agricultural by-products. Comparisons based on a production unit of 1 kg could potentially be further refined by comparing the biomass of edible parts, amounts of protein, content of essential nutrients or value of product units. The most desirable approach may depend on the objectives of the producer or intended beneficiaries of a particular program or innovation.

Cage-based aquaculture has been noted on Lake Albert, Lake Kivu, Lake Malawi, Lake Tanganyika and Lake Victoria in the African Great Lakes

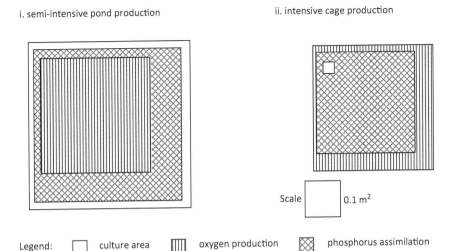

i. semi-intensive pond production

ii. intensive cage production

Scale ☐ 0.1 m²

Legend: ☐ culture area ▥ oxygen production ▨ phosphorus assimilation

Figure 5.3 Reassessed ecological footprints for (i) semi-intensive pond culture and (ii) intensive cage culture of 1 kg of tilapia based on ecological services appropriation per unit biomass production (m² kg^{-1}).

Source: Bunting (2001a).

region. Most operators on Lake Victoria are reportedly small-scale, producing 1–10 t y^{-1}; however, one larger operation is producing 7000 t y^{-1} and has created employment opportunities for several hundred people across associated food systems (Bunting et al., 2022). As with cage-based culture in shared waterbodies elsewhere, it is crucial that the cumulative demands for environmental services of both small- and large-scale installations be maintained with the carrying capacity of supporting ecosystem areas. Appropriate zoning and regulation and fish health management plans and surveillance strategies (for both established and emergent diseases) could be critical to protecting the environment and safeguarding fish production and employment generated by the sector (Bunting and Stentiford, 2014; Bunting et al., 2022).

Operators producing tilapia in cage-based systems may have choices to make when it comes to the strains of fish to select and whether to stock mono-sex of mixed-sex fish. Some producers, buyers and consumers may prefer the use of mixed-sex stocks as this can avoid the use of 17α-methyltestosterone which is needed for one procedure used to produce all male progeny. Trials in Thailand have indicated that mixed-sex stocks of tilapia may be commercially viable if the fish produced could command a price premium above 8 per cent for the final product and if at least 13 per cent of females could be sold as broodfish to hatcheries (Bostock et al., 2022). Smaller female fish of 100–200 g removed intermittently could also be utilised directly or through the formation of 'smallholder outgrower networks' to enhance nutrition and health outcomes in poorer communities (Bostock et al., 2022: 8).

Owing to the proliferation of cages for tilapia culture in Southern Vietnam (see example in Figure 5.2), problems of deteriorating water quality and disease outbreaks have been reported by producers in Tien Giang Province, where around 1600 cages with an average stocking density of 8.5 kg m^{-3} and typical dimensions of 4 m long, 8 m wide and 3 m deep were in operation (The Fish Site, 2011). Consequently, it was proposed that BMPs addressing cage design enhancement, seed quality checks, water quality monitoring, fish health inspection and proactive management, biosecurity and record-keeping should be devised to guide producers to limit losses and reduce environmental impacts. Production of appropriate BMPs could assist in improving the situation, but a more strategic assessment by the responsible authorities appears warranted to assess whether current production levels would be within the carrying capacity of the receiving environment, even if BMPs were widely adopted. Cage culture of striped catfish has largely ceased as pond culture is more economically viable, whilst regulation of water quality is possible in pond-based systems.

Coldwater and highland culture systems

Flow-through grow-out systems for brown trout (*Salmo trutta*) and rainbow trout (*Oncorhynchus mykiss*) have a long history in Europe, and current production levels appear sustainable. Continuity of production may be regarded as a proxy indicator for sustainability. Provenance associated with many producers and their products appears to offer credence to assertions that such systems have caused minimal negative social and environment impacts, whilst provisioning dependable supplies of high-quality products. Operating in highly developed regulatory environments and serving customers who are arguably amongst the most well-informed and environmentally concerned consumers of aquaculture products, operators have had to develop responsible husbandry practices to comply with statutory standards and meet expected product quality standards. Producer organisations such as the British Trout Association in the UK have been at the forefront of promoting better management practices amongst their members, coordinating responses to disease and husbandry challenges and marketing the quality attributes of produce from members to secure market share and premium prices.

Historically, trout eggs and husbandry techniques were exported around the globe under the purview and often direction of colonial powers. Trout transported from North America and Europe were used to establish populations in Africa, in South America and throughout Asia, principally for angling. Negative impacts of introduced salmonid species have been widely reported. Despite this, culture of introduced salmonid species is still commonly practiced in Asia and often proposed as a potential means of social and economic development in remote highland regions. Prospects for trout aquaculture were assessed during an Asian Development Bank–sponsored assistance programme to promote development in Papua New Guinea and the Indo-Norwegian Institutional Cooperation Programme which focused on rainbow trout culture in Himachal Pradesh, India. Although introduced fish species may well be regarded as established or naturalised in some quarters, it should be regarded as unacceptable to sanction re-stocking of introduced species where such activity could impact negatively on endemic populations or potentially globally threatened aquatic species. Restocking programmes are often proposed to safeguard endangered species or to enhance stocks of both native and introduced species to sustain capture fisheries (Box 5.6).

Box 5.6   Small-scale fisheries enhancement

Assessments of small-scale inland fisheries across Africa and in Bangladesh, China and Vietnam (Cai et al., 2010; Nguyen et al., 2010; Ahmed et al., 2013a, 2013b; Liu et al., 2018) have shown that they are an important component of local livelihoods and food and nutrition security. Globally such fisheries produce several million tonnes of catch annually and provide employment for millions of people. Surveys and co-monitoring with fishers on the Brahmaputra River, Mymensingh, Bangladesh, indicated that professional, seasonal and subsistence fishers went fishing on average 290, 179 and 187 days per year, respectively, and landed 1–1.3 kg per day (Ahmed et al., 2013b). Although landings by individual fishers may be relatively small, national-level estimates indicate that inland capture fisheries landings in Bangladesh total over 1 million tonnes in 2020 (FAO, 2022). Overfishing of what are perceived to be common property resources, habitat loss and pollution have led to significant declines in the once abundant freshwater fisheries of Bangladesh. Research findings indicate, however, that community-based management could enhance decision-making and enforcement of fisheries regulations, promote broader environmental awareness and contribute to better pollution control and planning and management of proposed conservation areas (Ahmed et al., 2013a).

Studies within the HighARCS project funded by the European Commission showed that capture fishery returns from the Beijiang River, China declined significantly between the 1950s and 2000 from around 8000 to 2000 tonnes per year (Cai et al., 2010). Reviewing the situation in northwest Vietnam indicated a decline of around 50 per cent in capture fisheries yields in the Da River since the 1980s (Nguyen et al., 2010). Declining fish stocks in China and Vietnam have been blamed on overfishing, pollution, sand mining and dam construction for irrigation purposes and hydro-electricity generation (Cai et al., 2010; Nguyen et al., 2010). Systematic assessment of potential biodiversity conservation and wise-use actions with applied economic modelling demonstrated that stringent pollution controls, and increased fish re-stocking could potentially benefit small-scale fishers (Liu et al., 2018).

Stock enhancement and aquaculture development of native species may be more desirable, but delineation of natural distribution ranges may be subjective or wrong, whilst systemic problems have been identified concerning stock enhancement programmes (Blankenship and Leber, 1995). Throughout Asia, artificial prorogation and re-stocking of endangered indigenous snow trout species is being implemented, but such actions constitute an inherent risk in terms of genetic dilution and disease and pest introductions.

Considering potential negative consequences of stock enhancement programmes in highland areas, a range of BMPs can be identified, including conformity to international treaties and national legislation; supportive policy and economic mechanism; species and location selection; quality and provenance of seed stocks; biosecurity with respect to diseases and pests; optimal release strategies; assessment of social, technical, environmental, political or institutional and sustainability conditions required for successful implementation; and selection of monitoring and evaluation criteria. Driving forces and associated pressures that resulted in native fish stock declines must also be addressed and benefits of stock enhancement must be shared equitably if programmes are to be effective and broadly supported.

Summary

Aquaculture in rural areas developed in response to various opportunities and needs but was often restricted to making use of locally available seed, production enhancing inputs and supplementary feed ingredients. Ponds, water bodies and wetlands where aquaculture was introduced were often being used for other purposes and consequently there was a danger that such multiple uses would be incompatible or result in tension and disputes. Ponds managed for semi-intensive production using a combination of waste resources and supplementary inputs were seen as an efficient way to produce a valuable fish crop, whilst conveying other benefits to producers and contributing to ecological processes in wider agroecosystems. Use of livestock waste in fishponds provided a sanitary route for disposal and effective means to recycle nutrients and energy through aquaculture and integrated agricultural crops. Areas with modified hydrological conditions developed principally for agricultural purposes presented opportunities for aquaculture development. However, intensification of agricultural production with increased agrochemical application has constituted a major constraint to integrated aquaculture. Notable examples of concurrent and sequential integrated rice–aquaculture persist and knowledge of these might inform the evolution of comparable systems in areas subject to similar environmental conditions, but prospects for wider adoption appear limited. The proliferation of reservoirs for hydroelectricity generation and irrigation appears to present opportunities for aquaculture development in remote and highland areas, but there are notable technical constraints to cage culture and risks to aquatic biodiversity associated with culture-based fisheries development. Well planned and managed stock enhancement programmes guided by appropriate BMPs could constitute an important source of income and employment and might be conceived to bolster stocks of endangered species and facilitate adaptation to climate change.

Note

1 Reclamation is often a misnomer for draining natural wetlands to increase the area of land under agriculture or to permit urban and industrial development.

References

Ahmed, N., Brown, J.H. and Muir, J.F. (2008) 'Freshwater prawn farming in *gher* systems in southwest Bangladesh', Aquaculture Economics and Management, vol 12, pp. 207–223.

Ahmed, N., Bunting, S.W., Rahman, S. and Garforth, C.J. (2014) 'Community-based climate change adaptation strategies for integrated prawn–fish–rice farming in Bangladesh to promote social–ecological resilience', Reviews in Aquaculture, vol 6, pp. 20–35.

Ahmed, N., Rahman, S. and Bunting, S.W. (2013a) 'An ecosystem approach to analyse the livelihood of fishers of the Old Brahmaputra River in Mymensingh region, Bangladesh', Local Environment, vol 18, pp. 36–52.

Ahmed, N., Rahman, S., Bunting, S.W. and Brugere, C. (2013b) 'Socioeconomic and ecological challenges of small-scale fishing and strategies for its sustainable management: a case study of the Old Brahmaputra River, Bangladesh', Singapore Journal of Tropical Geography, vol 34, pp. 84–100.

Armillas, P. (1971) 'Gardens on swamps', Science, vol 174, pp. 653–661.

Azim, M.E., Beveridge, M.C.M., van Dam, A.A. and Verdegem, M.C.J. (2005) 'Periphyton and aquatic production: an introduction' in M.E. Azim, M.C.J. Verdegem, A.A. van Dam and M.C.M. Beveridge (eds) Periphyton: Ecology, Exploitation and Management, CAB International, Wallingford, UK.

Bailey, R., Kaggwa, R., Kipkemboi, J. and Lamtane, H. (2005) 'Fingerponds: an agrofish polyculture experiment in East Africa', Aquaculture News, vol 32, pp. 9–10.

Beveridge, M.C.M. (2004) Cage Aquaculture, third edition, Blackwell Publishing, Oxford, UK.

Beveridge, M.C.M. and Muir, J.F. (1993) 'Environmental impacts and sustainability of cage culture in Southeast Asian lakes and reservoirs' in Proceedings of the First AADCP Workshop, Types of lakes and reservoirs and the ecological constraints on the enhancement of production, Melaka, Malaysia, 18–22 October 1993.

Beveridge, M.C.M. and Phillips, M.J. (1993) 'Environmental impact of tropical inland aquaculture' in R.S.V. Pullin, H. Rosenthal and J.L. Maclean (eds) Environment and Aquaculture in Developing Countries, ICLARM Conference Proceedings 31, International Centre for Living Aquatic Resources Management, Makati City, Philippines.

Blankenship, H.L. and Leber, K.M. (1995) 'A responsible approach to marine stock enhancement', American Fisheries Society Symposium, vol 15, pp. 167–175.

Bostock, J., Albalat, A., Bunting, S., Turner, W.A., Mensah, A.D. and Little, D.C. (2022) 'Mixed-sex Nile tilapia (*Oreochromis niloticus*) can perform competitively with mono-sex stocks in cage production', Aquaculture, vol 557, 738315.

Brugere, C., McAndrew, K. and Bulcock, P. (2001) 'Does cage aquaculture address gender goals in development? Results of a case study in Bangladesh', Aquaculture Economics and Management, vol 5, no 3–4, pp. 179–189.

Bunting, S.W. (2001a) 'Appropriation of environmental goods and services by aquaculture: a re-assessment employing the ecological footprint methodology and implications for horizontal integration', Aquaculture Research, vol 32, pp. 605–609.

Bunting, S.W. (2001b) A Design and Management Approach for Horizontally Integrated Aquaculture Systems, PhD thesis, Institute of Aquaculture, University of Stirling, UK.

Bunting, S.W., Bostock, J., Leschen, W. and Little, D.C. (2023) 'Evaluating the potential of innovations across aquaculture product value chains for poverty alleviation in Bangladesh and India', Frontiers in Aquaculture, vol 2, no 3, pp. 1–23.

Bunting, S.W., Karim, M. and Wahab, M.A. (2005) 'Periphyton-based aquaculture in Asia: livelihoods and sustainability' in M.E. Azim, M.C.J. Verdegem, A.A. van Dam and M.C.M. Beveridge (eds) Periphyton: Ecology, Exploitation and Management, CAB International, Wallingford, UK.

Bunting, S.W., Kundu, N. and Ahmed, N. (2017) 'Evaluating the contribution of diversified shrimp–rice agroecosystems in Bangladesh and West Bengal, India to social–ecological resilience', Ocean and Coastal Management, vol 148, pp. 63–74.

Bunting, S.W. and Little, D.C. (2005) 'The emergence of urban aquaculture in Europe' in B.A. Costa-Pierce, P. Edwards, D. Baker and A. Desbonnet (eds) Urban Aquaculture, CAB International, Wallingford, UK.

Bunting, S.W., Mishra, R., Smith, K.G. and Ray, D. (2015) 'Evaluating sustainable intensification and diversification options for agriculture-based livelihoods within an aquatic biodiversity conservation context in Buxa, West Bengal, India', International Journal of Agricultural Sustainability, vol 13, pp. 275–294.

Bunting, S., Pounds, A., Immink, A., Zacarias, S., Bulcock, P., Murray, F., et al. (2022) The Road to Sustainable Aquaculture: On Current Knowledge and Priorities for Responsible Growth, World Economic Forum, Cologny, Switzerland.

Bunting, S.W. and Pretty, J. (2007) Global Carbon Budgets and Aquaculture: Emissions, Sequestration and Management Options, University of Essex, Centre for Environment and Society Occasional Paper 2007-1, Colchester, UK.

Bunting, S. and Stentiford, G. (2014) 'Regulatory regimes and disease controls (strategies, frameworks and methods to ensure robust and transparent sustainability' in M. James (ed) United Kingdom–South East Asia Aquaculture Workshops – Fish & Shellfish | Health & Nutrition, MASTS, St Andrews, Scotland.

Cai, K., Chen, F., Cui, K., Gao, M., Fu, J., Gan, L., et al. (2010) HighARCS Situation Analysis Report – China Site, South China Agricultural University, Guangzhou, China.

Colman, J.A. and Edwards, P. (1987) 'Feeding pathways and environmental constraints in waste fed aquaculture: balance and optimization' in D.J.W. Moriarty and R.S.V. Pullin (eds) Detritus and Microbial Ecology in Aquaculture, ICLARM Conference Proceedings 14, International Center for Living Aquatic Resources Management, Manila, Philippines.

Costa-P Costa-Pierce, B.A. and Chopin, T. (2021) 'The hype, fantasies and realities of aquaculture development globally and its new geographies', World Aquaculture, vol 52, no 2, pp. 23–35.

Edwards, P., Little, D.C. and Demaine, H. (eds) (2002) Rural Aquaculture, CAB International, Wallingford, UK.

Ezban, M. (2020) Aquaculture Landscapes: Fish Farms and the Public Realm, Routledge, Abingdon, UK.

FAO (2010) The State of World Fisheries and Aquaculture 2010, FAO, Rome.

FAO (2022) The State of World Fisheries and Aquaculture 2022, FAO, Rome.

FAO (2023) *Eriocheir sinensis*, Cultured Aquatic Species Information Programme, FAO, Rome, https://www.fao.org/fishery/en/culturedspecies/eriocheir_sinensis/en

Faruque, G. Sarwer, R.H., Karim, M., Phillips, M., Collis, W.J., Belton, B. and Kassam, L. (2017) 'The evolution of aquatic agricultural systems in Southwest Bangladesh in response to salinity and other drivers of change', International Journal of Agricultural Sustainability, vol 15, no 2, pp. 185–207.

The Fish Site (2011) Better Management Practices for Tilapia Cage Farming in Viet Nam, http://www.thefishsite.com

Gnudi, S., Caputo, A. and Salomoni, C. (1991) 'Mass culture of waterfleas fed on microalgae grown on swine manure during the cold season' in N. De Pauw and J. Joyce (eds) Aquaculture and the Environment, EAS Special Publication 14, Bredene, Belgium.

Godfray, H.C.J., Crute, I.R., Haddad, L., Muir, J.F., Nisbett, N., Lawrence, D., et al. (2010) 'The future of the global food system', Philosophical Transactions of the Royal Society B: Biological Sciences, vol 365, no 1554, pp. 2769–2777.

Groeneweg, J. and Schlüter, M. (1981) 'Mass production of freshwater rotifers on liquid wastes II. Mass production of *Brachionus rubens* Ehrenberg 1838 in the effluent of high-rate algal ponds used for the treatment of piggery waste', Aquaculture, vol 25, pp. 25–33.

Gurung, T.B. (2003) 'Fisheries and aquaculture activities in Nepal', Aquaculture Asia, vol 8, no 1, pp. 14–19.

Halwart, M. and Gupta, M.V. (eds) (2004) Culture of Fish in Rice Fields, FAO, Rome, and The WorldFish Center, Malaysia.

Hem, S., Avit, J.B.L.F. and Cisse, A. (1995) 'Acadja as a system for improving fishery production' in J.J. Symoens and J.C. Micha (eds) The Management of Integrated Freshwater Agro-piscicultural Ecosystems in Tropical Areas, Seminar Proceedings, 16–19 May 1994, Technical Centre for Agricultural and Rural Co-operation (CTA), Royal Academy of Overseas Sciences, Brussels.

Irfanullah, H.M., Azad, M.A.K., Kamruzzaman, M. and Wahed, M.A. (2011) 'Floating gardening in Bangladesh: a means to rebuild lives after devastating flood', Indian Journal of Traditional Knowledge, vol 10, no 1, pp. 31–38.

Kautsky, N., Berg, H., Folke, C., Larsson, J. and Troell, M. (1997) 'Ecological footprint for assessment of resource use and development limitations in shrimp and tilapia aquaculture', Aquaculture Research, vol 28, pp. 753–766.

Kibria, G., Nugegoda, D., Fairclough, R., Lam, P. and Bradly, A. (1997) 'Zooplankton: its biochemistry and significance in aquaculture', Naga, vol 20, no 2, pp. 8–14.

Kibria, G., Nugegoda, D., Fairclough, R., Lam, P. and Bradly, A. (1999) 'Utilization of wastewater-grown zooplankton: nutritional quality of zooplankton and performance of silver perch *Bidyanus bidyanus* (Mitchell 1838) (Teraponidae) fed on wastewater-grown zooplankton', Aquaculture Nutrition, vol 5, pp. 221–227.

Knud-Hansen, C.F., Batterson, T.R. and McNabb, C.D. (1993) 'The role of chicken manure in the production of Nile tilapia, *Oreochromis niloticus* (L.)', Aquaculture and Fisheries Management, vol 24, pp. 483–493.

Korn, M. (1996) 'The dike–pond concept: sustainable agriculture and nutrient recycling in China', Ambio, vol 25, pp. 6–13.

Liang, Y., Melack, J.M. and Wang, J. (1981) 'Primary production and the fish yields in Chinese ponds and lakes', Transactions of the American Fisheries Society, vol 110, pp. 346–350.

Little, D.C. and Edwards, P. (1999) 'Alternative strategies for livestock–fish integration with emphasis on Asia', Ambio, vol 28, pp. 118–124.

Little, D.C. and Muir, J. (1987) A Guide to Integrated Warm Water Aquaculture, Institute of Aquaculture, University of Stirling, UK.

Little, D.C., Surintaraseree, P. and Innes-Taylor, N. (1996) 'Fish culture in rainfed rice fields of northeast Thailand', Aquaculture, vol 140, pp. 295–321.

Liu, Y., Bunting, S.W., Luo, S., Cai, K. and Yang, Q. (2018) 'Evaluating impacts of fish stock enhancement and biodiversity conservation actions on the livelihoods of small-scale fishers on the Beijiang River, China', Natural Resource Modeling, vol 32, no 1, e12195.

Lo, C.P. (1996) 'Environmental impact on the development of agricultural technology in China: the case of the dike–pond (*jitang*) system of integrated agriculture–aquaculture in the Zhujiang Delta of China', Agriculture, Ecosystems and Environment, vol 60, pp. 183–195.

Lopes, P.F.M., Carvalho, A.R., Villasante, S. and Henry-Silva, G.G. (2018) 'Fisheries or aquaculture? Unravelling key determinants of livelihoods in the Brazilian semi-arid region', Aquaculture Research, vol 49, pp. 232–242.

Lovatelli, A. and Holthus, P.F. (eds) (2008) Capture-Based Aquaculture. Global Overview, FAO, FAO Fisheries Technical Paper, No. 508, Rome.

Luu, L.T., Trang, P.V., Cuong, N.X., Demaine, H., Edwards, P. and Pant, J. (2002) 'Promotion of small-scale pond aquaculture in the Red River Delta, Vietnam' in P. Edwards, D.C. Little and H. Demaine (eds) Rural Aquaculture, CABI Publishing, Wallingford, UK.

Lynch, A.J., Baumgartner, L.J., Boys, C.A., Conallin, J., Cowx, I.G., Finlayson, C.M., et al. (2019) 'Speaking the same language: can the sustainable development goals translate the needs of inland fisheries into irrigation decisions?', Marine and Freshwater Research, vol 70, pp. 1211–1228.

Mboyerwa, P.A., Kibret, K., Mtakwa, P. and Aschalew, A. (2022) 'Greenhouse gas emissions in irrigated paddy rice as influenced by crop management practices and nitrogen fertilization rates in eastern Tanzania', Frontiers in Sustainable Food Systems, vol 6, 868479.

Melack, J.M. (1976) 'Primary productivity and fish yields in tropical lakes', Transactions of the American Fisheries Society, vol 105, pp. 575–580.

Muir, J.F. (2003) The Future for Fisheries: Livelihoods, Social Development and Environment, Fisheries Sector Review and Future Development Study, Commissioned by the World Bank, DANIDA, USAID, FAO, DFID with cooperation of the Department of Fisheries and Ministry of Fisheries and Livestock, Government of Bangladesh, Dhaka.

Munro, J.L., Iskandar, A. and Costa-Pierce, B.A. (1990) 'Fisheries of the Saguling reservoir and a preliminary appraisal of management options' in B.A. Costa-Pierce and O. Soemarwoto (eds) Reservoir Fisheries and Aquaculture Development for Resettlement in Indonesia, ICLARM Technical Report 23, Manila, Philippines.

Murray, F., Little, D.C., Haylor, G., Felsing, M., Gowing, J. and Kodithuwakku, S.S. (2002) 'A framework for research into the potential for integration of fish production in irrigation systems' in P. Edwards, D.C. Little and H. Demaine (eds) Rural Aquaculture, CAB International, Wallingford, UK.

Newton, R., Zhang, W., Xian, Z., McAdam, B. and Little, D.C. (2021) 'Intensification, regulation and diversification: the changing face of inland aquaculture in China', Ambio, 50, no 9, pp. 1739–1756.

Nguyen, H.S., Bui, A.T., Nguyen, D.Q., Truong, D.Q., Le, L.T., Abery, N.W. and De Silva, S.S. (2005) 'Culture-based fisheries in small reservoirs in northern Vietnam: effect of stocking density and species combinations', Aquaculture Research, vol 36, pp. 1037–1048.

Nguyen, T.H.T., Nguyen, T.T., Nguyen, H.D., Do, V.T. and Nguyen, T.D.P. (2010) Situation Analysis Report on Highland Aquatic Resources Conservation and Sustainable Development in Northern and Central Vietnam, HighARCS Project Report to the European Commission, Research Institute for Aquaculture No. 1, Bac Ninh, Vietnam.

Norberg, J. (1999) 'Periphyton fouling as a marginal energy source in tropical tilapia cage farming', Aquaculture Research, vol 30, pp. 427–430.

Nwachukwu, V.N. (1999) 'Periphyton fauna as an alternative live food in the rearing of *Clarias gariepinus* (Burchell) fry', Bamidgeh, vol 51, pp. 169–171.

Pounds, A., Islam, F.U., Barman, B.K., Haque, M.M., McAdam, B. and Little, D.C. (2023) 'Cash or crop? On-going adoption of rice–fish fingerling production contributes towards food and nutrition security in northwest Bangladesh'. (in prep).

Pretty, J., Sutherland, W.J., Ashby, J., Auburn, J., Baulcombe, D., Bell, M., et al. (2010) 'The top 100 questions of importance to the future of global agriculture', International Journal of Agricultural Sustainability, vol 8, no 4, pp. 219–236.

Schlüter, M. and Groeneweg, J. (1981) 'Mass production of freshwater rotifers on liquid wastes I. the influence of some environmental factors on population growth of *Brachionus rubens* Ehrenberg 1838', Aquaculture, vol 25, pp. 17–24.

Schwartz, M.F. and Boyd, C.E. (1995) 'Constructed wetlands for treatment of channel catfish pond effluents', The Progressive Fish-Culturist, vol 57, pp. 255–266.

Shaalan, M., El-Mahdy, M., Saleh, M. and El-Matbouli, M. (2018) 'Aquaculture in Egypt: insights on the current trends and future perspectives for sustainable development', Reviews in Fisheries Science and Aquaculture, vol. 26, pp. 99–110.

Shrestha, M.K. and Knud-Hansen, C.F. (1994) 'Increasing attached microorganism biomass as a management strategy for Nile tilapia (*Oreochromis niloticus*) production', Aquaculture, vol 13, pp. 101–108.

Sindilariu, P.D., Schulz, C. and Reiter, R. (2007) 'Treatment of flow-through trout aquaculture effluents in a constructed wetland', Aquaculture, vol 270, pp. 92–104.

Tidwell, J.H., Coyle, S., Van Arnum, A. and Weibel, C. (2000) 'Production response of freshwater prawns *Macrobrachium rosenbergii* to increasing amounts of artificial substrates in ponds', Journal of the World Aquaculture Society, vol 31, pp. 452–458.

van Beijnen, J. and Yan, G. (2019) Local heroes: embracing the culture of native species in Brazil and beyond, The Fish Site, https://thefishsite.com/articles/local-heroes-embracing-the-culture-of-native-fish-species-in-brazil-and-beyond

Wahab, M.A., Azim, M.E., Ali, M.H., Beveridge, M.C.M. and Khan, S. (1999a) 'The potential of periphyton-based culture of the native major carp calbaush, *Labeo calbasu* (Hamilton)', Aquaculture Research, vol 30, pp. 409–419.

Wahab, M.A., Mannan, M.A., Huda, M.A., Azim, M.E., Tollervey, A.G. and Beveridge, M.C.M. (1999b) 'Effects of periphyton grown on bamboo substrates on growth and production of Indian major carp, rohu (*Labeo rohita* Ham.)', Bangladesh Journal of Fisheries Research, vol 3, no 1, pp. 1–10.

Wenhua, L. and Qingwen, M. (1999) 'Integrated farming systems an important approach towards sustainable agriculture in China', Ambio, vol 28, pp. 655–662.

Wong Chor Yee, A. (1999) 'New developments in integrated dike–pond agriculture–aquaculture in the Zhujiang Delta, China: ecological implications', Ambio, vol 28, pp. 529–533.

Yusoff, F.M. and McNabb, C.D. (1989) 'Effects of nutrient availability on primary productivity and fish production in fertilised tropical ponds', Aquaculture, vol 78, pp. 303–319.

Zajdband, A.D. (2011) 'The introduction of aquaculture into family farming systems in subtropical northeast Argentina', Paper presented at World Aquaculture 2011, 7–10 June, Natal, Brazil, World Aquaculture Society.

6 Sustainable urban and peri-urban aquaculture

Key points

The aims of this chapter are to:

- Review the diversity and characteristics of aquaculture practices in urban and peri-urban settings and assess prospects for continued operation and new developments.
- Critically evaluate prospects for urban and peri-urban culture-based fisheries enhancement and shellfish and seaweed culture for enhanced ecosystem services.
- Describe large-scale aquatic plant cultivation around major cities in Southeast Asia and consider prospects for aquaponics for crop and livelihoods diversification.
- Reflect on the development of aquaponics systems, notably in the USA, and consider prospects for such systems elsewhere.
- Assess prospects for urban aquaculture sector growth, focusing on multi-functional recirculating systems in Europe and North America and ornamental production in Asia.
- Review factors prompting the emergence of small-scale commercial aquaculture in Africa and assess the contribution it could make to livelihoods, local economic development and resilient food systems drawing on SWOT and STEPS framework assessment by Bunting and Little (2015).
- Assess the current status of wastewater-fed aquaculture, including recent developments in the East Kolkata Wetlands, India, which accommodate the largest area of wastewater-fed aquaculture ponds in the world and assess prospects for future developments.
- Consider alternative strategies to convert waste resources to valuable aquatic products, notably using intermediaries and producing biomass for non-food and industrial purposes, including the DPSIR (drivers, pressures, state, impacts and responses) framework analysis by Bunting and Edwards (2018).

DOI: 10.4324/9781003342823-6

Production systems characteristics

Traditional urban aquaculture systems ranging from fish culture in fort moats in Europe and Asia and fishponds as integral components of monastic granges to expansive wastewater-fed fishpond complexes developed during the early twentieth century on the fringes of major cities in China, India and Vietnam have been reviewed extensively elsewhere (Edwards, 1992; Bunting and Little, 2005). Often motivations behind urban and peri-urban aquaculture development were fuelled by opportunity or necessity or, more pragmatically, a combination of the two. Wastewater flowing from urban drainage systems was often the only source of nutrient-rich water for aquaculture in low-lying areas around the peripheries of cities; as it was a readily accessible source of water, strategies could be foreseen for its productive use (see resilience discussion in chapter 1). Aquaculture systems devised to promote resource use efficiency are presented in Table 6.1.

Earthworks left by excavations for infill material or brickmaking present a viable, if not ideal, situation for aquaculture development. Abandoned business premises and industrial units offer a suitably located if not correctly serviced or configured site for recirculating aquaculture systems conceived with combined food production and education goals in mind. Elsewhere, rural aquaculture has been engulfed by the rapid urbanisation being witnessed around the globe, and often ponds, lakes and diminished wetlands remain as the last vestiges of rural landscapes owing to costs associated with developing such areas.

Despite being in urban and peri-urban settings where it might be assumed that only more intensive production modes and associated income generation might be able to compete for space with alternative urban activities and land-use practices, extensive and semi-intensive aquaculture is routinely instigated. Such developments occur when prevailing environmental conditions permit, associated risks are deemed acceptable and alternative uses are prohibited, for example, by legislation or physical and hydrological constraints. Pertinent instances of intensive, semi-intensive and extensive aquaculture were reviewed comprehensively by Bunting and Little (2003); typical characteristics of aquaculture systems managed at these different levels of intensity are summarised in Table 6.2.

Stocking urban and peri-urban waterbodies

Urban lakes and reservoirs throughout the tropics are routinely stocked with fish, either by government agencies, community groups or companies that have acquired the necessary rights. Such operations are at risk from pollution, theft and conflict with other stakeholder groups. Cages and pens can be deployed in large water bodies, offering greater control over production processes but potentially amplifying associated risks to aquaculture in urban environments. Cage-based aquaculture is practiced in a wide array of urban and peri-urban lakes, reservoirs and rivers. An early example

Table 6.1 Integration of aquaculture practices in urban and peri-urban settings to optimise resource use efficiency

Integration strategy	Management practices	Constraints and conditions	Potential water-use efficiency outcomes
Urban and peri-urban aquaculture	• fish cages in canals and lakes in Bangladesh and Vietnam • fish culture in canals, lakes, ponds and borrow-pits in peri-urban areas throughout Asia • macrophyte cultivation in drainage canals and low-lying water bodies; e.g. Bangkok, Hanoi, Phnom Penh • aquaculture exploiting food and drink production and processing by-products	• multiple use of urban and peri-urban water bodies may mean that hydrology is out of the control of aquaculture producers and associated operational constraints result in sub-optimal management • risks from pollution and poaching may constrain aquaculture development • insecure land tenure and pressure from urban residential and industrial development may constrain investment in aquaculture systems	• floodwater storage and groundwater recharge associated with extensive wastewater-fed aquaculture operations can contribute to stabilising local hydrological conditions • vigilance of aquaculture producers helps in monitoring pollution and safeguarding water quality for other users
Wastewater-fed aquaculture	• intentional use of wastewater to supply water and nutrients for aquaculture • lagoon-based sewage treatment systems incorporating fishponds developed under the Ganges Action Plan initiative, India • fish culture in 3900 ha of ponds in the East Kolkata Wetlands, West Bengal, India • duckweed cultivation on wastewater for processing to biofuel	• health risks posed by wastewater use for aquaculture demand that appropriate treatment and control measures be adopted • consumer perceptions, prevailing beliefs and institutional barriers may constrain development • land area required for combined wastewater treatment and reuse through aquaculture may prohibit development	• management of wastewater promoted by integration of aquaculture could help operators meet statutory discharge standards and help to safeguard public health • wastewater reuse through aquaculture can help protect the quality of water bodies receiving discharge from the system • exploitation of wastewater flows for biomass production could help alleviate pressure on finite freshwater resources

(Continued)

Table 6.1 (Continued)

Integration strategy	Management practices	Constraints and conditions	Potential water-use efficiency outcomes
Aquaculture in thermal effluents and cooling water	• production of juvenile fish in cooling water effluents from nuclear power stations in France • farming marine worms in thermal effluents in the UK • exploitation of excess heat from data centres to raise the culture water temperature above ambient and enhance production	• chemicals used to clean power station and variations in water temperate may affect growth and product quality • farming species for human consumption may pose unacceptable health risks or not gain consumer acceptance • greenhouse gas emissions from electricity generation to supply data centre must be considered	• retention of thermal effluents for aquaculture production could facilitate heat dissipation and contribute to meeting statutory discharge standards • exploitation of thermal effluents could help avoid greenhouse gas emissions associated with heating water for culturing cold-intolerant species

Table 6.2 Characteristics of urban aquaculture systems managed at different intensities

Characteristic	Management intensity		
	Extensive	Semi-intensive	Intensive
Feed source	• natural production enhanced indirectly through nutrient-rich surface runoff and drainage water	• exploitation of waste resources and fertiliser applications to enhance natural production and/or the provision of basic supplementary feed	• dependence on externally supplied high-protein feed, which in some cases may have been produced using by-products; e.g. fly larvae, tubifex worms
Access, ownership and tenure	• open access, common property resources	• private, cooperatives, leaseholders, community-based management	• private, commercial, research and development, vertically integrated
Markets	• subsistence, local retail markets	• subsistence, local and regional wholesale and retail	• high-value food and ornamental species, regional and export orientated, food products processed to add value
Constraints	• variable productivity; access may be denied to poorer community members and new entrants; urban sprawl; competition with other user groups; theft and poaching	• contamination of waste resources and pollution may inhibit production and affect consumer sentiment; urban sprawl; limited control over environmental perturbations	• high capital costs; inherent financial risks; susceptible to disease outbreaks, technical failures, changing market conditions and competition
Opportunities	• poorer community members may benefit through continued access or cheaper food from low-investment systems	• where hazards can be minimised, local production of fish and plants from urban systems can contribute to food security, enhanced livelihoods and environmental protection	• investment opens up access to new and larger markets; possibility of higher returns from money and resources invested

Source: Bunting et al. (2006).

from Indonesia was carp culture in modest bamboo cages in drainage channels where stocked fish could freely feed on invertebrates that proliferated in the nutrient rich water (Edwards, 1992). Cage culture proliferated more recently in the Saguling Reservoir downstream of Bandung, Indonesia, with an estimated 4425 cages producing 6000t of tilapia (*Oreochromis* sp.) annually (Hart et al., 2002), but large fish-kills occurred and self-pollution was implicated. Elsewhere, cages were outlawed in drinking water reservoirs in Shaoguan City, Guangdong Province, China, owing to concerns over water quality and potential contamination. Operators of fishponds on the periphery of larger towns and cities or found within the confines of provincial towns can exert a high degree of control over stocking and production-enhancing inputs but may still find themselves subject to the vagaries of surface water quality and threats from poaching. Cognisant of potential problems, operators may well adopt socially orientated mitigating management strategies (see Box 6.1), and such measures, it might be surmised, are central to resilient aquaculture practices in urban and peri-urban settings.

Box 6.1   Socially orientated management strategies for aquaculture systems

Interviews conducted with pond operators in the town of Saidpur in northwest Bangladesh provided a valuable insight into the management of small-scale aquaculture systems in the dynamic peri-urban environment. Certain pond operators had adopted socially orientated management strategies to minimise the risks associated with their aquaculture system. Mr Rhaman operated a single pond that was used to culture a variety of fish, including tilapia (*Oreochromis* spp.), rohu (*Labeo rohita*), catla (*Catla catla*), mrigal (*Cirrhinus mrigala*), silver carp (*Hypophthalmichthys molitrix*) and common carp (*Cyprinus carpio*). In addition, several species of wild fish such as snakehead (*Chana* spp.) and liver catfish (*Heteropneustes fossilis*) were transferred into the pond with floodwater during the rainy season. During the floods associated with the rainy season, Mr Rhaman suspected that fish were escaping from his pond; however, poaching was also blamed for reducing the productivity of the pond. Poachers from other wards in Saidpur and from outside of the town had previously visited the pond; the local authorities were informed but no action was taken. The employment of a guard was seen as one possible solution; unfortunately, the guard employed last year was beaten up and fish were taken anyway.

In contrast, Mr Haque, the operator of two ponds, did not consider poaching to be a problem. The reason given for this was that each year at harvest time, some fish were distributed to the people living around the pond, apparently reducing the tendency for local people to steal the fish. Another socially orientated management strategy adopted by

Figure 6.1 People removing earth from the bottom of a pond to raise their
dwellings above the expected level of the seasonal flood associated
with the monsoon; dwellings positioned on the pond's embankments can be seen in the background.

Photo credit: Stuart Bunting.

Mr Haque was to allow landless people to settle on the embankments
surrounding one of the ponds, which resulted in several positive effects.
The physical presence of these people helped to reduce the incidence of
poaching and predation. In addition, when the floodwaters threatened
people's houses, they raised the embankment upon which they built
their dwellings, thereby protecting both their property and the pond
(Figure 6.1), which meant that Mr Haque did not need to engage
contractors to deepen his pond and maintain its perennial nature.

Source: Bunting et al. (1999).

Urban and peri-urban culture-based fisheries enhancement

Culture-based fisheries were introduced in chapter 5 in the context of rural
development, supporting fishing activity in lakes and reservoirs and associated livelihoods in remote and often highland areas. Elsewhere, notably in
coastal areas affected by overfishing and urban and peri-urban areas affected
by pollution and urban encroachment, culture-based fisheries enhancement
programmes are being widely proposed.

Aquaculture development has been proposed to supplement fish supplies in urban areas of Shaoguan City, China, and regions of Northern Vietnam affected by dams constructed for hydroelectric power generation, but attendant environmental and social concerns have been noted. Culture-based fisheries are being promoted with several million common carp (*Cyprinus carpio*) and crucian carp (*Carassius carassius*) being stocked annually from fry release platforms moored in Shaoguan City. Other species such as grass carp (*Ctenopharyngodon idella*), Chinese sturgeon (*Acipenser sinensis*), *Spinibarbus hollandi* and black Amur bream (*Megalobrama terminalis*) are released intermittently. Bioeconomic modelling has contributed to a better understanding of the socio-economic impact of such restocking programmes and alternative livelihood strategies and conservation plans (Liu et al., 2019). Action is being taken to conserve and restore wild fish stocks, and aquatic conservation areas have been established under Shaoguan City with objectives including protection of spawning areas and conservation of rare species. Insights from the planning and management of these aquatic conservation areas could potentially be useful in guiding the establishment of proposed protected freshwater areas throughout Asia.

The success of such schemes will, however, depend on first addressing underlying fisheries management and environmental protection failings. For example, evaluating the potential of Florida Bay scallop culture to enhance stocks in Tampa Bay, Florida, and reinstate commercial aquaculture production, it was noted that despite growing to market size in 9–10 months, water quality criteria required to produce whole scallops for sale would mean that scallops produced in Bayboro Harbor would have to be relayed elsewhere for a suitable depuration period or producers would be restricted to selling extricated adductor muscles (Blake, 2005). Aquaculture might be undertaken as a means to achieve ecosystem services enhancement in degraded urban and peri-urban waterways; potential water quality improvement as a consequence of scallop culture in Tampa Bay was alluded to by Blake (2005: 193), including the potential contribution of such filter feeders to 'increasing water clarity and lowering organic and nutrient loadings'. Elsewhere, seaweed cultivation to sequester nutrients derived from urban runoff and sewage treatment plants has been proposed as a means of bioremediation (Fish Information and Services, 2012); whilst the biomass produced may not be acceptable for direct human consumption, it may have potential as feedstock for industrial processes and biofuel production.

Aquatic plant cultivation in Southeast Asia

Aquatic macrophytes are cultivated close to urban markets throughout Southeast Asia, traditionally exploiting low-lying land that is unsuited to other uses and relatively costly to develop for residential or commercial purposes. Moreover, urban authorities may be reluctant to sanction development of such land as it probably functions as a receptacle for floodwater that

---

Box 6.2   Aquatic vegetable production in Southeast Asian cities

Aquatic vegetable production in semi-intensive and intensive systems was widespread and commercially significant around many cities in Southeast Asia. According to Nguyen and Pham (2005) in Hanoi, water spinach (or morning glory, *Ipomoea aquatica*) was produced throughout the year, whilst water mimosa (*Neptunia oleracea*) was cultivated only in the summer (April to August) and water dropwort (*Oenanthe stolonifera*) and water cress (*Rorippa nasturtium-aquaticum*) were produced in the winter (September to March). Most production occurred in flooded fields, some of which were converted from rice production to generate a higher income; water spinach was also cultivated floating on canals within the city. Water mimosa and water spinach production were reported from peri-urban provinces around Bangkok (Yoonpundh et al., 2005). Around Ho Chi Minh City, Vietnam, many farmers in Binh Chanh District combined water mimosa cultivation with fish production in separate ponds; mimosa provided a daily income, whilst fish consumed the duckweed that grew alongside the mimosa (Hung and Huy, 2005). Duckweed (*Lemna* and *Wolffia* spp.) were commonly removed from aquatic vegetable crops in both Vietnam and Thailand for use locally as fish feed. Water spinach was grown in converted rice-fields in Thu Duc District utilising wastewater, and in some cases water spinach shoots were used to feed cultured fish species such as giant gourami (*Osphronemus goramy*) that can readily digest and benefit from them.

Source: Bunting et al. (2006).

---

might otherwise inundate built-up areas, resulting in major financial losses and adverse public opinion. Assessments of aquatic vegetable cultivation were carried out as part of the European Commission–sponsored PAPUSSA project around major urban centres in Bangkok, Thailand; Hanoi and Ho Chi Minh City, Vietnam; and Phnom Penh, Cambodia. The study revealed widespread and productive practices (Box 6.2). But rather than being benign, many producers applied heavy doses of agrochemicals, spraying cut stems every time new growth was cropped, presenting significant environmental, animal and public health risks.

Comparative advantages associated with aquatic vegetable cultivation in urban and peri-urban Asia were summed up by Leschen et al. (2005: 5): 'aquatic vegetables are far less vulnerable to theft and chemical contamination; generally more land efficient, involve lower entry costs and normally require lower value inputs'.[1] However, it was also noted by the authors that aquatic vegetable growers faced a number of problems: limited access to extension, training and technology transfer compared to fish producers; poor representation in urban planning, with uncertainty concerning which authority had responsibility for their productive practices; and an absence of groups or associations to protect their interests, although in certain instances urban authorities such as those in Ho Chi Minh City, Vietnam, had allocated land on the periphery of the city for aquatic plant and agricultural production.

Aquaponics

Aquaponics, comprising the hydroponic cultivation of macrophytes to condition, process and produce saleable crops in recirculating aquaculture facilities has been demonstrated and proposed for wider adoption. Lettuce and basil were grown in water being reused to culture rainbow trout, and assuming a feed conversion ratio of 1.2, it was estimated that thirteen to eighteen lettuce heads could be produced with waste nutrients resulting from 1 kg of fish feed (Adler, 1998). Waste streams produced by mechanical and biological filtration of aquaculture process water can contain reasonable levels of nitrogen and phosphorus as well as the majority of micronutrients required by greenhouse or salad crops. Pioneering studies indicated that waste produced from culturing 100t of European eel (*Anguilla anguilla*) contained sufficient nitrogen to grow 690t of tomatoes annually (Jungersen, 1991). Preliminary economic assessment of an aquaponics system for the U.S. Virgin Islands noted that producing 30t of tilapia and 8424 cases of lettuce annually would make a financially attractive investment opportunity (Rakocy and Bailey, 2003). Aquaponics is also being implemented elsewhere to promote socio-economic development; for example, in Phnom Penh, Cambodia, and Kathmandu, Nepal.

Although the technical viability of aquaponics has been widely demonstrated, constraints to commercial-scale development have been noted, including balancing available nutrient levels and states with plant nutrient requirements and uptake mechanisms, preventing salt accumulation and controlling plant and fish diseases and pests (van Rijn, 1996).

Hydroponic cultivation of horticulture crops has evolved into a precise, closely controlled large-scale commercial enterprise producing high-value perishable crops destined for multiple retailers. Growing conditions and culture medium composition are carefully monitored and adjusted, and this degree of refinement, necessary to achieve a financially viable operation, is indicative of constraints to wider uptake of aquaponic principles in conjunction with commercial aquaculture operations. Husbandry and management skills required for aquaculture and hydroponics are different, thus placing an added burden on potential operators. Prospective markets and associated food safety and quality assurance scheme compliance requirements will demand separate approaches to sales and implementation of sector-specific monitoring, record-keeping and protocols for appropriate remedial actions.

Aquaponics has potential in certain locations and when targeting specific ethnic or niche markets where prices for fresh fish and perishable salad and herb crops are comparatively high but perhaps demand levels and costs and constraints associated with access do not favour imports.

Aquaponics development in the USA

Diverse aquaponic systems have been developed in the USA and internationally (Love et al., 2014, 2015). In a survey of 809 practitioners in 2013 it was noted that over half (53 per cent) cultured tilapia and 48 per cent raised ornamental species (Love et al., 2014). A wide array of plant varieties were being cultivated, with 70, 69 and 64 per cent of respondents having produced basil, tomatoes and salad greens in the preceding 12 months. Aquaponics systems were generally being operated as a 'niche' activity (Love et al., 2014: 6) by entrepreneurs or had been developed in schools, colleges, universities and not-for-profit organisations. Over half of those surveyed (57 per cent) were using renewable energy and 50 per cent were using alternatives such as live feed or aquatic plants to supplement pelleted fish feeds. Producing their own food was a priority for many respondents.

Aquaponics may be perceived as an ecologically balanced and synergistic production paradigm, and this makes it an engaging system to use in educational settings. Survey findings mentioned above have, however, highlighted that such systems still require nutrient and energy inputs. Consequently, care must be taken to ensure that sourcing these does not have negative environmental or unintended social and food security impacts elsewhere. When commercial operators can meet the expectations of consumers regarding aquaponics as a sustainable means of production, large-scale commercial facilities may become more commonplace.

Urban RAS for food and food-for-thought

Recirculating aquaculture systems (RAS) integrating fish and plant culture have been central to several initiatives in Europe and North America to facilitate urban regeneration, through combining local food production, employment, recreation and education. Life cycles of cultured organisms, ecosystem process and functions involved in maintaining a healthy environment and principles of integrated production provide important insights for students concerning ecology and water and nutrient cycles. Broader environmental and socio-economic considerations and concerns are often invoked when proposing urban and peri-urban aquaculture developments to planners and investors. Local food production where consumers can witness the means of production and perhaps discuss this with producers, and where transport costs are minimised, is often espoused as inherently more desirable than centrally processing and distributing or importing products. Production in intensively managed RAS systems is, however, energy hungry, whilst alternative but responsible aquaculture systems in developing countries supplying export markets may actually convey significant socio-economic benefits to local communities.

Several pilot-scale urban aquaculture operations have been established in North America, primarily for education and community development; for example, recirculating aquaculture systems in Boston Harbor, Massachusetts (Goudey and Moran, 2005), and Brooklyn, New York (Schreibman and Zarnoch, 2005), and as part of the urban aquaculture education programme, New Haven, Connecticut (Roy, 2005). Aims cited for the Massachusetts Institute of Technology (MIT) Sea Grant programme in Boston Harbour by Goudey and Moran (2005) encompassed culture protocols for commercial marine finfish species, evaluating growth and environmental impacts, demonstration of recirculating aquaculture and outreach and education activities. Unfortunately, the authors noted that water quality problems, a legacy of contaminated sediments associated with former naval dockyards and increasing competition for property with harbour access, constitute major barriers to commercial aquaculture development.

Reeling from what Prain (2006: 308) called the 'implosion of the Soviet-dependent Cuban economy', 'households and political structures' supported by 'local social organisations' instigated innovative livelihoods responses, including 'patio agriculture', integrated crop–livestock systems, vermiculture and household aquaculture. Establishment of an enabling institutional environment was critical, with the constitution of people's councils as grassroots organisations focused on encouraging the interactive participation of local households in joint assessment and problem-solving. Pressures leading to the reform of agricultural policy throughout Cuba included declining average daily per capita calorie intake, nutritional deficiencies experienced by pregnant women, falling birth weights and a general deterioration in public health (Prain, 2006). Prain (2006) noted that urban agriculture can perform 'multiple roles for the urban community, contributing to food security, nutritional well-being, income supplements for low-income families, child and youth education about natural processes and resources and the role of science, family recreation, and to the sustainability of cities' (310). But this was tempered by highlighting tension between urban agriculture development and other economic, natural and social elements of urban environments, notably the interests of other stakeholder groups.

Adoption of recirculation has potential to increase the production capacity of an aquaculture facility within prescribed discharge consent limits and without the need to appropriate more water. Water consumption and discharge volumes could be substantially reduced, thus avoiding associated abstraction costs and effluent disposal charges, although water recirculation and process water treatment would constitute major new costs. Recirculation could facilitate enhanced stock control and optimal feeding strategies, prevent escapes and restrict disease transmission between farms and wild animals (Muir, 1996). Recirculation could conserve heat and nutrients, but

Table 6.3 Parameters characterising European eel, African catfish, seabass and turbot production in recirculation aquaculture units in Europe

Parameter	European eel	African catfish	Seabass	Turbot
Stocking density (kg m^{-3})	80–250[a,b,c]	170[a]	50[d]	35[a]
Hydraulic retention time (h)	0.25–0.5[a]	1[a]	–	1[a]
Specific growth rate (% d^{-1})	0.75[¥]	1.7[a]	–	0.42[a]
Feed conversion ratio	1.3–1.4[a,b]	0.85[a]	0.8[d]	1.1[a]
Production rate (kg m^{-3} y^{-1})	173–260[a]	740–1110[a]	118[d]	41[a]

Source: Bunting and Little (2005).

Notes
[a]   Eding and Kamstra (2002);
[b]   Schmidt-Puckhaber (2000);
[c]   Torres (2000);
[d]   Rigby (2000).

carbon dioxide, inorganic nitrogen (notably total ammonia nitrogen) and organic matter can accumulate in the culture system, thus necessitating close monitoring and advanced water treatment. Ponds and lagoons have been advocated as potential biofilters, facilitating the incorporation of biological processes in process water conditioning in recirculating systems. Combined intensive–extensive culture unit configurations have been demonstrated in varied settings, but, as with other integration strategies, it is often difficult to reconcile optimal growth condition requirements in intensive culture units with the conditioning and treatment capacity of extensive components.

Considering prospects for further development of recirculating aquaculture systems in urban and peri-urban settings in Europe, it is important to note that culture parameters for several promising species, notably European eel, African catfish, seabass (*Dicentrarchus labrax*) and turbot (*Scophthalmus maximus*), are well defined (Table 6.3).

Artisanal and commercial ornamental production

Ornamental fish production in urban and peri-urban environments is common. However, owing to the often small-scale of operations and indoor nature of production, it may be less apparent and perhaps not as likely to factor in policymaking and planning compared with higher profile, more extensive production systems occupying large land areas. Clusters of ornamental fish producers have been established in peri-urban Kolkata, West Bengal, India. In Udayrampur Village it was noted that income generation through small-scale ornamental fish production in concrete tanks for the domestic market and formation of a cooperative society including women members had contributed to better nutrition for household members, housing improvements and enhanced opportunities for children to attend school (Mukherjee et al., 2004). Elsewhere, in Bangkok, Singapore and Thailand, for example, commercial hatcheries, intermediary traders and quarantine facilities are well established, with operators supplying ornamental fish, invertebrates and coral to the international trade through highly evolved marketing networks; these enterprises constitute a significant source of revenue and employment.

There are numerous concerns surrounding the international trade in aquatic animals for ornamental purposes. These centre on destructive and unsustainable collection methods for wild juveniles and broodstock; risks associated with disease transfer and introduction of invasive species outside of their native range; animal welfare during collection, transport and husbandry in the tanks of traders and hobbyists; and ethical and emotional arguments over catching wild animals and keeping them in captivity. Trade organisations representing actors in the ornamental sector are cognisant of such concerns. Thus, they have entered into dialogue with policymakers to allay fears and promote awareness of positive socio-economic and environmental outcomes of responsible ornamental trade. They have also developed codes of conduct and ethics accordingly to enable their members to demonstrate that they are committed to sustainable and fair trade, as well as law-abiding (Bassleer, 2007).

Commercial tank culture in Africa

Small-scale, commercially orientated urban and peri-urban aquaculture has emerged in several countries in Africa as an innovative sector producing fish to meet demand from burgeoning numbers of urban, middle-class consumers. African catfish (*Clarias arietinous*) farming in Nigeria presents an interesting case study concerning the rapid development of this sector. Reviewing growth in the sector, Miller and Atanda (2011) reported that since its inception in 2000, production of 30,000t achieved by 2642 producers recorded in 2003 grew to around 120,000t at the time of their assessment. Furthermore, the authors noted that typical production units were comparatively small but intensively managed and that one operator might have concrete tanks with a surface area of $50m^2$ yet be able to produce 1.5t of fish annually.

Continued success of peri-urban catfish farming in Nigeria has been attributed to a number of factors: fast-growing demand for fish, juxtaposed with declining wild-capture fisheries; development of locally appropriate production systems for African catfish; investment to ensure sound management, including operators attending training course in Europe and recruitment of international aquaculture managers; establishment of well-run hatcheries with quality imported broodstock and selective breeding programmes; and local production of quality feeds specifically formulated for catfish production. 'Fish farming villages' constitute another interesting innovation whereby groups of investors acquired access to land close to potential urban markets and oversaw setting up and staffing of intensive tank-based production units, with some developments reportedly having 800 concrete tanks (Miller and Atanda, 2011).

Aggregation of production units raises concerns over disease transfer and environmental degradation as cumulative releases of culture water from large numbers of intensively managed production units could cause local environmental problems. However, the Environmental Protection Agency, Nigeria, has been actively monitoring the situation, whilst aquaculture development policy in the country calls for environmental and biodiversity protection and adoption of best management practices. Cost–benefit analysis has indicated that producers may be able to realise a profit of US$1 per kilogram of fish produced; however, rising input and transport costs mean such margins are likely to be eroded. Furthermore, early signs of consolidation may result in fewer employment opportunities in the future and more restricted distribution of benefits. Some producers operating peri-urban catfish culture units in Nigeria have reportedly begun to culture tilapia commercially to meet demand from more affluent consumers and supply export markets. Elsewhere in Africa (e.g. Dar es Salaam, Tanzania), tilapia are being cultured preferentially in urban and peri-urban areas in ponds, but water quality concerns and institutional barriers to further development have been identified (Rana et al., 2005).

Opportunities to overcome barriers to urban aquaculture

Weaknesses constraining urban aquaculture globally, including pollution of surface waters, were summarised in a strengths, weaknesses, opportunities and threats (SWOT) analysis (Bunting and Little, 2015; Table 6.4). Others identified were insecurity of tenure, theft and indiscriminate dumping and in-filling; lack of recognition as a legitimate land-use practice; and contributions to food security, livelihoods and economic development being overlooked. Existing and possible future threats were access to land and inputs being denied or disrupted, growing competition, changing patterns of consumer demand, conflict with other groups and excessive development overwhelming the capacity of supporting ecosystems. Key strengths were cited, notably strong demand for aquaculture products, income and employment opportunities generated, the diversity of systems and strategies that can be deployed to capitalise on niches in urban environments and growing recognition in national and international policies. Market accessibility constituted an important opportunity, as did increasing demand from both affluent and poorer consumers and the existence of underutilised resources that urban and peri-urban aquaculture could capitalise on.

The existence of urban aquaculture development is testament to the value of such practices. Given the dynamic nature of urban and peri-urban areas, however, such systems may be expected to be transitory, and operators may need to be particularly resilient and adaptable. Other social, technical, environmental, political and sustainability (STEPS) conditions that could support or promote urban aquaculture are summarised in Table 6.5.

Table 6.4 SWOT assessment of urban aquaculture in developing countries

Strengths: *existing or potential resources or capability*	**Weaknesses**: *existing or potential internal force that could be a barrier to achieving objectives/results*
• strong demand for aquaculture products	• deficiencies with urban environmental management lead to widespread pollution of surface waters
• broad range of aquaculture production systems and strategies suited to niches found in urban environments	• insecurity of tenure constrains investment in urban aquaculture
• urban aquaculture recognised in national and international policy and supported by donors and development agencies	• difficulties in monitoring aquaculture systems in urban settings can lead to thefts and indiscriminate dumping and in-filling
• aquaculture in urban areas can be a good indicator of environmental health and help restore degraded urban ecosystems	• dispersed and often transient nature of small-scale urban aquaculture means support from government institutions and access to service providers are lacking
• urban aquaculture provides income and employment opportunities in production phase and across associated value chains and produces fish and plants that can be an affordable and important sources of protein and nutrients for the poor	• urban authorities may not recognise aquaculture as a legitimate land-use practice
	• lack of information on extent of peri-urban aquaculture means its contribution to livelihoods, economic development and food security is overlooked

Opportunities: *existing or potential factors in the external environment that, if exploited, could provide a competitive advantage*	**Threats**: *existing or potential force in the external environment that could inhibit maintenance or attainment of unique advantage*
• opportunities to access markets locally with fresh produce on a timely basis	• access to land and inputs (water and nutrient sources) denied or disrupted owing to competition or development plans that do not consider or recognise the claims of aquaculture producers
• increasing demand for high-value aquatic products amongst burgeoning middle classes in many urban areas in developing countries	• improved transport links and communications mean urban aquaculture must compete with production in rural areas with lower capital and operating costs
• rising demand for affordable aquatic products for nutrition and food security amongst poor urban communities	• demand for fish from urban aquaculture systems declines owing to negative media coverage resulting from animal, environmental or public health concerns
• underutilised resources (low-lying areas, nutrients in organic waste streams, wastewater flows) that urban aquaculture could exploit	• inappropriate urban aquaculture development results in conflict with other resource users or local residents
• international agreements and guidelines that support responsible aquaculture development and the use of waste resources	• excessive urban aquaculture development overwhelms capacity of supporting ecosystem areas, resulting in self-pollution and environmental degradation
• national policies that advocate and recognise urban agriculture (encompassing aquaculture) as a legitimate urban activity	

Source: Bunting and Little (2015).

Table 6.5 STEPS assessment of condition needed to support and promote urban aquaculture

STEPS elements	Conditions
Social	• acceptance and support for aquaculture as a legitimate and worthwhile urban activity • demand for products from urban aquaculture continues and grows • urban aquaculture is able to generate sufficient financial returns to make it viable and a continued and novel employment and income-generating activity
Technical	• access to appropriate spaces within urban environments is possible and sufficient periods of tenure guaranteed to safeguard investments made by producers • inputs to establish and sustain aquaculture are readily available • haulage providers and processing and marketing facilities willing and able to accept products from urban aquaculture • transaction costs and overheads are reasonable given the production volumes and financial returns generated by urban aquaculture systems
Environmental	• responsible authorities implement and enforce policies and laws that prevent pollution and environmental degradation in urban areas • city planning and infrastructure development (including green infrastructure) safeguards areas where urban aquaculture can be practised against encroachment and shocks (floods, drought, disruption to electricity and water supplies) and includes provisions for aquaculture as potential element of multifunction urban water management plans • wastewater aquaculture included as a legitimate element in establishing and upgrading sewage treatment systems in accordance with WHO (2006) guidelines to help protect receiving water bodies and facilitate productive reuse of waste resources
Political (institutional)	• national and international polices explicitly support urban aquaculture for employment and income generation and associated ecosystem services and food security benefits • municipal authorities recognise and encourage aquaculture as a legitimate urban activity • land-use policy and tenure agreements provide sufficient security so as to encourage prospective producers and reassure investors and credit providers • government extension and private sector aquaculture support services cover aquaculture in urban environments
Sustainable (long-term viability)	• policies supporting urban aquaculture and production system management strategies are adaptable given the rapidly changing urban environment • urban aquaculture producers join forces to promote knowledge sharing, raise awareness and lobby for greater support, negotiate for cheaper inputs and coordinate sales and marketing • producers develop links with business advisors, development agencies and researchers to enhance efficiency and capitalise accessible resources and income-generating opportunities

Source: Bunting and Little (2015).

Wastewater-fed aquaculture

Wastewater-fed production systems reached their zenith around the middle of the twentieth century, with swollen wastewater flows from post-colonial sewerage systems in several countries in Asia being widely used to culture fish and aquatic plants, solid waste from pit latrines and sullage from septic tanks serving burgeoning urban populations being deposited routinely in fishponds and productive wastewater use in lagoon-based systems instigated extensively in impoverished, post-conflict Central and Eastern Europe. Aquaculture systems fertilised with solid and liquid waste of domestic origin were comprehensive reviewed by Edwards (1992), and accounts from twenty-one countries were presented and classified according to the mode of faecal material delivery, ranging from overhung latrines to cartage of urban nightsoil and intentional and unintentional flows of faecally polluted surface water and sewage. Inadvertent use of faecally polluted surface water for aquaculture was noted for China, Indonesia, Japan, Sri Lanka and Taiwan. Systematic exploitation of wastewater to fertilise fishponds through purposeful appropriation of faecally polluted surface water was recounted from China, Indonesia, India and Israel (see Edwards and Pullin, 1990). Generally, however, it was noted that the magnitude, extent and acceptability of wastewater-fed aquaculture was declining (Edwards, 2000, 2005).

Thanh Tri District, Hanoi, was the epicentre for wastewater-fed aquaculture development in Vietnam. During the 1960s, rivers were used to convey wastewater away from the city and subsequently fishponds were established in low-lying land to the south of the city. Wastewater was pumped from the rivers to fishponds, and production of silver carp, rohu and tilapia species ranged from 5 to 8 t ha^{-1} y^{-1}. Total average annual production from wastewater-fed aquaculture was estimated at 4500t, equivalent to 40–50 per cent of the fish supplied to Hanoi, and constituted an important and affordable source of food within poor communities (Edwards, 1996). Comprehensively reviewing peri-urban aquaculture production around Hanoi, Nguyen and Pham (2005: 10) noted that 'aquatic production in peri-urban Hanoi has been increasing in both overall area and yield' and that over the 'past 13 years, the land area used to produce fish increased from 2061 ha to 3348 ha, with the associated yield more than doubling from 4207 to 8972 [metric] tons'.

Rapid urbanisation, governed by comprehensive planning by the city authorities, has meant, however, that fishponds managed for wastewater-fed aquaculture in much of peri-urban Hanoi have been reclaimed for housing, commercial premises and urban infrastructure. Aquaculture continues elsewhere around the city in urban lakes, although it is a secondary consideration with respect to water management and amenity concerns. Nguyen and Pham (2005) noted that aquatic plant cultivation remained widespread (see below); integrated rice–fish culture continued in Dong Anh, Gia Lam and Tranh Tri Districts; and VAC farming practices (see chapter 5) combining horticulture, livestock husbandry and aquaculture were being practiced in Dong My Commune. Concerns have been voiced over perceived risks from integrated aquaculture–livestock systems, especially those supplied with wastewater, and emergent pandemic flu strains. An account of discourse on this issue is presented in Box 6.3.

Box 6.3   Integrated aquaculture and avian flu

A possible relationship between outbreaks of H5N1 avian flu with poultry waste–fed integrated aquaculture systems in Southeast Asia has been suggested. This has serious implications for both small- and large-scale integrated fish farming systems in urban and peri-urban areas, particularly those utilising wastes from their own chickens, ducks and geese or bringing in commercial poultry wastes from outside sources. The Convention on the Conservation of Migratory Species of Wild Animals (CMS, 2023) has called for the prevention of using untreated poultry products and poultry faeces as fish feeds in open aquaculture systems at times of heightened risk from highly pathogenicity avian influenza.

Scholtissek and Naylor (1988: 215) identified a similar hazard earlier in a letter titled 'Fish Farming and Influenza Pandemics' published in *Nature* and on the promotion of integrated aquaculture systems in Asia stated that:

> the result may well be creation of a considerable potential human health hazard by bringing together the two reservoirs of influenza A viruses, generating risks that have not hitherto been considered in assessment of the health constraints of integrated animal–fish farming.

In integrated systems where poultry and pigs are reared in proximity, there is the possibility of mutation of the virus within pigs into a more virulent strain which can more readily be transmitted from human to human, thus leading to fears of a pandemic. However, as Edwards et al. (1988: 506) noted in their reply, pigs and poultry have been raised together on farms in Asia and Europe for centuries, and they pointed out that 'co–location of pigs and poultry to supply manure for fish culture is neither prevalent nor likely to become so', noting that most integrated livestock–fish farms combine a single terrestrial species with fish.

A more recent newspaper article from Ho Chi Minh City stated that considerable quantities of chicken manure were used as feed for fish in Tri An lake, Dong Nai Province, and that 'the practice of using chicken excrement to feed fish in southern Vietnam is threatening millions of people with bird flu in Ho Chi Minh City and should be stopped' (Thanh Nien News, 2005). Despite such reports, there is no evidence for the transmission of the virus to humans mediated by such integrated fish farming. In this regard, there is a clear need for further research that could support the much-needed risk assessments concerning the possible role of integrated aquaculture systems as reservoirs for the transmission of the virus and whether the joint rearing of poultry, pigs or other livestock could lead to the mutation of the virus into a more virulent strain.

Considering the future of urban aquaculture, this issue highlights more general food safety concerns regarding how fish and aquatic plants are produced using recycled wastes – for example, wastewater – and findings from past research, notably at the Asian Institute of Technology, Thailand, on wastewater-fed aquaculture may be critical in addressing such issues. The PAPUSSA project assessed heavy metal levels in aquatic plants and fish raised on wastewater in Phnom Penh and Ha Noi, as well as the biological and chemical water treatment capacity of peri-urban aquaculture systems in Phnom Penh, Ha Noi, Ho Chi Minh City and Bangkok. If peri-urban aquaculture is to be sustained and deliver the potential benefits attributed to it in the future, such concerns would require greater attention and targeted human health- and food safety–orientated research so policymakers, city planners, potential investors, entrepreneurs and, perhaps most impor-tant, consumers would have the necessary and pertinent information available to them to feel more reassured.

Source: Bunting et al. (2006).

Several reviews concern the origins and evolution of wastewater aquacul-ture practice in the East Kolkata Wetlands (EKW; see Figure 6.2). Ghosh and Sen (1987) provided a comprehensive historical account of key moments and developments, Kundu (1994) approached the subject from an insightful urban planning perspective and Bunting et al. (2010) endeavoured to assess whether the persistence of productive wastewater-fed systems in the EKW against unenviable odds from urban development pressure revealed anything about how livelihoods and cultures elsewhere could develop sustainably and become resilient. With most Asian wastewater-fed aquaculture systems having ceased to operate or now considerably reduced, the concept and practices are in danger of being consigned to history books – the prospects for wastewater-fed aquaculture in the EKW seemed bleak (Edwards, 2003; Box 6.4). However, more recently, protection has been afforded to the area owing to its designation as a Ramsar site, constitutional commitments have been made to protect the wetlands with the passing of 'The East Kolkata Wetlands (Conservation and Management) Act, 2006' (Government of West Bengal, 2006) and investment by the Government of West Bengal to

Figure 6.2 Captain Bhery on the edge of the East Kolkata Wetlands has been used for wastewater aquaculture for over a century; the Eastern Metropolitan Bypass (EMB) can be seen in the background.

Photo credit: Stuart Bunting.

rehabilitate infrastructure serving the system appears to have signalled a reprieve. Such interventions were not undertaken lightly and depended to a large extent on proponents within and outside government championing the case of the EKW, its unique heritage, the livelihoods of those it supports and broader ecosystem services sustained by this iconic wetland agroecosystem.

Wastewater-fed fishponds east of Kolkata totalled around 7300 ha in 1945 but had declined to 3900 ha at the time of drafting the East Kolkata Wetlands (Conservation and Management) Act, 2006. However, since then, canals conveying wastewater to most fish culture areas have been rehabilitated. Several fishponds, the largest of which were 60–70 ha, have been systematically desilted to provide infill material for urban development to the north and defunct siphons feeding wastewater to the southern regions of the EKW have been cleared. Consequently, 1000 ha of former fishponds that had been converted to rice farming could potentially be reinstated for more productive and profitable wastewater-fed aquaculture.

Practices in the EKW have been widely cited as a model for productive and sustainable wastewater management through aquaculture (Edwards and Pullin, 1990). A series of lagoon-based treatment systems combined with fishponds, inspired by knowledge of the EKW system, were commissioned under the Ganges Action Plan to reduce wastewater flows from towns and

Box 6.4   Apparent decline of wastewater aquaculture to the east of 'Calcutta' (Kolkata)

Interviews conducted with key informants in the wetlands to the east of Calcutta highlighted a range of technical factors constraining the operation of the existing wastewater aquaculture system. Competition with the irrigation system regarding the equitable distribution of the wastewater was an important constraint. Regulation of the supply of wastewater to the *bheries* (local name for the wastewater-fed fishponds) was also constrained owing to the absence of suitable outlet channels allowing the water level within the *bheries* to be manipulated as desired.

Blue-green algae blooms have been observed within certain fishponds, and as the presence of these blooms has been blamed for inhibiting the growth of fingerlings, they have been purchased from external sources. This not only constituted an additional cost to the operators but also constrained the development of vertically integrated systems. Furthermore, reliance on externally produced fry was a consequence of the fishers being preoccupied with the problems of today rather than planning for a hatchery in the future. Production was also being constrained by a decrease in the quality and therefore survival rate of fry purchased from external sources. The cause of the decline in fry quality was believed to be due to the use of broodstock of only 200 g for spawning as opposed to previous years when fish of 500 g were used.

Accumulation of sediments within the wetland system was considered to be a problem, although not an insurmountable one. Contaminants introduced into the fishponds with the wastewater were also considered to be potential constraints to production within the existing wetland system. Labour problems represented another important concern; *bheri* operators had been forced to employ large numbers of fishers who had been displaced by the reclamation of wetlands for development, resulting in an artificially high level of employment within the remaining wetland system. Consequently, harvesting was being conducted too frequently to create work, resulting in the sub-optimal management of the wetland system with the production of small, lower value fish. The increasing cost of inputs required by the wastewater aquaculture system – that is, labour and seed – combined with the decreasing survival of fry stocked in the fishponds was blamed for the decline in profits generated by the *bheries*. However, the major constraint facing the existing wastewater aquaculture system was considered by some operators to be poaching, sometimes forcible poaching by armed gangs. In addition, theft was blamed for stifling innovation within the wetland system; the theft of ducks and flowers that were being cultured on the embankments of one *bheri* had prevented the potential benefits of integrated systems from being realised.

Source: Bunting et al. (1999).

cities being discharged untreated to the river. Elsewhere, building on highly evolved engineering designs for lagoon-based wastewater treatment to meet statutory discharge standards, a rational design approach was formulated by Mara et al. (1993) to optimise fish production through better nutrient utilisation whilst maintaining adequate wastewater treatment.

Rational design approach for wastewater-fed aquaculture

Cognisant of constrains threatening the continued operation of wastewater-fed aquaculture in the EKW (Edwards, 2003; Bunting, 2004; Bunting et al., 2005) and the importance of ensuring that practices were in accordance with revised international standards for wastewater use in aquaculture (WHO, 2006), bioeconomic modelling was employed to test the potential of the rational approach to wastewater-fed aquaculture devised by Mara et al. (1993). Formulated to simultaneously optimise wastewater treatment and fish production, the rational design approach consisted of anaerobic and facultative lagoons to treat incoming wastewater prior to reuse in fishponds (Table 6.6). The design is in contrast to conventional lagoon-based treatment systems that routinely include maturation ponds following facultative lagoons to achieve statutory quality levels prior to productive reuse through fish culture and which therefore would result in limited fish production

Table 6.6 Design criteria for conventional and rational lagoon-based wastewater treatment and fish culture

Component	Design criteria	Conventional	Rational
Anaerobic pond	Retention (d)	1	1
	Depth (m)	4	2
	Volumetric BOD loading for validation	<300 g m^{-3} d^{-1}	<300 g m^{-3} d^{-1}
Facultative pond	Retention (d)	4	5
	Depth (m)	1.5	1.5
	Aereal BOD loading for validation	<350 kg ha^{-1} d^{-1}	<350 kg ha^{-1} d^{-1}
Maturation ponds	Depth (m)	1.5	–
	Number of ponds (n) where: faecal coliform count (100 ml^{-1}) required in the maturation pond effluent is N_e	$N_e = N_i(1+k_T\emptyset_m)^n$	–
	Minimum retention (d)	3	–
	Aereal BOD loading for validation	<350 kg ha^{-1} d^{-1}	–
Fishpond	Depth (m)	1	1
	Optimal N loading rate	4 kg ha^{-1} d^{-1}	4 kg ha^{-1} d^{-1}

Source: Bunting (2007).

because most of the nutrients would be removed in the wastewater treatment process before the effluent flowed into fishponds.

The omission of maturation ponds in the rational design was intended to reduce the amount of nitrogen removed from the wastewater stream, thus enhancing its value as a pond fertiliser. After retention for 1 and 5 days in anaerobic and facultative lagoons, respectively, it was asserted that this system constituted a 'design method for the minimal treatment of wastewater and the maximal production of microbiologically safe fish that effectively resolves the dilemma of simultaneously optimizing wastewater treatment and fish production in a practical way' (Mara et al., 1993: 1797). Details of the design criteria are presented in Box 6.5. Estimates arrived at by these authors put the average production of fish at 13 t ha^{-1} y^{-1}, with more than one crop of tilapia being cultured per year, and financial assessment appeared to indicate that the system would be commercially viable in India. Reviewing prospects for transforming practices in the EKW to align with the rational design approach, notable constraints identified included the need for additional land area, capital costs and practical limitations to reconfiguring the system (Bunting, 2007), although nutrient retention, fish production and financial returns would all be increased in the medium to long term.

Box 6.5   Design criteria for rational lagoon-based waste-water treatment and fish culture

Multiplying the flow volume (m^3) by retention time (1d) and dividing by depth (2 m) gives the anaerobic lagoon mid-depth area; volumetric biochemical oxygen demand (BOD) loading (λ_V, g m^{-3} d^{-1}) is used to validate the design (Mara, 1997); thus:

$$\lambda_V = L_i Q/V_a,$$

where L_i is influent BOD concentration (g m^{-3}), Q is the flow rate (m^3 d^{-1}) and V_a is the anaerobic lagoon volume (m^3). At 25°C the maximum permissible load is 300 g m^{-3} d^{-1} (Mara and Pearson, 1986, as cited in Mara et al., 1993).

Facultative lagoon mid-depth area is found by multiplying flow volume (m^3) by retention time (5d) and dividing by depth (1.5 m); following Mara et al. (1993), the design was validated using the BOD loading. At 25°C the maximum permissible loading is 350 kg ha^{-1} d^{-1}. Mara et al. (1993) discounted evaporation as the facultative lagoon had a relatively short retention time and surface scum significantly reduces losses from anaerobic lagoons under normal operating conditions. However, evaporation from facultative lagoons should be taken into account where the treatment of larger wastewater flows is under consideration.

Fishponds were dimensioned using an optimal nitrogen loading of 4 kg ha^{-1} d^{-1} (Edwards, 1992). Assuming no removal occurs in the anaerobic pond (Mara et al., 1993), total nitrogen (TN) removal in the facultative pond was estimated following Reed (1985; cited in Mara et al., 1993):

$$C_e = C_i \exp\{-[0.0064(1.039)^{T-20}][\emptyset + 60.6(\text{pH} - 6.6)]\}$$

where C_e is effluent TN (mg l^{-1}), C_i is influent TN (mg l^{-1}), T is temperature (°C) and Ø is retention time (d).

Faecal coliform numbers in fishponds were estimated using the relationship established by Marais (1974, as cited in Mara et al., 1993):

$$N_p = N_i/(1 + k_T \emptyset_a)(1 + k_T \emptyset_f)(1 + k_T \emptyset_p),$$

where N_p is faecal coliforms (100 ml^{-1}) in fishpond, N_i is faecal coliforms (100 ml^{-1}) in untreated wastewater, k_T is the rate constant for faecal coliforms removal per day $(2.6(1.19)^{T-20})$, $\emptyset_a$ is anaerobic retention time (d), $\emptyset_f$ is facultative retention time (d) and $\emptyset_p$ is fishpond retention time (d). Mara et al. (1993) recommended that an additional land area equivalent to 55 per cent of that required for lagoons and fishponds be allocated for access and infrastructure.

Source: Bunting et al. (2011).

Productive wastewater reuse in both agriculture and aquaculture is seldom acknowledged or accounted for in planning and policymaking, and the extent of such practices may actually be downplayed to allay public concerns and to present a more sanitised profile to onlookers. Against this backdrop, the Hyderabad Declaration on Wastewater Use in Agriculture (IWMI, 2002: 3) urged 'policy-makers and authorities in the fields of water, agriculture, aquaculture, health, environment and urban planning, as well as donors and the private sector' to 'safeguard and strengthen livelihoods and food security, mitigate health and environmental risks and conserve water resources by confronting the realities of wastewater use in agriculture ...' whilst instigating 'cost-effective and appropriate treatment suited to the end use of wastewater'.

Internationally recognised standards have been established by WHO (2006) based on sound scientific principles to promote the wider recognition of wastewater-fed aquaculture as a legitimate approach to the productive management of wastewater discharges. Current wastewater-fed aquaculture practices, whether intentional or not, must, however, be upgraded to employ appropriate strategies to safeguard public, animal and environmental health.

As more than 1 billion people are without access to even basic sanitation, productive reuse might constitute a much-needed motivation for sanitary waste collection and disposal management. Conventional wastewater treatment results in sludge requiring disposal, and this is costly and wasteful when it is destined for landfill. Otherwise, nutrients are not necessarily readily available when application to agricultural land and forests is undertaken; following application, the sludge may result in diffuse pollution. Energy consumption and greenhouse gas emissions by conventional treatment plants are increasingly coming under scrutiny and were included in assessments for the Asian Development Bank and Government of West Bengal, India, for example, concerning whether to commission a conventional sewage treatment plant or invest in rehabilitating the EKW system.

Indirect waste exploitation through aquaculture

Concerns over direct wastewater exploitation to produce aquatic products destined for human consumption centre on worries over potential animal, environmental and public health risks and the capacity of aquatic ecosystems to consistently attain statutory discharge standards and assimilate nutrients over a prolonged period. Considering the global significance of how best to capture nutrients form waste streams before they are lost to the biosphere (Willett et al., 2019), the DPSIR framework was used to analyse this pressing resource recovery conundrum (see Figure 6.3 for further details; Bunting and Edwards, 2018). The assessment demonstrated that the challenge is complex and multi-layered and that corrective action is needed across several sectors and at different scales.

Potential responses to address high-level driving forces (analogous to those in the food system framework introduced in chapter 1 and discussed in chapter 7; HLPE, 2020) were noted and included commitments of responsible authorities to meet international obligations to tackle the United Nations' Sustainable Development Goals by 2030, notably 'SDG 6. Ensure availability and sustainable management of water and sanitation for all' and 'SDG 11. Make cities and human settlements inclusive, safe, resilient and sustainable' (United Nations, 2015: 16); policies and planning guidelines that stipulate that safe wastewater reuse is feasible; instigating legislation to legitimise safe wastewater reuse; and introducing safeguards to protect and reassure consumers and others across value chains and food systems (Bunting and Edwards, 2018).

Immediate and serious concerns regarding the state of the system and impacts on public health and receiving ecosystems could be addressed with appropriate wastewater reuse strategies with significantly lower capital and operating costs compared with conventional treatment plants. Appropriate polices and regulations (e.g. preventing industrial effluents contaminating waste resources, ensuring operators have security of tenure and planning for a managed retreat from urban expansion) could be devised to encourage investment in safe wastewater reuse through aquaculture.

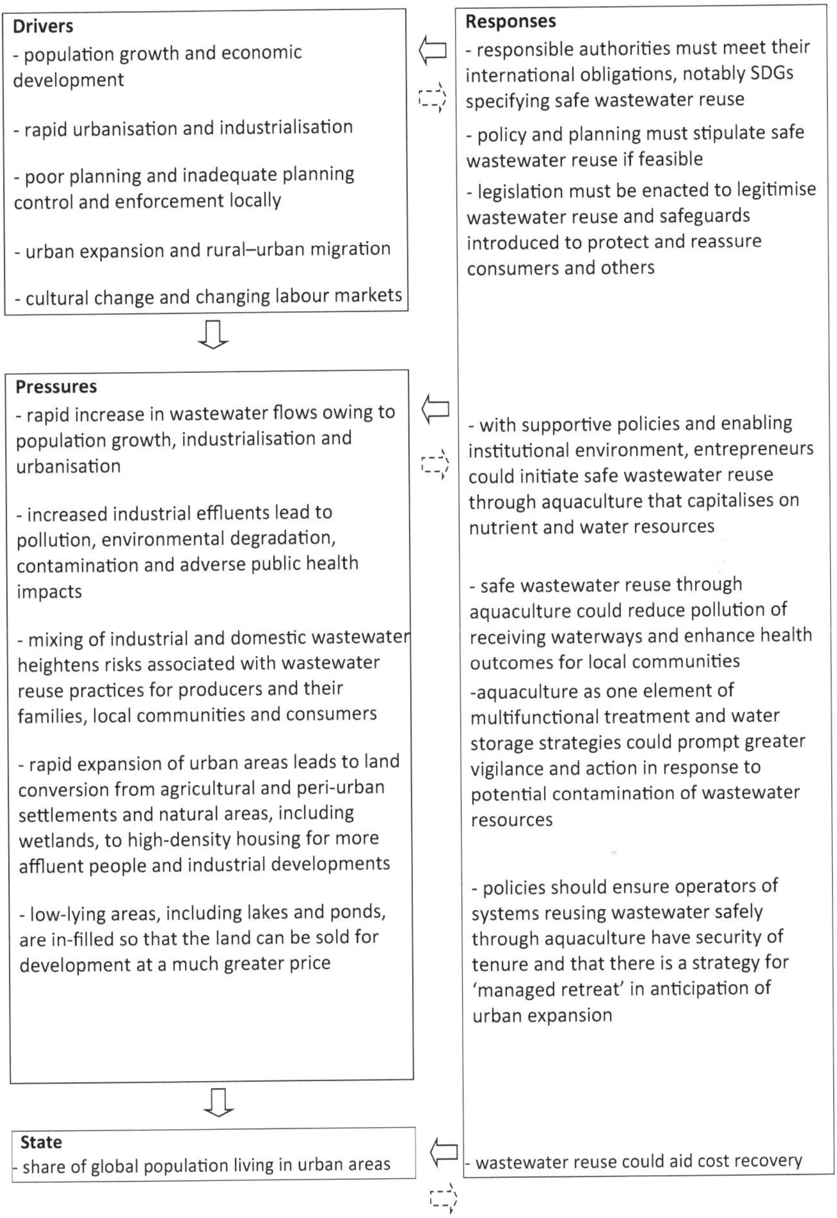

Drivers
- population growth and economic development

- rapid urbanisation and industrialisation

- poor planning and inadequate planning control and enforcement locally

- urban expansion and rural–urban migration

- cultural change and changing labour markets

Responses
- responsible authorities must meet their international obligations, notably SDGs specifying safe wastewater reuse

- policy and planning must stipulate safe wastewater reuse if feasible

- legislation must be enacted to legitimise wastewater reuse and safeguards introduced to protect and reassure consumers and others

Pressures
- rapid increase in wastewater flows owing to population growth, industrialisation and urbanisation

- increased industrial effluents lead to pollution, environmental degradation, contamination and adverse public health impacts

- mixing of industrial and domestic wastewater heightens risks associated with wastewater reuse practices for producers and their families, local communities and consumers

- rapid expansion of urban areas leads to land conversion from agricultural and peri-urban settlements and natural areas, including wetlands, to high-density housing for more affluent people and industrial developments

- low-lying areas, including lakes and ponds, are in-filled so that the land can be sold for development at a much greater price

- with supportive policies and enabling institutional environment, entrepreneurs could initiate safe wastewater reuse through aquaculture that capitalises on nutrient and water resources

- safe wastewater reuse through aquaculture could reduce pollution of receiving waterways and enhance health outcomes for local communities

- aquaculture as one element of multifunctional treatment and water storage strategies could prompt greater vigilance and action in response to potential contamination of wastewater resources

- policies should ensure operators of systems reusing wastewater safely through aquaculture have security of tenure and that there is a strategy for 'managed retreat' in anticipation of urban expansion

State
- share of global population living in urban areas

- wastewater reuse could aid cost recovery

Figure 6.3 Drivers, pressures, state, impacts and responses framework analysis for safe wastewater management through intermediaries and innovative biorefinery processes.

Source: Bunting and Edwards (2018).

exceeded 50 per cent in 2007, and this trend continues, with many hundreds of millions or poor people suffering as they lack access to basic services, such as sanitation, and have to live in insecure and polluted environments

- surface water quality in many urban settings and downstream environments is below acceptable international standards (and adversely affects human health)

- vast amounts of nutrients are discharged to receiving waters (causing eutrophication and the loss of a valuable resource)

- extent of low-lying areas that could previously accommodate runoff and flood water has declined significantly (leading to substantial financial costs associated with need to build drainage infrastructure and heightened risk of flooding in established and newly urbanised areas)

to subsidise appropriate collection and treatment and provide income opportunities for poor and marginal groups

- appropriate reuse systems could avoid capital and O&M costs for conventional sewage treatment systems

- wastewater reuse could achieve better environmental protection and biodiversity conservation and restoration of ecosystem services that sustain economic and socio-cultural activities

- wastewater reuse could facilitate nutrient recycling and added value from water appropriated for human use

- production of affordable aquatic plants and animals could enhance nutrition and food security amongst poor communities

- water bodies established for wastewater reuse could provide additional capacity to store urban runoff and floodwater

Impacts

- poor and marginal groups suffer owing to limited employment and income, insecure living conditions and food supplies, absence of health care and sanitation, air and water pollution, and flooding

- population pressure leads to informal settlements on accessible areas that are not served by sewers and consequently discharge wastewater indiscriminately and lead to open defecation and overhung latrines along canals and over ponds

- discharge of untreated wastewater negatively impacts biodiversity, undermines stocks and flows of ecosystems services, and increases animal and public health risks associated with direct and unintentional wastewater reuse through aquaculture

- conflicts between land-use activities, with poor communities and agriculture and aquaculture generating modest financial returns displaced or further marginalised

-financial costs associated with living in urban settings force younger generation to avoid farm work and seek urban employment

Figure 6.3   (Continued)

Two primary strategies can be envisaged to help allay health concerns, namely, waste-based aquaculture employing intermediaries destined for feed formulation, described as 'lengthening the food chain' (Edwards, 1990: 210), who comprehensively assessed producing either tilapia or duckweed as high-protein animal feed and aquatic production of non-food products to enhance ecosystem services. Culturing intermediary products can break pathogen life-cycles, thus preventing disease transmission, permitting optimisation of nutrient assimilation and helping distance waste use both physically and psychologically from aquaculture making use of intermediary products. Intermediaries assessed for inclusion in integrated aquaculture systems have included fish, invertebrates, zooplankton, microalgae and aquatic macro-phytes. An account of cultivating duckweed using wastewater for use in fish and chicken feed in Bangladesh is presented in Box 6.6, but prospects for wider application in the country were considered limited (Edwards, 2005).

Box 6.6   Wastewater reuse to produce duckweed to feed fish

The wastewater treatment system developed by PRISM Bangladesh received the domestic wastewater from the resident population of between 2000 and 3000 individuals at Kumudini Hospital, Mirzapur, Bangladesh. The treatment system occupied an area of 2 ha, the duckweed-covered lagoon a surface area of 0.6 ha and the remainder of land consisted of embankments and an access road. The lagoon channel was reticulated in nature and had a length of 500 m; the depth of the lagoon increased from 0.4 m at the inflow to 0.9 m at the outflow. Alaerts et al. (1996) described the operation of the system in detail and reported that on a weekly basis, the lagoon received approximately 2000m^3 of sewage water that had been allowed to pre-settle in an anaerobic lagoon. The total retention time within the lagoon was reported to be 20.4 days and the surface loading of BOD$_5$ ranged between 48 and 60 kg ha^{-1} per day (Alaerts et al., 1996).

The duckweed-covered lagoon removed between 95 and 99 per cent of the BOD$_5$, between 90 and 97 per cent of the chemical oxygen demand and between 74 and 77 per cent of both the Kjeldahl nitrogen and total phosphorus (Alaerts et al., 1996). During the study period, duckweed production within the system was reported to range between 58 and 105 kg (dry weight) ha^{-1} per day; at this level of production, the removal of nitrogen and phosphorus from the system in the harvested duckweed was estimated to be 0.26 and 0.05 g m^{-2} per day, respectively. Annual duckweed yields from this lagoon-based system can be extra-polated to between 21,170 and 38,325 kg (dry weight) ha^{-1}, equivalent to the production of between 260,975 and 438,000 kg ha^{-1} of fresh duckweed biomass per year.

Figure 6.4 Duckweed-covered lagoon system developed by PRISM Bangladesh to treat wastewater from Kumudini Hospital, Mirzapur, Bangladesh; embankments planted with banana trees are visible.

Photo credit: Stuart Bunting.

Duckweed produced in the system was used to culture fish in adjacent ponds, with the fish produced using this system readily accepted by the local community. Commercial outlets for the duckweed existed in both local and regional markets; in Jessore, duckweed commanded a price of between 0.5 and 1 taka kg^{-1} and was used to feed fingerlings in local hatcheries. In the past, duckweed would have been collected from the wild to supply these hatcheries; however, wild stands of duckweed were increasingly hard to find, and those that did exist were probably already being exploited to provide feed for either fish or livestock.

An important initiative developed by PRISM Bangladesh has been to cultivate banana trees on the lagoons' embankments (Figure 6.4); producing crops on the embankments of the lagoon not only enhanced the economic performance of the overall system but also provided environmental and social benefits. The physical presence of the banana trees enhanced the micro-climate surrounding the lagoon, providing shade for the duckweed and reducing wind velocities over the lagoon. From a social perspective, the labour requirements of the integrated system provided employment for local individuals, many of whom were landless farmers or net-less fishermen.

Source: Bunting et al. (1999).

Intermediary aquatic plants and animals cultured employing waste resources represent what could be increasingly important, locally sourced and nutritionally important constituents of sustainably sourced formulated feeds for aquaculture (Bunting and Edwards, 2018). Essential amino acid and polyunsaturated fatty acid concentrations in chironomid larvae cultured on palm oil mill effluent were higher when compared to larvae produced using algae cultures (Habib et al., 1997). Numerous pilot-scale assessments concerning the optimal productivity of potential intermediaries were conducted in the 1980s, but relatively cheap fish meal and fish oil arguably thwarted commercial exploitation of even the most promising findings. This unedifying era is hopefully drawing to a close and could well usher in renewed interest in aquaculture as a means to produce feed constituents for more valuable species. Potential yields of Nile tilapia (*O. niloticus*) cultured on microalgae suspensions emanating from high-rate stabilisation ponds treating sewage were extrapolated to 20 t ha^{-1} y^{-1} (Edwards and Sinchumpasak, 1981). Apparent constraints to recycling human wastewater, including faecal coliform concentrations and issues of consumer acceptance, might be addressed through processing tilapia for inclusion in formulated feed. Alternative strategies to integrate intermediaries in wastewater-fed aquaculture systems were reviewed comprehensively by Edwards (1990) and may warrant renewed assessment given evolving environmental imperatives and financial conditions.

Drawing on knowledge of fishing practices involving fish aggregating brush parks such as *katha* fisheries in Bangladesh and *acadja* fisheries in Benin, West Africa, research in Bangladesh and India has assessed the potential contribution of various substrates in ponds for promoting attached periphyton growth to enhance the conversion of nutrients to fish biomass (see chapter 2). Elsewhere, stoneroller minnows (*Campostoma anomalum*) and tilapia (*Oreochromis mossambicus*) have been assessed as potential candidates to graze periphyton grown in wastewater to sequester nitrogen and phosphorus (Drenner et al., 1997). Observed nutrient removal through harvesting fish and faeces removal was less than anticipated and phosphorus removal was significantly lower at 48 mg m^{-2} d^{-1} than levels of 104–139 mg m^{-2} d^{-1} obtained through mechanical periphyton removal. Potential explanations centred on three propositions:

- fish excrete phosphorus, making them less effective at sequestration compared with mechanical periphyton removal;
- nitrogen concentrations employed may have limited periphyton growth and consequently phosphorus uptake;
- periphyton growth may have been retarded by suboptimal system configuration.

Low temperatures were cited as causing high fish mortality rates during the study, indicating that the species selected were inappropriate and raising questions over whether any cold tolerant, grazing species might be identified, when low periphyton production at higher altitudes seems to preclude this.

Aquaculture has potential to convert waste resources and untapped nutrient streams to products with value. Increasingly, aquaculture has been developed to produce animals and plants not destined for direct human consumption but to produce seed and feed for other culture operations, juveniles for restocking programmes and culture-based fisheries and feedstock for industrial processes and biofuel production. Unlike wastewater-fed aquaculture in which animals are integrated in the system, with a need to maintain water quality conditions conducive to growth, macrophytes and microalgae could be raised on more concentrated waste streams.

However, plant-based (macrophytes and algae) systems are prone to problems: dense stands and blooms can result in self-shading, whilst rapid growth can deplete potentially limiting nutrients, thus resulting in senescence and crashes; pests and disease agents still constitute potential problems to continuous cultivation; and technical failures and environmental perturbations are attendant hazards. The problems of maintaining algae cultures at higher latitudes and as part of integrated land-based marine aquaculture systems in France were discussed in chapter 4.

Summary

Urban and peri-urban aquaculture developed around cites in several Asian countries to exploit wastewater emanating from centralised sewerage systems. As these cities have expanded rapidly, areas supporting such systems have been developed for residential, commercial and other urban uses. Remnants persist in several cases, but the water bodies are increasingly valued for their contribution to flood control and aesthetic role in otherwise built-up areas. Although international standards have been agreed to facilitate productive wastewater use through aquaculture, it is probably not generally considered a viable option. Production of intermediaries on concentrated waste streams for subsequent use in formulated feeds and production of biomass as feedstock for energy production may signal a renaissance in wastewater reuse through aquaculture. Small-scale and intensive aquaculture production systems in urban and peri-urban areas have emerged in both developed and developing counties to meet specific requirements. Urban catfish farming has evolved in Africa to meet growing demand from more affluent consumers, whilst in Europe and North America recirculating and aquaponic systems have been conceived primarily as learning and teaching resources. Urban rivers and lakes in many cities have been polluted, and as action to rehabilitate them progresses, there appears to be another role for aquaculture in supplying fish and other aquatic animals for restocking. As concern over the loss of ecological knowledge amongst urban and more affluent communities grows, knowledge and experience of such initiatives may help redress this.

Note

1 Aquatic vegetables referred to here include water spinach (or morning glory, *Ipomoea aquatica*), water mimosa (*Neptunia oleracea*), water dropwort (*Oenanthe stolonifera*) and water cress (*Rorippa nasturtium-aquaticum*).

References

Adler, P.R. (1998) 'Phytoremediation of aquaculture effluents', Aquaponics Journal, vol 4, pp. 10–15.

Alaerts, G.J., Mahbubar, R. and Kelderman, P. (1996) 'Performance analysis of a full-scale duckweed-covered sewage lagoon', Water Research, vol 30, pp. 843–852.

Bassleer, G. (2007) 'Ornamental Fish International: an international organization for the ornamental aquatic industry', Aquaculture Europe, vol 32, no 1, p. 15.

Blake, N.J. (2005) 'Aquaculture of the Florida Bay Scallop, *Argopecten irradians concentricus*, in Tampa Bay Florida, USA, an urban estuary' in B.A. Costa-Pierce, P. Edwards, D. Baker and A. Desbonnet (eds) Urban Aquaculture, CAB International, Wallingford, UK.

Bunting, S.W. (2004) 'Wastewater aquaculture: perpetuating vulnerability or opportunity to enhance poor livelihoods?', Aquatic Resources, Culture and Development, vol 1, pp. 51–75.

Bunting, S.W. (2007) 'Confronting the realities of wastewater aquaculture in peri-urban Kolkata with bioeconomic modelling', Water Research, vol 41, no 2, pp. 499–505.

Bunting, S.W. and Edwards, P. (2018) 'Global prospects for safe wastewater reuse through aquaculture' in B.B. Jana, R.N. Mandal, and P. Jayasankar (eds) Wastewater Management through Aquaculture, Springer, New York.

Bunting, S.W., Edwards, P. and Kundu, N. (2011) Environmental Management Manual East Kolkata Wetlands, MANAK Publications, New Delhi.

Bunting, S.W., Edwards, P.E. and Muir, J.F. (1999) Constraints and Opportunities in Wastewater Aquaculture, Working Paper, UK Government's Department for International Development, Aquaculture Research Programme, Institute of Aquaculture, University of Stirling, UK.

Bunting, S.W., Kundu, N. and Mukherjee, M. (2005) 'Peri-urban aquaculture and poor livelihoods in Kolkata, India' in B.A. Costa-Pierce, P. Edwards, D. Baker and A. Desbonnet (eds) Urban Aquaculture, CAB International, Wallingford, UK.

Bunting, S.W. and Little, D.C. (2003) 'Urban aquaculture' in RUAF (ed) Annotated Bibliography on Urban Agriculture, ETC, Leusden, The Netherlands.

Bunting, S.W. and Little, D.C. (2005) 'The emergence of urban aquaculture in Europe' in B.A. Costa-Pierce, P. Edwards, D. Baker and A. Desbonnet (eds) Urban Aquaculture, CAB International, Wallingford, UK.

Bunting, S.W. and Little, D.C. (2015) 'Urban aquaculture for resilient food systems' in H. de Zeeuw and P. Drechsel (eds) Cities and Agriculture: Developing Resilient Urban Food Systems, Routledge, Abingdon, UK.

Bunting, S.W., Little, D. and Leschen, W. (2006) 'Urban aquatic production' in R. van Veenhuizen (ed) Cities Farming for the Future: Urban Agriculture for Green and Productive Cities, RUAF Foundation, Leusden, The Netherlands, IDRC, Ottawa, Canada and IIRR, Cavite, Philippines.

Bunting, S.W., Luo, S., Cai, K., Kundu, N., Lund, S., Mishra, R., et al. (2016) 'Integrated action planning for biodiversity conservation and sustainable use of highland aquatic resources: evaluating outcomes for the Beijiang River, China', Journal of Environmental Planning and Management, vol 59, pp. 1580–1609.

Bunting, S.W., Pretty, J. and Edwards, P. (2010) 'Wastewater-fed aquaculture in the East Kolkata Wetlands: anachronism or archetype for resilient ecocultures?', Reviews in Aquaculture, vol 2, pp. 138–153.

CMS (2023) Scientific Task Force on Avian Influenza and Wild Birds Statement On: H5N1 High Pathogenicity Avian Influenza in Wild Birds – Unprecedented Conservation Impacts and Urgent Needs, Convention on the Conservation of Migratory Species of Wild Animals, Bonn, Switzerland.

Drenner, R.W., Day, D.J., Basham, S.J., Smith, J.D. and Jensen, S.I. (1997) 'Ecological water treatment system for removal of phosphorous and nitrogen from polluted water', Ecological Applications, vol 7, pp. 381–390.

Eding, E. and Kamstra, A. (2002) 'Netherlands farms tune recirculation systems to production of varied species', Global Aquaculture Advocate, vol 5, no 3, pp. 52–55.

Edwards, P. (1990) 'An alternative excreta-reuse strategy for aquaculture: the production of high-protein animal feed' in P. Edwards and R.S.V. Pullin (eds) Wastewater-Fed Aquaculture, Proceedings of the International Seminar on Wastewater Reclamation and Reuse for Aquaculture, Calcutta, India, 6–9 December 1988, Environmental Sanitation Information Center, Asian Institute of Technology, Bangkok.

Edwards, P. (1992) Reuse of Human Waste in Aquaculture, a Technical Review, UNDP–World Bank Water and Sanitation Program, World Bank, Washington.

Edwards, P. (1996) 'Wastewater reuse in aquaculture: socially and environmentally appropriate wastewater treatment for Vietnam', Naga, vol 19, no 1, pp. 36–37.

Edwards, P. (2000) 'Wastewater-fed aquaculture: state-of-the-art' in B.B. Jana, R.D. Banerjee, B. Guterstam and J. Heeb (eds) Waste Recycling and Resource Management in the Developing World, Ecological Engineering Approach, University of Kalyani, India, and International Ecological Engineering Society, Switzerland.

Edwards, P. (2003) 'Peri-urban aquaculture in Kolkata', Aquaculture Asia, vol 8, no 2, pp. 4–6.

Edwards, P. (2005) 'Demise of wastewater-fed duckweed-based aquaculture in Bangladesh', Aquaculture Asia, vol 10, no 1, pp. 9–10.

Edwards, P., Lin, C.K., Macintosh, D.J., Leong Wee, K., Little, D. and Innes-Taylor, N.L. (1988) 'Fish farming and aquaculture', Nature, vol 333, pp. 505–506.

Edwards, P. and Pullin, R.S.V. (eds) (1990) Wastewater-fed Aquaculture, Proceedings of the International Seminar on Wastewater Reclamation and Reuse of Aquaculture, Calcutta, India, 6–9 December 1988, Environmental Sanitation Information Center, Asian Institute of Technology, Bangkok.

Edwards, P. and Sinchumpasak, O.A. (1981) 'The harvest of microalgae from the effluent of a sewage fed high rate stabilisation pond by *Tilapia nilotica*. Part 1: description of the system and the study of the high rate pond', Aquaculture, vol 23, pp. 83–105.

Fish Information and Services (2012) 'Seaweed could be used to clean up polluted waters', www.fis.com/fis/worldnews

Ghosh, D. and Sen, S. (1987) 'Ecological history of Calcutta's wetland conversion', Environmental Conservation, vol 14, pp. 219–226.

Goudey, C.A. and Moran, B.M. (2005) 'Four years of recirculating aquaculture in urban Boston Harbor' in B.A. Costa-Pierce, P. Edwards, D. Baker and A. Desbonnet (eds) Urban Aquaculture, CAB International, Wallingford, UK.

Government of West Bengal (2006) The East Kolkata Wetlands (Conservation and Management) Act, 2006, Government of West Bengal, Kolkata, India.

Habib, M.A.B., Yusoff, F.M., Phang, S.M., Ang, K.J. and Mohamed, S. (1997) 'Nutritional values of chironomid larvae grown in palm oil mill effluent and algal culture', Aquaculture, vol 158, pp. 95–105.

Hart, B.T., van Dok, W. and Djuangsih, N. (2002) 'Nutrient budget for Saguling Reservoir, West Java, Indonesia', Water Research, vol 36, pp. 2152–2160.

HLPE (2020) Food Security and Nutrition: Building a Global Narrative towards 2030, High Level Panel of Experts, Rome.

Hung, L.T. and Huy, H.P.V. (2005) 'Production and marketing systems of aquatic products in Ho Chi Minh City', Urban Agriculture vol 14, pp. 16–19.

IWMI (2002) The Hyderabad Declaration on Wastewater Use in Agriculture, Hyderabad, India, 14 November 2002, International Water Management Institute, Hyderabad, India.

Jungersen, G. (1991) 'Environmental benefit from integrating recirculation fish breeding systems with production of greenhouse crops' in N. De Pauw and J. Joyce (eds) Aquaculture and the Environment, European Aquaculture Society, Special Publication 14, Bredene, Belgium.

Kundu, N. (1994) Planning the Metropolis: A Public Policy Perspective, Minerva Associates, Calcutta, India.

Leschen, W., Little, D., Bunting, S. and van Veenhuizen, R. (2005) 'Urban aquatic production', Urban Agriculture, vol 14, pp. 1–7.

Liu, Y., Bunting, S.W., Luo, S., Cai, K. and Yang, Q. (2019) 'Evaluating impacts of fish stock enhancement and biodiversity conservation actions on the livelihoods of small-scale fishers on the Beijiang River, China', Natural Resource Modeling, vol 32, e12195.

Love, D.C., Fry, J.P., Genello, L., Hill, E.S., Frederick, J.A., Li, X., et al. (2014) 'An international survey of aquaponics practitioners', PLoS ONE, vol 9, no 7, e102662.

Love, D.C., Fry, J.P., Li, X., Hill, E.S., Genello, L., Semmens, K. and Thompson, R.E. (2015) 'Commercial aquaponics production and profitability: findings from an international survey', Aquaculture, vol 435, pp. 67–74.

Mara, D.D. (1997) Design Manual for Waste Stabilization Ponds in India, Lagoon Technology International, Leeds, UK.

Mara, D.D., Edwards, P., Clark, D. and Mills, S.W. (1993) 'A rational approach to the design of wastewater-fed fishponds', Water Research, vol 27, pp. 1797–1799.

Miller, J.W. and Atanda, T. (2011) 'The rise of peri-urban aquaculture in Nigeria', International Journal of Agricultural Sustainability, vol 9, pp. 274–281.

Muir, J.F. (1996) 'A systems approach to aquaculture and environmental management' in D.J. Baird, M.C.M. Beveridge, L.A. Kelly and J.F. Muir (eds) Aquaculture and Water Resource Management, Blackwell Science, Oxford, UK.

Mukherjee, M., Banerjee, R., Datta, A., Sen, S. and Chatterjee, B. (2004) 'Women fishers in peri-urban Kolkata', Urban Agriculture, vol 12, p. 40.

Nguyen, T.D.P. and Pham, A.T. (2005) 'Current status of peri-urban aquatic production in Hanoi', Urban Agriculture, vol 14, pp. 10–12.

Prain, G. (2006) 'Integrated urban management of local agricultural development: the policy arena in Cuba', in R. van Veenhuizen (ed) Cities Farming for the Future:

Urban Agriculture for Green and Productive Cities, RUAF Foundation, Leusden, The Netherlands, IDRC, Ottawa, Canada and IIRR, Cavite, Philippines.

Rakocy, J.E. and Bailey, D.S. (2003) 'Initial economic analysis of aquaponic systems' in T. Chopin and H. Reinertsen (eds) Beyond Monoculture, European Aquaculture Society, Special Publication 33, Bredene, Belgium.

Rana, K., Anyila, J., Salie, K., Mahika, C., Heck, S. and Young, J. (2005) 'Aquafarming in urban and peri-urban zones in sub Saharan Africa', Aquaculture News, vol 32, pp. 6–8.

Rigby, M. (2000) 'Sea bass farming in Europe: recirculation system technology, a viable alternative' in Abstracts from the Third Annual C-Mar PESCA Aquaculture Workshop, Portaferry, Northern Ireland, 7–8 September 2000, Centre for Marine Resources and Mariculture, Portaferry, Northern Ireland.

Roy, J.J. (2005) 'Growing a future crop of aquaculturists: creating an urban aquaculture education programme in New Haven, Connecticut, USA' in B.A. Costa-Pierce, P. Edwards, D. Baker and A. Desbonnet (eds) Urban Aquaculture, CAB International, Wallingford, UK.

Schmidt-Puckhaber, B. (2000) 'Intensive fish culture in recirculation systems in Germany: fish farming at the highest level of productivity', Global Aquaculture Advocate, vol 3, no 3, pp. 52–53.

Scholtissek, C. and Naylor, E. (1988) 'Fish farming and influenza pandemics', Nature, vol 331, p. 215.

Schreibman, M.P. and Zarnoch, C.B. (2005) 'Urban aquaculture in Brooklyn, New York, USA' in B.A. Costa-Pierce, P. Edwards, D. Baker and A. Desbonnet (eds) Urban Aquaculture, CAB International, Warrington, UK.

Thanh Nien News (2005) 'Chicken excrement for fish stirs concern in Vietnam', Thanh Nien News Online, http://www.thanhniennews.com/2005/pages/200511717 10180.aspx

Torres, J. (2000) 'Fish culture in Spain: recirculating systems', Global Aquaculture Advocate, vol 3, no 3, pp. 77–78.

United Nations (2015) Transforming Our World: The 2030 Agenda for Sustainable Development, United Nations, New York.

van Rijn, J. (1996) 'The potential for integrated biological treatment systems in recirculating fish culture: a review', Aquaculture, vol 139, pp. 181–201.

WHO (2006) Guidelines for the Safe Use of Wastewater, Excreta and Greywater, Volume 3: Wastewater and Excreta Use in Aquaculture, World Health Organization, Geneva, Switzerland.

Willett, W., Rockström, J., Loken, B., Springmann, M., Lang, T., Vermeulen, S., et al. (2019) 'Food in the Anthropocene: the EAT–Lancet Commission on healthy diets from sustainable food systems', Lancet, vol 393, no 10170, pp. 447–492.

Yoonpundh, R., Dulyapurk, V. and Srithong, C. (2005) 'Aquatic food production systems in Bangkok', Urban Agriculture, vol 14, pp. 8–9.

7 Enhancing prospects for sustainable aquaculture development

Key points

The aims of this chapter on enhancing prospects for sustainable aquaculture development are to:

- Promote systems thinking and practice in support of sustainable aquaculture development and highlight the advantages of adopting a food systems framework approach.
- Assess approaches to stakeholder, institutional and market assessment in support of equitable and appropriate aquaculture development and aquatic resources management.
- Present and evaluate a toolbox of approaches, drawing on practical examples and published accounts that could be deployed to support and foster widespread adoption of sustainable aquaculture practices.
- Describe findings from an assessment on participatory approaches to selecting desirable traits and discuss prospects for gene editing innovations.
- Critically review application of participatory action planning, focusing on peri-urban wetland management in West Bengal, India, to facilitate joint assessment and decision-making.
- Introduce the seven principles identified to support participatory action planning (see Bunting et al., 2016) and discuss opportunities and constraints to wider adoption.
- Describe the STEPS (social, technical, environmental, political and sustainability) approach to testing feasibility with stakeholders.
- Review calls for consensus-based aquaculture sector planning and development and assess the potential of the stakeholder Delphi approach for consensus-building and testing the strength of agreement.
- Introduce preliminary findings from a stakeholder Delphi assessment on the use of small pelagic fish species from sub-Saharan Africa as feed ingredients (see Thiao and Bunting, 2022a, 2022b) and discuss the potential social, economic and environmental implications.

DOI: 10.4324/9781003342823-7

- Highlight promising approaches to joint financial and economic appraisals of aquaculture practices and review prospects for enhanced assessments of environmental impacts with stakeholders.

Systems thinking and practice for sustainable aquaculture development

Research and development activities commonly focus on technical aspects of aquaculture implementation, engineering design, management parameters, feed conversion rates and production efficiency, process water conditioning and treatment performance. Operators deciding whether to adopt such strategies, however, must consider a wide array of possible constraints and potential opportunities. Resource use efficiency and potential contributions to livelihood portfolios are major concerns for subsistence and small-scale producers, whilst financial viability and market demand are critical for commercial operators.

Most rural and peri-urban livelihoods are typified by a diversity of income-generating, subsistence and coping activities, often including common property resource exploitation.

Owing to investment costs and potential revenues associated with selling aquatic products versus the cost of staple foods, it might be assumed wrongly that subsistence-level households are not engaged in aquaculture and that small-scale producers are not motivated by commercial interests. A continuum of practice, in constant transition and flux exists, however, with subsistence producers contemplating small-scale commercial activities orientated towards expanding and increasingly accessible markets and commercial producers faced with disease problems or poor prices switching to traditional and extensive production modes, as demonstrated by shrimp farmers in the Mahakam Delta, East Kalimantan, Indonesia (Bunting et al., 2013). Decisions over where to invest household time and resources and whether to focus on meeting subsistence needs or labouring and market opportunities will be moderated by the prevailing or foreseen poise in the 'equilibrium of survival' discussed by Professor Amartya Sen (2000: 164). In extreme cases, income generated from livelihood activities such as hunting and gathering, cultivating cash crops or small-scale business may be insufficient to pay for adequate supplies of staple food. Fishermen, for example, were singled out by Professor Sen as a group that faced extreme hardship during the famine of 1943 in Bengal, India.

Checkland and Poulter (2006: 7) noted that 'the complexity of real situations is always to a large extent due to the many interactions between different elements in human situations' and that given 'systems ideas concern interaction between parts which make up a whole', it is 'not surprising that systems ideas have some relevance to dealing with real-world complexity'. These authors identified four traits that characterise conventional hard systems. To persist over time and adapt to a changing environment, systems must exhibit 'communication processes' (trait 1) to discern change and 'control processes' (trait 2) to adapt accordingly; be composed of sub-systems or be considered a sub-system of a larger entity, thus comprising a 'layered structure' (trait 3); and display 'emergent properties' (trait 4) relating to the functioning of the system as a whole.

Systems thinking was central to conventional agricultural development in the second half of the twentieth century. Spedding (1988: 18) described a system as:

> a group of interacting components, operating together for a common purpose, capable of reacting as a whole to external stimuli: it is unaffected directly by its own outputs and has a specified boundary based on the inclusion of all significant feedback.

Aquaculture systems were defined as 'interdependent elements of environment, physical structure, rearing process, resource use and socio-economic context which form the entity from which farmed aquatic organisms are produced' (Muir, 1996: 19). Both definitions testify to the fact that systems should be regarded as dynamic and complex entities, composed of sub-systems and having a discernible boundary.

Principles of a 'systems approach to management' for aquaculture were identified by Muir (1996), who provided a comprehensive overview of essential elements requiring consideration. Components, or systems features, deemed to interact within aquaculture systems included ethical, cultural, socio-political, economic, climatic, agro-ecological, production, farming, resource use, mass balance, ecological and biological factors. Preliminary steps in applying a systems-based approach include defining culture practices to be assessed and delineating systems boundaries based on physical areas occupied, comprehensive ecological footprint assessments or cradle-to-grave product evaluations, including inputs, by-products and waste disposal. Subsequent steps include exploring relationships between intra-system elements, leading to context and explanation, assessing functions and controls governing systems behaviour and describing implications and impacts of outcomes (Muir, 1996). Subsequent synthesis of these elements should lead to greater comprehension of a complete systems context, permitting aspects subject to prioritisation and choice to be elaborated and associated outcomes predicted.

Conceptualising aquatic food systems

To help operationalise a systems-based approach, the High-Level Panel of Experts convened to promote sustainable food systems globally have developed a conceptual framework (HLPE, 2020). This presents a comprehensive and systematic scheme visualising interconnected sub-systems and aims to transform thinking and practice away from a production centric approach to focus instead on what motivates consumers and governs the

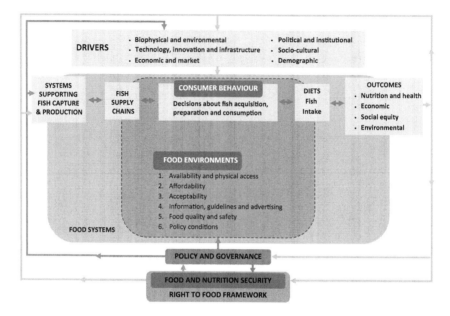

Figure 7.1 An adapted version of a food systems framework (HLPE, 2020), positioning fish supply chains and the contribution of fish to diets within the wider food system. This review collates evidence of aspects of the food environment which influence fish acquisition and consumption by populations in the African Great Lakes Region.

Source: de Bruyn et al. (2021).

types and timing of food items that different groups of people have access to geographically and temporally and what influences the diets of individuals, households, communities and society. High-level drivers including bio-physical, environmental, demographic, infrastructure, innovation, market, economic, socio-cultural, political and institutional factors largely govern food system configuration and operations. Analysis at a local or household level demonstrates, however, that the choices made about what aquatic foods people buy and consume depends on what items are available and accessible and an array of factors within their external and internal food environments (see Figure 7.1; de Bruyn et al., 2021).

Drawing on the sustainable livelihoods framework (DFID, 2001; chapter 1) greater appreciation of the vulnerability context could help account for adverse shocks and negative trends that might exacerbate food and nutrition insecurity. Cross-cutting food system issues to be borne in mind include health and safety at work considerations, animal welfare, feed and food safety, food and nutrition security outcomes, human rights, environmental impacts and economic and social outcomes of food systems. It is also imperative that food systems are governed, regulated, organised, configured and supported to maximise the contribution of aquatic foods to sustainable and healthy diets globally (see chapter 1; Willett et al., 2019; Golden et al., 2021). Innovations to manufacture alternative feed ingredients and develop novel industrial closed-loop and bioeconomy sectors – for example, the 'shell biorefinery' (Yan and Chen, 2015; chapter 2) – could potentially have unintended negative outcomes (e.g. economic, environmental, equity, health or nutrition) for poor and marginal groups and, consequently, careful analysis and social safeguards may be needed to protect the human rights and legitimate interests of vulnerable stakeholder groups (de la Caba et al., 2019).

Stakeholder analysis

Acknowledging the right of stakeholders to participate in joint assessment of and decision-making and planning for the management of aquatic resources and aquaculture development, a logical progression is to query how to identify the concerned stakeholders. First, it is necessary to define and delineate system boundaries. These might be established geographically as a watershed or irrigation command and control area or be based on administrative and national boundaries. Systems boundaries might even extend to a global frame and consider international treaties, export markets and global biodiversity concerns and non-use and existence values. Next, it is useful to consider the roles and responsibilities of people who may be considered as stakeholders. Guidance provided by the UK Government's Department for International Development (DFID, 2001) identified two stakeholder categories: primary stakeholders who may be impacted positively or negatively by a proposed intervention or development activity and key stakeholders in a position to influence or facilitate implementation and affect outcomes.

Assessment of stakeholders associated with the East Kolkata Wetlands, India, through brainstorming during a joint stakeholder workshop, permitted identification of twenty-nine stakeholder groups; this led to a great deal of discussion concerning their respective roles and interactions (Bunting et al., 2001). Subsequent analysis employing the DFID conception of primary and key stakeholders revealed a larger number of key stakeholder groups (Table 7.1). However, such a categorisation is not straightforward as it revealed that stakeholders can have several positions, and within key or primary categories there are hierarchies and multiple perspectives, motivations and agendas; individual stakeholder groups are not homogenous.

The wives of fishermen were considered the poorest group during subsequent assessment exercises by workshop participants, emphasising the need for disaggregated and gender-, age-, wealth-, class- and ethnicity-sensitive assessments. Intra-household differences in power, control of resources and decision-making are often ignored or overlooked in routine appraisals, but these can have significant consequences for the success or otherwise of proposed aquaculture development.

Table 7.1 Groups associated with the East Kolkata Wetlands categorised as primary or key stakeholders

Primary stakeholder groups	Key stakeholders
Poor	Advocates for the poor
Land owners	NGOs
Fish, vegetable, rice and livestock producers	Department of Irrigation and Waterways
Tannery operators and polluting activities	Kolkata Metropolitan Authority
Kolkata city/society[a]	West Bengal Pollution Control Board
Wholesale markets, market owners and retailers[a]	Department of Fisheries
Processors	Department of Environment
Seed traders	Bankers and money lenders[a]
People in flood-prone areas	Developers; e.g. West Bengal Industrial Development Corporation
Consumers[a]	Scientists
Poachers	Taxpayers[a]
Country liquor producers	Police
Farmers and growers downstream	Wildlife stakeholders; e.g. Ramsar Convention
	Politicians
	Planners; e.g. Housing Infrastructure Development Corporation

Note
[a] Groups that could reasonably be classified as either primary or key stakeholders.

Central to soft systems methodology (SSM) thinking and practice is the CATWOE mnemonic elaborated by Checkland and Poulter (2006: 41) as a framework around which to structure a 'generic model of any purposeful activity':

- C – customers who would be affected positively or negatively by a proposed transformation;
- A – actors who would undertake or facilitate the purposeful transformation activity;
- T – transforming process;
- W – worldviews held by people based on learning, beliefs and assumptions which are varied and can conflict;
- O – owners of the process, who could influence, curtail or stop purposeful activity;
- E – environment in which the transformation takes place.

This more nuanced approach to assessing the different roles and responsibilities of people in a given situation could assist in understanding what governs transformations and influences worldviews and how these affect purposeful activity. SSM is a marked departure from traditional systems-based thinking and practice, however, with the focus instead placed on establishing systems of enquiry to gain insights into perceived problem situations. CATWOE-guided assessment of purposeful activity provides a richer picture or 'root definition' of the 'activity system' giving rise to observed or proposed transformations (Checkland and Poulter, 2006: 40). The authors proposed several performance measures that might be applied to the process, centred on the efficacy, efficiency and effectiveness of the transformation, whilst noting that elegance, ethicality or other criteria might be applicable in a given context.

Institutional assessment

Knowledge and understanding of the position, remit and influence of institutions as stakeholders in aquaculture development are crucial, yet they are often overlooked. The influence of policies, institutions and processes (PIPs) was highlighted by DFID within the livelihoods assessment framework (DFID, 2001: section 4.11), but it was noted that 'current institutional appraisal methodologies and checklists tend to focus on national and sector levels and are not particularly well suited to being used with a specific group of stakeholders, or in a defined geographic area'. Whether policies, institutions and processes 'promote social sustainability and create an overall enabling environment for *sustainable* livelihoods' was seen as fundamental and a prime focus for any institutional assessment (DFID, 2001: section 4.11). The importance of 'an enabling framework of support' provided by governments and NGOs for 'self-mobilisation' in support of sustainable agricultural development was noted earlier by Pretty (1995: 1252). He pointed out that, for rural and agricultural institutions to meet the challenge of becoming 'learning organizations', they should 'improve learning by encouraging systems that develop a better awareness of information' (Pretty, 1995: 1258) and that such organisations would by necessity be decentralised and respond to farmers needs with multidisciplinary and heterogeneous outputs.

It is sensible to try and build constructive rapport with key and primary stakeholder representatives early in the institutional assessment process to jointly review the situation. Meeting face to face often proves most rewarding, but a good deal of preparation is required to get the most from such opportunities. Secondary data, supplemented with new knowledge garnered from meetings and discussion, can be used initially to better define the roles, responsibilities, relationships and influence of selected institutions. Legislation, planning frameworks and principles that govern decision-making should be identified and reviewed with regard to the proposed focus for development activity. Concerning the current status and future prospects of peri-urban farming, notably aquaculture, a series of supplementary areas for enquiry were identified to guide a process of institutional assessment (Edwards, 2002), including:

- classify peri-urban farming activity, zoning, relevant legislation and references to peri-urban farming in official documentation;
- discern the priority of peri-urban farming with respect to other activities;
- elicit perceived constraints to peri-urban farming and investigate initiatives to address these problems;
- assess perceived benefits of peri-urban farming;
- discuss historical land-use change that has impacted peri-urban farming and future prospects in light of planning initiatives and state-level development objectives and foreign investment.

Synthesis of findings, mainly derived from face-to-face meetings and semi-structured interviews, consisted of describing key stakeholder roles in managing peri-urban natural resources and relationship to the formal land-use planning process, including their jurisdiction, sphere of influence and position in the overall hierarchy. Committees and institutions governing planning policy were of particular interest, whilst assessment of the management and planning hierarchy extended to landowners, leaseholders, cooperative management bodies and employees and their power, influence and roles. Prospects for strengthened producer associations and more informed advocates for the poor and local planners in formulating appropriate management and development strategies for peri-urban natural resources were considered. Information on the knowledge needs of key stakeholders was used subsequently to identify appropriate communication pathways and materials (chapter 8).

The SWOT (strengths, weaknesses, opportunities and threats) framework has been applied liberally to aquaculture development to guide assessments concerning the performance of project concepts, innovations, strategy plans or business models. Guidelines for aquaculture sector planning and policy development published by FAO (2008: 13) under 'Theme 1: Policy Formulation Process' specifically note that to enable the aquaculture sector to 'develop optimally and sustainably', it may be advisable to 'proactively identify opportunities and constraints to aquaculture development resulting from policy gaps and failures (e.g. via a strengths, weaknesses, opportunities and threats (SWOT) analysis, risk assessment or other analytical method) and address these through policy change'. Subsequently, SWOT analysis was conducted on aquaculture policy and national planning process in several Southeast Asian counties and for the region generally (FAO, 2010).

It was noted that political stability, a robust government framework and a strongly supported aquaculture development plan constitute major strengths in Vietnam (FAO, 2010). Conducive environmental conditions and well-developed human capacity, combined with international market potential and existing collaboration, present major opportunities. Important weaknesses identified were limited financial and infrastructure provision, poor information access to inform policy formulation, low levels of producer education and limited stakeholder participation in policy formulation. Threats were identified, several of which were common to aquaculture policy and planning procedures in other Southeast Asian counties, notably climate change, environmental degradation and competition in global markets.

SWOT analysis of production systems, management practices and strategies and aquatic food systems could also be informative and guide future research, innovation and decision-making. An assessment of diversified shrimp–rice agroecosystems in Bangladesh and West Bengal, India, highlighted specific opportunities concerning established markets and value chains, existing strong social networks to mediate information exchange and changing hydrological regimes associated with climate change that could extend the area suited to culture (see Figure 7.2; Bunting et al., 2017). Key

- prevailing typography and hydrology across large areas suited to shrimp–rice culture
- knowledge of shrimp–rice culture is widespread
- private sector input suppliers and service providers established
- institutions and mechanisms for credit provision established
- education and extension services cover integrated aquaculture-agriculture systems
- national-level policies support integrated aquaculture–agriculture systems development

Strengths: existing or potential resources or capability

- inadequate planning and lack of policy support may prevent expansion of shrimp–rice culture in promising areas
- economic and institutional barriers may prevent reconfiguration of existing infrastructure
- failure to counter saline intrusion may force producers to convert to less diverse, high-salinity cropping strategies
- sector-specific research, development and innovation may undermine synergistic effects and not match producer needs

Weaknesses: existing or potential internal force that could be a barrier to achieving objectives/results

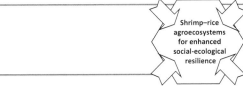

Shrimp–rice agroecosystems for enhanced social-ecological resilience

Opportunities: existing or potential factors in the external environment that, if exploited, could provide a competitive advantage

- established markets and value chains with capacity to handle increased supply
- international agreements and national policies support responsible aquaculture development
- climate change–induced hydrological regime shift may extend area for shrimp–rice culture
- evidence that strong social networks and efficient information exchange promote resilience should inform policy and practice to support existing and new producers
- promotion of better management practices from existing systems could aid successful adoption in promising new locations

Threats: existing or potential force in the external environment that could inhibit maintenance or attainment of unique advantage

- land-use policy and adaptation (climate change, salinisation) strategies fail to capitalise on potential benefits of diversified shrimp–rice agroecosystems
- irrational objections to converting low-yielding cereal crop land to more productive and financially rewarding diversified shrimp–rice agroecosystems
- reticence by development agencies to sponsor initiatives involving shrimp culture as it may be perceived as only occurring in a damaging monoculture

Figure 7.2 SWOT assessment of diversified shrimp–rice agroecosystems development.
Source: Bunting et al. (2017).

weaknesses include inadequate planning and policy support in anticipation of expansion into promising new areas, barriers (economic and institutional) to systematic reconfiguring, failure to counter saline intrusion inland where it could restrict the diversity of crops grown and sector-specific development, innovation and research that could fail to capitalise on synergies and not meet producer needs (Bunting et al., 2017).

Indigenous groups, traditional resource use systems and community-based organisations are often assumed to be endowed with an automatic ability to establish and maintain sustainable and wise-use practices, but arguably they suffer from the same deficiencies as other institutions. Moreover, they are often poorly resourced and not attuned to meeting contemporary and emerging challenges of industrialisation, globalisation, population growth and movements (seasonal and permanent migration), escalating multiple use, invasive species introductions and climate change. Furthermore, novel threats constitute major challenges for all institutions on multiple scales and transcend administrative, sector and geographical boundaries.

New institutions are being widely promoted, notably self-help groups, clusters of farmers, producer associations and co-management bodies. Noteworthy opportunities and constraints associated with emerging institutions tasked with managing common property resources and overseeing conformity with product certification standards, for example, were reviewed in chapter 3 and chapter 8. Institutions, including newly formed ones, do not exist in isolation. It is vitally important that institutional analyses address the hierarchical nature of institutions, especially within government systems, and identify areas of common ground and mutual support, replication of effort and inefficiency and competing agendas and conflict. Many institutions, and by default senior individuals within institutions, have political, religious or ideological affiliations which may consequently prohibit joint working, even preventing people from attending the same meeting or sharing a platform with others. Patronage, personal relationships and affiliations to common groups can further complicate and influence the institutional landscape. Furthermore, illegal and immoral activity and relationships perhaps play an often-underestimated role in governing what happens. Venn diagrams, such as the one in Figure 7.3 depicting stakeholder groups associated with mangrove management in the Mahakam Delta, East Kalimantan, Indonesia, help to visualise institutional arrangements, affiliations and support networks and gaps where stakeholders fail to communicate or avoid working together.

Institutional analysis applied to international agreements and treaties has proved revealing and instructive. A landmark case was presented through the Millennium Ecosystem Assessment (MEA, 2005) in which it was clearly shown that there are inconsistencies and trade-offs inherent in implementing some of the most significant global development and environmental protection legislation and policy. Two examples from aquatic resources: development of hydro-electricity schemes to achieve renewable energy targets considering climate change is thought to be having major impacts on provisioning services, thus

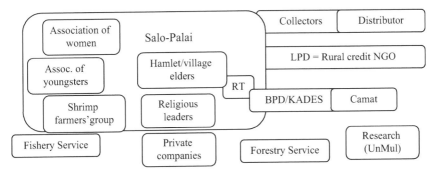

Figure 7.3 Venn diagram showing institutional relationships between mangrove affili-
ated stakeholders in Salo-Palai village, Mahakam Delta, East Kalimantan,
Indonesia.

Source: Bosma et al. (2007).

threatening food security, and measures to address food insecurity through
aquaculture development predicated on imported species and technology will
have negative consequences for aquatic biodiversity. Thus, the matrix developed for the MEA (2005) assessment is a useful approach to tease out
unintended consequences of policy and legislation developed for aquaculture
and to identify inconsistencies and contradictions.

Market demand assessment and network analysis

Secondary data concerning national or regional production rates and trends for
selected species, combined with information on supply from other areas and
capture fisheries landings, can provide a great deal of insight concerning
prospects for further aquaculture development or opportunities to culture new
species. Reviewing data on species with potential for inclusion in integrated
production systems in Europe assisted in identifying high-potential candidate
species for culture (Table 7.2). Such an analysis also permits areas of
uncertainty requiring further assessment to be highlighted. Subsequent enquiry
may be required, for instance, to explain observed production trends that may
be associated with disease problems or to understand the consequences of
changing capture fisheries landings on demand for aquaculture products.

Focus groups are often convened to test consumer reaction to novel
products and gauge opinion amongst specific demographic groups concerning environmental and ethical questions. Seafood eaters were purposefully selected to form focus groups to deliberate the relative merits of
aquaculture products from horizontally integrated systems within the EU-
funded GENESIS project. It transpired that these consumers might be willing
to pay 20–30 per cent extra for products from GENESIS-type integrated
systems (Ferguson et al., 2005). Focus group discussions yield a wealth of
information and insights concerning more general concerns and opinions of

Table 7.2 Trends in European aquaculture and capture fishery returns for species shown and an indication of production/market potential

Species	Aquaculture* ($t\,y^{-1}$; 2001)	Trend (% y^{-1}) (1992–2001)	Ex-farm value* (€ kg^{-1}; 2001)	Fishery* ($t\,y^{-1}$; 2001)	Trend (% y^{-1}) (1992–2001)	Production/ market potential
Primary species						
Seabass (*Dicentrarchus labrax*)	42,750	20	6.36	8318	3	++++
Seabream (*Sparus aurata*)	65,770	27	5.35	5709	1	+++
Turbot (*Scophthalmus maximus*)	4856	12	9.86	7253	–5	+++++
Sole (*Solea solea*)	60	14	11.84	40,524	–5	+++++
Tiger prawn (*Penaeus semisulcatus*)	0	–	–	–	–	++++
Biofilter I						
Phytoplankton	–	–	–	–	–	++
Biofilter II						
Sea urchins	0	–	–	1869[a]	–13	+++
European flat oyster	6714	4	4.3	953[b]	–19	+++
Japanese carpet shell	56,405	9	2.8	3193[c]	–17	++++
Pacific cupped oyster	133,957	0	1.85	124[d]	–24	++
Hard clam	0	–	–	11,460[e]	1	++
Polishing unit						
Samphire (*Salicornia* spp.)	–	–	–	–	–	+++
Seaweed	3518	–6	3.76	343,035[f]	–1	++

Source: Bunting (2005).

Primary data source:
*   Fishstat+ (FAO, 2003)
[a]   European edible, stony and miscellaneous sea urchins;
[b]   European & miscellaneous flat oyster;
[c]   grooved and pullet carpet shell;
[d]   Pacific and miscellaneous cupped oyster;
[e]   donax, hard, razor, solid surf, Venus and miscellaneous clams;
[f]   brown, green and red seaweed, kelp, eelgrass and miscellaneous.
Production/market potential: + very poor; ++ poor; +++ reasonable; ++++ good; +++++ very good

consumers and help to highlight contradictory or contrary points of view. Major challenges are how to capture the richness of exchanges in such discussions, to ensure that participants are representative of the required groups and that the findings are analysed and represented in a systematic manner. Logistics and costs to organise and facilitate focus groups may constitute further constraints. Similar explicit and concealed forces that preclude and interfere with interactive stakeholder participation such as deference, subjugation, bias and variation in inter-personal and social interaction will distort and restrict focus group discussions and temper confidence in outcomes.

Analysis of marketing arrangements with stakeholders throughout the processing, distribution and retail chain can reveal important information, perhaps explaining why certain types of culture or aquatic resources exploitation emerge or persist. Mapping market networks can help to understand the geography of market arrangements and identify the ultimate destination for products. Discussion prompted by such activities can give greater insight concerning patron–client relationships, motivations concerning species selection or product quality and opportunities for producers to capture a greater share of the true product value. Mapping market networks for mud crab (*Scylla serrata*) caught by members of three communities in Nakhon Si Thammarat, Thailand (Figure 7.4), demonstrated that a proportion of the catch was destined for provincial and export markets and consequently there might be opportunities for capturing a price premium for fishers through sustainable or responsible labelling schemes. However, major barriers to small-scale producers include the complex nature of the marketing network, combined with limited experience with traceability and

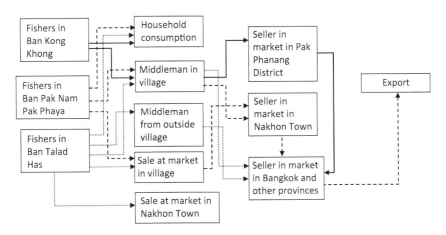

Figure 7.4 Market network maps for mud crab (*Scylla serrata*) originating from three communities in Nakhon Si Thammarat, Thailand.

Source: Dulyapurk et al. (2007).

knowledge of certification schemes locally and transaction costs associated with entering such schemes.

Differing perceptions of trait selection and genetic improvement strategies

Identification and selection of desirable traits in cultured animals and plants is crucial for successful and sustainable aquaculture development. Analysis using state-of-the-art choice experiment software in Bangladesh and India demonstrated that female and male smallholders culturing rohu (*Labeo rohita*) may have different preferences, but generally they were similar for the most important traits tested (Mehar et al., 2022). Variations were noted between producers in different countries. In Bangladesh, preference was given to enhanced growth rates and shorter culture periods, whilst in India the preference was given to producing 'larger size fish whether by length, weight and value' (Mehar et al., 2022: 9).

Certain traits may be appropriate for inclusion in genetic improvement programmes, whilst others (e.g. freshness) might be enhanced most effectively through careful pre-harvest strategies and post-harvest handling. Broodstock can be chosen based on the identification of a selectable trait. Formal breeding programmes may, however, need to assess and manage trade-offs among several desirable traits. Genetic markers and genomic technologies can contribute to designing and monitoring of breeding programmes to promote the expression of desirable traits and enhance production efficiency and disease resistance in cultured organisms (Nguyen et al., 2022). Application of genome editing has been used to produce fast-growing fish, but this may be at the expense of other desirable traits and consumer preferences.

Joint assessment and decision-making

The following sections present a review of participatory approaches to developing action plans for enhanced management of aquatic resources, a complementary approach to feasibility assessment and a commentary on consensus building, centred on application of the stakeholder Delphi. Striving for interactive participation where 'people participate in joint analysis, development of action plans and formation or strengthening of local institutions' (Pretty, 1995: 1252) is considered critical to ensure that development is sustainable in situations where 'uncertainties are high' and 'problems are open to interpretation' (Pretty, 1995: 1247).

Participatory action planning and implementation

Developed as a workshop-based consensus building and planning approach, PAPD (participatory action plan development) works by building mutual awareness of livelihoods and concerns amongst different stakeholder groups.

PAPD consisted of a fixed sequence; first, a problems census across different stakeholder groups; second, problem-solving in separate groups and plenary sessions; third, feasibility analysis; and fourth, commitment to act and develop an implementation committee. Intense facilitation is required during the workshop. Influential stakeholders such as local politicians and community leaders are invited at decisive points to witness the proceedings, adding gravitas to the planning process. Further information regarding the evolution of PAPD and situations where it has been used is presented elsewhere (Lewins et al., 2007). Application in Bangladesh centred on promoting a supportive activity within larger projects dealing with natural resources management involving predefined objectives and activities. According to Bunting and Lewins (2006: 21), 'Consensual planning of this type aims to highlight the inter-connectedness of floodplain stakeholders and the opportunity for simple interventions that cross-cut their needs and concerns'.

Planning water resources management in the East Kolkata Wetlands (EKW) necessitated PAPD adaptation as the system is imbued with the complex physical, social and institutional character of peri-urban[1] environments and communities. Conventional, comprehensive and technocratic planning processes are not suited to identify and address the needs of poor and vulnerable groups, especially in complex interface settings. Moreover, limited knowledge and awareness of alternative planning approaches and fear amongst authorities that 'wider involvement is less controllable, less precise and so likely to slow down planning processes' (Pretty, 1995: 1252) more than likely perpetuated continued reliance on established, yet unrepresentative approaches of poor repute. Given such concerns, research was conducted in the EKW to generate new knowledge of participatory action planning in peri-urban, land–water interface settings that would help to reconcile multiple demands and benefit poor groups. Contrasting with previous PAPD application in Bangladesh, both the context and objectives were markedly different. A brief overview of how PAPD was adapted to the EKW context is presented in Box 7.1.

Disaggregating the EKW into eleven regions based on discernible physical, environmental and social–political characteristics permitted better representation in the planning process. Joint meetings were convened initially in each region to elicit and prioritise constraints and discuss potential solutions. Further meetings with representatives of key stakeholder groups, notably government departments, producer associations, labour unions and political parties, raised awareness of the initiative and led to constructive dialogue on desirable and feasible water management enhancements. Periods in the planning process when representatives of political parties and producer associations could consult with their members and seek broad-based support, consolidate their position and work to refine proposals were critical. Having ascertained three to four potential water management–related interventions per region, three preliminary development activities were identified that had potential to benefit people from different sectors and across several regions.

Box 7.1   Adaptation of the PAPD process to the EKW context

Rather than operating horizontally to develop understanding and agreement between local stakeholders, the intention was to explore a planning process that could result in feasible pro-poor actions with government support and facilitation. This necessitated a mechanism to 'report back' local-level issues and suggestions to intermediaries such as farmer associations, community-based organisations and government institutions. Contributions to planning by the poor, as primary stakeholders, were provided at distinct periods within the process but not continuously. The planning process was not conducted specifically to build consensus but to develop feasible pro-poor plans with the required backing and to test the suitability of the pilot process to Kolkata and other peri-urban interface settings. The scale and complexity of the EKW production system provided special problems regarding proper representation and the identification of potential actions that could benefit the range of poor stakeholders simultaneously, without significant negative impacts on other users or livelihood functions of the system.

Discussions during initial project team meetings ensured that the project aimed to address a key issue identified by previous work as a significant constraint to sustaining poor livelihoods in the EKW; that is, water supply to peri-urban farming systems (Bunting, 2002). Poor households and individuals in the EKW engaged in farming-related activities experienced seasonal vulnerability, which appeared to be most closely related to insufficient access to water for irrigation and fish culture during the dry winter months, despite the continuous discharge of wastewater from municipal Kolkata (Punch et al., 2002). Siltation in the primary and secondary feeder canals compounded the problem, and coordinated action on behalf of all stakeholder groups, including rural and urban government agencies and poor communities, was urgently required. Acknowledging the need to consider the varying demands of the main primary stakeholder groups in a holistic approach to planning, it was decided to develop action plans that would address the main concerns and expectations of those groups dependent on aquaculture, horticulture and paddy farming. However, it was also acknowledged that there were differing problems, priorities and agendas between stakeholders in different geographical regions of the EKW; therefore, the wetland was divided into eleven distinct regions based on physical, livelihoods, social and institutional attributes (see Figure 7.5).[2]

Source: Bunting and Lewins (2006).

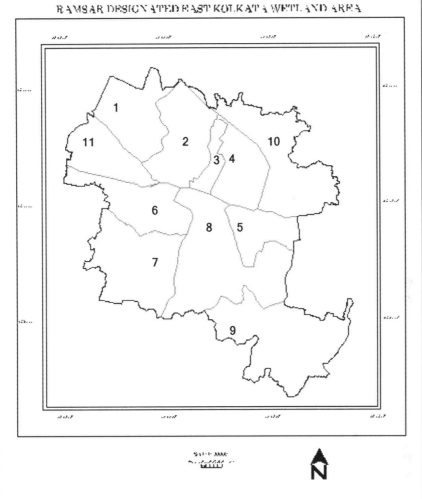

Figure 7.5 Extract from the East Kolkata Wetlands water management action plan.

Source: Bunting et al. (2005b).

The feasibility of these activities was assessed in a workshop convened by local stakeholders, described in further detail below. Supplementary focus groups were organised with women and other groups, notably *panchayat* or village council members, who had not participated in workshops to verify possible impacts.

Participatory action planning described here resulted in a widely supported water management action plan for the EKW that addressed constraints faced by communities across eleven wetland regions and highlighted promising solutions identified by stakeholders. Working in this way permitted various resource user groups (i.e. fish, rice and vegetable farmers and relevant community-based organisations and government agencies) to jointly identify major constraints to water resources management in the wetlands, propose potential solutions and formulate a plan of action to implement change. Both the East Kolkata Wetlands Management Authority and ADB and DFID-sponsored Kolkata Environmental Improvement Programme capitalised on the resulting action plan to guide investment and action to renovate and reinstate the primary and secondary canal network serving the EKW.

Feasibility assessment

Given the multitude of worldviews surrounding proposed interventions in any real-world situation, Checkland and Poulter (2006: 11) noted that this necessitates 'finding possible changes which meet two criteria simultaneously', namely, being both desirable and feasible. Assuming that outcomes of constructing models of purposeful activity point to desirable transformations for the stakeholders involved, it was stated that these 'must also be culturally feasible for these particular people in this particular situation with its unique history and the unique narrative which its participants will have constructed over time in order to make sense of their experience' (Checkland and Poulter, 2006: 11).

The STEPS (social, technical, environmental, political, sustainability) framework was used for feasibility testing of action plans formulated with stakeholders of floodplain fisheries enhancement in Bangladesh and water resources management in the EKW, India (Bunting et al., 2005b; Lewins et al., 2007). Critical elements of the workshop-based STEPS method included arranging joint meetings at convenient and neutral places, inviting all those with a stake in the proposed activity. Workshops commenced with reiterating objectives for the meeting. For the EKW this included demonstrating to government departments and potential investors that stakeholder groups were active and committed to working together to solve problems. Moreover, by focusing initially on feasible activities, it was anticipated that it would be possible to demonstrate immediate and beneficial progress, thus building confidence and trust and establishing mechanisms and a way of working to tackle intransient and difficult problems.

Detailed objectives outlined to participants at action planning meetings for water management in the EKW included testing the feasibility of proposed pilot activities (using STEPS); developing a better knowledge and database to assist implementation; reaching agreement on what happens next (when, how, why, who is responsible) and determining how to monitor progress and revise the plan as required. Care was taken to ensure that the role of the research team to facilitate and document the process was explained. Participants were invited to introduce themselves and to state their interest in the proposed water management intervention. An important verification step is to request participants to identify stakeholder groups that are not represented at the meeting, for whatever reason. Guidelines developed for STEPS-based assessment in the EKW included an outline for group work during the workshop that it was envisaged would be applicable to test the feasibility of future activities proposed by stakeholders (Box 7.2). It was expected that participants would be in a position towards the end of the meeting to agree on what needed to be done urgently and what was required

---

Box 7.2   Description of PAPD workshop group working sessions

Request the stakeholders to form groups based on the region(s) in which they have a stake and then their interest in the pilot activity. During the group discussion make sure to get information on past planning initiatives on this topic (especially failures and conflicts), existing plans or ongoing work regarding this topic and associated data that could be used to support implementation. Remember to keep the discussion focused on the pilot activity and related matters. Then, mapping in groups, ask participants to:

- draw an outline map to show where the planned work is to be carried out and provide notes to annotate;
- include physical change required and technical/financial inputs needed;
- identify institutions and social groups that need to be involved or consulted or will be positively or negatively impacted and explore why and where to locate them;
- specify the location of 'biodiversity'-rich areas that should be considered or monitored during implementation and other potential positive and negative environmental impacts as a direct or indirect result of implementation;
- identify the extent of threats to continued usefulness and how to overcome them.

Make sure that each group gets a chance to present a summary of their deliberations or, using the map, provide feedback to the other groups on their behalf. Only permit others to make comments at the end of the group feedback session. Note any discussions and try to resolve any differences. Highlight areas of agreement or breakthroughs. Note new issues arising that should be added to the STEPS analysis.

Source: Bunting et al. (2005b).

in the short and long term. Associated reasoning should be documented and potential constraints and bottlenecks highlighted. Finally, it should be decided how to proceed with the required actions identified; agreement must be reached on who should take responsibility for individual tasks. Ideally, all those concerned will be represented at the meeting, but targeted follow-up may be required for certain points. Agreement should be reached on when tasks will be completed, and a process to monitor progress and revise the plan should be implemented as deemed necessary.

Ensuring participatory processes are inclusive, efficient and trustworthy

Building on experiences of facilitating integrated action planning and implementation in upland areas of China, India and Vietnam, it was possible to identify seven principles to guide comprehensive, inclusive, equitable and efficient participatory processes (see Table 7.3; Bunting et al., 2016). The primary objective of action planning here was to promote aquatic biodiversity conservation in vulnerable highland areas and wise use of aquatic resources and wetlands in support of sustainable livelihoods; hence, the first three principles focus on these. In another context, these principles could be modified to better represent the core objectives the process.

Principles 4 and 5 address 'stakeholder participation and representation' and 'gender and age', respectively, and aim to ensure that participatory processes are fully inclusive. This is crucial as women and children from poor and marginal communities are often missing from decision-making process that can affect them greatly (Bunting et al., 2016) yet could benefit disproportionally from enhanced access to nutritious aquatic foods to help avert micronutrient deficiencies and promote better health, developmental and life-course outcomes (Thilsted et al., 2016; Cohen et al., 2021; Simmance et al., 2021). The next principle concerns making the process efficient and timely, to encourage people to engage and minimise attrition, and ensuring that implementation can adapt to changing circumstances. Trustworthiness and accountability are covered by the last principle, and this is crucial when, for example, actions being planned for multifunctional floodplains and wetlands or shared waterbodies could have implications for disparate groups of stakeholders with very different status and power. A common understanding of the issues at stake and broad-based consensus on what can be done to address problems and enhance the situation can be crucial to achieving improvements and sustained change. Despite the best of intentions, potential barriers to achieving fully inclusive and representative participatory process have been noted (chapter 3; Bunting, 2010). It may be that targeted activities are required to engage with poor and marginal groups across food systems or to safeguard their interests (de la Caba et al., 2019; Thiao and Bunting, 2022a, 2022b; Bunting et al., 2023).

Table 7.3 Seven principles for integrated action planning and implementation

Domain	Principle
Biodiversity conservation	1 Ensure that aquatic biodiversity conservation is central to integrated action plan formulation and implementation; avoid over-harvesting, the introduction of invasive and non-native species and other threatening activities; and adopt the precautionary principle, integrated wetland assessment and management and better management practices when appropriate
Wise use of natural resources	2 Principle of wise use is respected with appropriate measures instigated to avoid exceeding the environmental carrying capacity for ecosystem services or disrupting environmental stocks and flows
Sustainable livelihoods and resilience	3 Promotion of sustainable livelihoods and well-being based on equitable and rational use of ecosystem services whilst maintaining environmental stocks and flows necessary to ensure continued ecosystem functioning and promote social–ecological resilience
Stakeholder participation and representation	4 Integrated action plan formulation and implementation founded on promoting self-organisation and adaptive management, with good stakeholder representation and interactive participation assured
Gender and age	5 Ensure joint assessment of needs and priorities with heterogeneous social groups disaggregated by age, gender and wealth
Efficiency, timeliness and adaptability	6 Integrated action plan implementation process should be as efficient as possible and adhere closely to mutually agreed time frames and deadlines and appropriate mechanisms to enable adaptive management and appropriate integrated action plan reformulation to be adopted
Trustworthiness and accountability	7 Implementation of actions should be consistent with integrated action plan and reflect adaptations, and the process should be transparent and trustworthy, with appropriate measures to promote accountability, whilst respecting the rights of groups and individuals

Source: Developed from Bunting et al. (2016).

Consensus building and testing

Consensus is often implied by the absence of dissenting voices or contradictory opinions. More refined and nuanced assessments, however, of the nature and strength of consensus between and within groups could potentially contribute to better targeted interventions, research or development assistance and point to issues demanding further explanation, practices requiring modification or perceptions, social norms and beliefs that constitute major barriers to aquaculture development. Joint assessment, in which different stakeholder groups come together to evaluate alternative development options and pathways, has been advocated in several settings. Such approaches are, however, susceptible to overt and hidden processes. The views of the more powerful or eloquent are likely to hold sway and those of the poor and reticent are unlikely to be expressed, although these may ultimately govern prospects for progress and sustainable development.

The Delphi method, developed by the RAND Corporation, USA, was proposed as a means of consensus building amongst experts concerning strategic military planning. Acknowledging problems with unequal power relationships and possible bias, the original Delphi approach was founded on four main assumptions:

- expert opinion is a valid input to inexact research areas;
- consensus amongst experts is more valid than the opinion of an individual;
- joint meetings amongst experts induce a follow-the-leader bias;
- ensuring the anonymity of participants compensates for inherent opinion biases.

Apparent advantages of the Delphi approach seemed to offer potential solutions to problems that constrain the interactive participation of stakeholders (chapter 3, Table 3.5). Moreover, it is proposed that such an approach has potential for facilitating joint assessment in a development context, given the iterative nature of the process and scope to facilitate interactive participation and sharing of information and perceptions. Assessments should aim to cut across social, sector and physical divides; possibly contribute to building consensus; and, perhaps more important, identify differences in opinion and areas demanding further assessment.

Consequently, a process guided by the principles of the original Delphi method has been proposed in which stakeholders should participate in characterising problems and joint assessment of potential solutions. A critical difference compared with restricting this proviso to expert opinion is the assumption that the knowledge and experience of stakeholders is a valid input to the research process. The resulting *stakeholder Delphi* offers a structured and iterative process that permits the degree of consensus to be qualified and areas of weak agreement to be highlighted for further investigation. Given the nature of the intended study, relevant stakeholders should be identified, and individuals and representatives for specific groups should be invited to participate as panel members for the duration of the assessment. The stakeholder Delphi process can be depicted as a sequence of activities and rounds (Figure 7.6).

Considering the range and geographic distribution of potential participants, a preliminary schedule should be developed to structure the assessment. The first phase of this is to draft the first-round questionnaire and pretest it with a selection of stakeholders. Attention should focus on ensuring that questions are easy to comprehend given the non-technical background or low literacy and numeracy levels of some stakeholder groups yet be specific enough to elicit meaningful and insightful responses. Ensuring that interdisciplinary perspectives are included when drafting the questionnaire helps

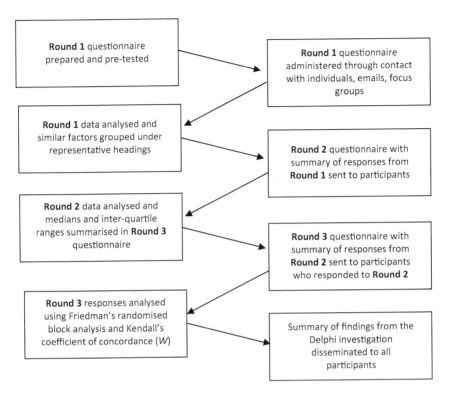

Figure 7.6 Schematic for the stakeholder Delphi process.

to ensure that a broad range of responses are elicited. A comprehensive array of potential areas for joint assessment were identified through the application of the stakeholder Delphi as part of the European Commission–funded HighARCS (Highland Aquatic Resources Conservation and Sustainable Development) project and invoking the DPSIR (drivers, pressures, state, impacts and responses) framework, including:

- threats to aquatic biodiversity (separated out as drivers of change and pressures);
- resulting status of highland aquatic resources;
- impacts resulting from pressures on highland aquatic resources;
- ongoing and proposed responses with potential to conserve aquatic biodiversity or permit wise use.

Open questions to elicit a wide range of responses on a number of strategic issues should be included in Round 1. Open questions encourage participants to specify constraints and potential solutions and areas for improvement from their particular perspective. Subsequent assessment should have heightened significance, encouraging continued engagement, whilst portraying the views of all relevant stakeholder groups. Moderation of Round 1 involves aggregating together similar responses (trying to retain nuances and descriptive elements) and recording the frequency with which similar (aggregated) responses occur. Round 2 involves presenting a summary of Round 1 outcomes to the stakeholder panel with their frequency of occurrence to give an indication of the relative importance of responses and request participants to rate them (e.g. on a scale of 1–10).

Analysis of Round 2 responses involves calculating the median and interquartile range for rates for each response and preliminary assessment of consensus employing Kendall's coefficient of concordance.[3] If strong agreement is found across all fields, there is no need to continue. If deemed necessary, during Round 3, medians and quartiles for rates are reported back to the panel and they are asked to either agree with the median rate or provide an alternative. When this lies outside the inter-quartile range, they are requested to provide a written justification concerning their reasons. Three rounds are generally regarded as sufficient to elicit a good measure of consensus. Following reassessment of the strength of agreement, a final report is presented back to the stakeholder panel. Critical areas demanding attention throughout the assessment include clearly specifying the schedule, reiterating deadlines for responses and actively following up non-respondents to try and minimise attrition.

Quasi-anonymity, when the identity of respondents is known only to those facilitating the process, should overcome barriers to participation such as subjugation, deference and leader and opinion bias. But this means that a requisite degree of trust is essential in terms of the integrity of those conducting the study and processes put in place to ensure anonymity. The iterative nature of the assessment encourages participants to agree with the median response or, where specific evidence or strongly held beliefs prohibit this, to provide statements qualifying responses that may in turn inform and influence other participants. Knowing the source of such statements may make them more or less influential. An essential component in assessing the significance of the assessment is to provide a summary of stakeholders involved, but care should be taken to maintain the anonymity of individual respondents where this has been specified. Ensuring participants' consent to have their names or roles disclosed would assist interested parties in evaluating the trustworthiness of the process and the degree of representation achieved. It may be difficult, however, to persuade certain stakeholders to engage in the process or agree to any sort of disclosure concerning their participation when the issue under consideration is location specific or politically sensitive or where prospective participants are wary of misrepresenting an official or received position or stance.

Conducting a Delphi study with stakeholders concerning prospects for enhanced wastewater management and horizontally integrated aquaculture, prospective participants were identified based on publications addressing the

Table 7.4 Friedman's $X^2{}_F$ at probability levels *(p)* indicated and Kendall's *W* for rank patterns in rates assigned to alternative strategies following Round 3

Subcategory for alternative strategies	Friedman's $X^2{}_F$	*(p)*	Kendall's W	Agreement	Confidence
Technological	33	<0.001	0.786	strong–unusually strong	high–very high
Managerial	32.1	<0.001	0.765	strong	high
Institutional	39.1	<0.001	0.697	strong	high
Socioeconomic	32.4	<0.001	0.579	moderate–strong	fair–high

Source: Bunting (2010).

issue of aquaculture wastewater or integration, farm management or aquaculture consultancy experience or responsibility for monitoring and regulating discharges. After pre-testing the Round 1 questionnaire with colleagues, it was sent to potential participants, of whom twenty-four replied. Notable outcomes included eighteen strategies for better managing aquaculture wastewater proffered by participants. The four highest-ranked strategies were distributed across managerial, institutional and technological subcategories. The highest socio-economic strategy was ranked eighth and agreement was strong to unusually strong concerning potential technological strategies following Round 3 (Table 7.4). Agreement over potential socio-economic strategies to enhance aquaculture wastewater management was fair to high, reflecting perhaps different social settings and economic realities encountered by participants in different locations or dealing with specific aquaculture systems.

The number of rounds required to reach a satisfactory degree of agreement or demonstrate a consistent degree of disagreement will vary depending on the nature of the study. It is important to inform participants at the start what conditions will be used when deciding to conclude the study, whether it is the degree of agreement, level of confidence or number of rounds, for example. Possible limitations include not having representatives from all stakeholder groups; there are logistical barriers and potential pitfalls to facilitating the inclusion of poor and vulnerable groups in the structured manner presented here. The degree of representation achieved and timing of investigation will influence outcomes, the nature of strategies proposed and subsequent ranking and degree of agreement achieved. Studies should commence with a preliminary stakeholder analysis to ensure as far as possible that the array of stakeholders invited to participate is comprehensive. Where necessary, supplementary steps should be included to elicit or verify the priorities and concerns of marginal groups.

Mechanisms to ensure accountability and transparency are indispensable. These include a neutral or trustworthy facilitator, peer review and sufficient information about participants to assess the degree of representation achieved and relevance of people involved. Demands and expectations of participants may become more personal and value laden, and prospects for high levels of agreement diminish as studies become more focused. The stakeholder Delphi process should make such views explicit, with people having to justify their position, thus giving rise to greater understanding regarding the differences between stakeholder groups. When groups are in open conflict, more specialist reconciliation and conflict resolution approaches will be required to initiate constructive dialogue. Although the stakeholder Delphi might have a role to play here, it has not been tested in such circumstances.

Knowledge concerning application of various Delphi-based approaches in aquaculture development is expanding and evolving. Conventional Delphi studies with experts assessed sustainability indicators for aquaculture in the southern USA (Caffey, 1998) and perceptions concerning the accuracy of screening for viruses (Bruneau et al., 1999). Prospects for stakeholder participation in Delphi studies have been reviewed (Bunting, 2001). Subsequently, indicators for enhanced policies concerning aquaculture service provision in India were assessed (Haylor et al., 2003). Drawing on these examples and knowledge of studies concerning aquaculture wastewater management and horizontal integration (Bunting, 2008, 2010), stakeholder Delphi studies have assessed threats to highland aquatic resources in Asia and opportunities for biodiversity conservation and wise use (Lund et al., 2014) and opportunities to better manage greenhouse gases emissions across product value chains for the UK marine aquaculture sector (Bunting, 2014a).

Priorities were identified for policymakers and future research to promote the rational use of small pelagic fish species (SPFS) in sub-Saharan Africa across eight countries in the African Great Lakes region (Malawi and Uganda), Central-West Africa (Congo, Ghana and Sierra Leone) and North-West Africa (Gambia, Mauritania and Senegal) (Thiao and Bunting, 2022a, 2022b). In this instance, the stakeholder Delphi was used to gather insights from sub-panels from each of the countries. The motivation for the study was to better understand the role of SPFS in the livelihoods of small-scale fisheries and artisanal processors of fish for direct human consumption (most of whom are women) and for animal feeds and to promote the rational use of this valuable natural resource (Thiao and Bunting, 2022a, 2022b). Over two rounds of the assessment, the strength of agreement amongst participants increased. This was encouraged by requesting participants to agree with median responses from the first round or to specify an alternative score. When this was outside the inter-quartile range, participants were requested to share a brief justification with the other panel members. The study was conducted in both English and French, and a French translation of the findings is available (Thiao and Bunting, 2022a).

Through the stakeholder Delphi process, it was possible to prioritise actions for decision-makers and for future research, and these are summarised in Table 7.5. In the context of the assessment, priority was assigned to recommendations that received a mean score above eight in the second

Table 7.5 Priority recommendations that received a mean score above eight from participants in Round 2 of the stakeholder Delphi assessment across eight countries in Central and West Africa and the African Great Lakes region

Recommendation
Implement an environmental audit for existing fish meal plants to check and monitor their capacity and level of enforcement of national norms/standards (DM)
Assist and train local fish and livestock farmers so that they can formulate and produce alternative and efficient feeds (DM)
Promote environmentally friendly and healthy/safe (for workers) fish meal production technologies (DM)
Ensure regular assessment of key-stocks of fish and effective monitoring of harvest and post-harvest activities/operations of the fisheries sector at a national level (R)
Conduct research to assess the chemical properties of all types of wastes from fish meal plants and their environmental and health effects (R)
Prohibit fish meal factories from dumping toxic wastes into the sea and inland waterbodies (e.g. lakes, rivers and wetlands) (DM)
Make sure that fish meal factories are constructed far away from towns and villages to avoid adverse impacts on residents (DM)
Promote the use of plant-based and/or insect-based protein as feed alternatives in national aquaculture and livestock sectors (DM)
Promote national research programs to identify alternatives to fish-based feed and assess their feasibility, viability, efficiency and profitability (R)
Assess and monitor fish (categorised by, e.g. size, species, source of production and means of processing/preservation) consumption, affordability and importance for food security and nutrition (R)
Implement and effectively enforce the policies and norms/standards specific to the fish-based feed industry (DM)
Ensure regular assessment of key stocks of fish and effective monitoring of harvest and post-harvest activities/operations of the fisheries sector at a regional level (African Great Lakes region or West Africa coastal zone) (R)
Promote better fish harvesting and post-harvesting methods to reduce by-catches being directed away from human consumption and used instead for fish meal and fish-based animal feeds production (DM)
Regulate and limit the number, capacity and production of fish meal factories based on the status of fish stocks and need for fish for human consumption (DM)
Assess the national/regional demand/need and affordability of fish-based feed for aquaculture and livestock sectors (R)
Define and introduce minimum price controls for fish that can be purchased by fish meal and fish-based animal feed producers to ensure more income for fishers and encourage fish availability for local consumers and processors (DM)

Source: Developed from Thiao and Bunting (2022a, 2022b).

Note: (DM) denotes recommendation for decision-makers and (R) denotes recommendation for future research for development investment.

study round. The highest priorities for decision-makers were to implement environmental audits of existing fish meal plants, assist local farmers in formulating alternative and efficient to use fish and livestock feeds and promote environmentally sound and safe to use fish meal production strategies. Considering research, the top priority was to ensure that key stocks of SPFS are assessed regularly and that associated harvest and post-harvest activities are monitored effectively.

Recognition is growing amongst practitioners and policymakers of the potential of Delphi-based approaches to:

* achieve interactive stakeholder participation;
* promote consensus-building, often amongst hierarchical and sometimes antagonistic stakeholders;
* facilitate shared learning to aid the reconciliation of multiple demands and better characterise areas of competition or conflict;
* improve planning and policy formulation for aquaculture development;
* assess stakeholder perceptions of emerging threats, notably invasive species impacts and climate change risks and adaptation options, when limited information precludes conventional assessments.

Brugere and Ridler (2004: 32) concluded that planning will 'be key to the sustainable development of aquaculture' in FAO Fisheries Circular No. 1001 titled *Global Aquaculture Outlook in the Next Decades: An Analysis of National and Aquaculture Production Forecasts to 2030*. They recommended adoption of a planning framework underpinned by application of the Delphi method. Guidelines resulting from the *Report of the Expert Consultation on Improving Planning and Policy Development in Aquaculture* (FAO, 2008: 14) specified that 'policy development based on consensus is desirable', in which case the stakeholder Delphi may have a role to play as these guidelines become more widely adopted. Although in some instances where joint assessment and participatory planning processes are established, application of the stakeholder Delphi approach may be seen as retrograde and cumbersome.

Joint financial and economic assessment

Promising approaches to joint financial and economic assessment of small- and large-scale sustainable aquaculture practices with stakeholders are reviewed below.

Bioeconomic modelling

Bioeconomic modelling is an approach to combine the assessment of production activities (e.g. aquaculture, horticulture, agriculture and fishing) with financial analysis of capital and operating costs and output values.[4] Bioeconomic models matching this description have been developed to assess various commercial and prototype aquaculture systems: combined culture of salmon and seaweed in Canada (Petrell et al., 1993); wastewater-fed aquaculture in the East Kolkata Wetlands, India (Bunting, 2007); horizontally integrated, land-based, marine aquaculture in Europe (Bunting and Shpigel, 2009); intensive shrimp culture with and without green-water technology, extensive shrimp–fish culture and mangrove–shrimp systems for the Mahakam Delta, Indonesia (Bosma et al., 2012); integrated mangrove–shrimp culture in East Kalimantan, Indonesia (Bunting et al., 2013); strategies across food systems to reduce greenhouse gas emissions from the UK marine aquaculture sector (Bunting, 2014b); diversified agroecosystems combining horticulture, fishponds and the System of Rice Intensification (SRI) (National Consortium on SRI, 2012; Uphoff, 2012) for enhanced resilience and agrobiodiversity in Buxa Tiger Reserve, West Bengal, India (Bunting et al., 2015); opportunities for culture-based fisheries enhancement in the Beijiang River, China (Liu et al., 2019); and cage culture of mixed-sex as compared to mono-sex stocks of Nile tilapia in Thailand (Bostock et al., 2022). A framework for bioeconomic modelling is presented in Figure 7.7. This builds on the above studies and is cognisant of the need for refined approaches in which conventional financial assessments might be better informed through incorporating assessment of livelihoods diversification opportunities and assets, ecological footprints and values and potential payments associated with ecosystem services.

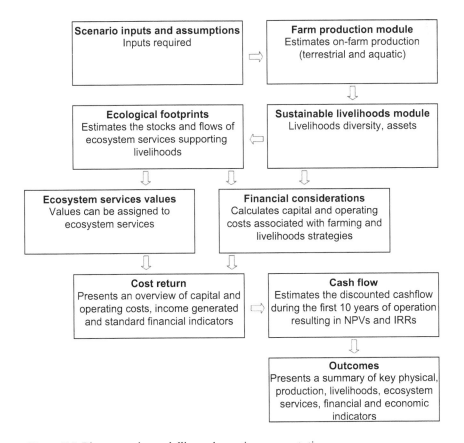

Figure 7.7 Bioeconomic modelling schematic representation.

Application of bioeconomic modelling to assess the potential financial performance of horizontally integrated, multi-trophic, land-based, marine aquaculture systems in Europe constituted a crucial aspect of the GENESIS (Development of a Generic Approach to Sustainable Integrated Marine Aquaculture for European Environments and Markets) project (Bunting and Shpigel, 2009). Bioeconomic modelling outcomes indicated that it would not be financially viable to establish integrated land-based culture of fish, phytoplankton and shellfish on the Atlantic coast of France (Table 7.6). Where producers are engaged in what might be considered traditional practices of fattening shellfish in earthen ponds, it might be financially viable to integrate phytoplankton production and overwintering of fish in this system. Capitalising on existing infrastructure and market opportunities, it

Table 7.6 Financial parameters and key performance indicators for the temperate prototype

Parameter		Baseline	No labour cost	No land or basin costs	Premium on production
Capital costs (€)					
Land		1267	1267	0	0
Basin construction		18,388	18,877	0	0
Infrastructure and materials		29,303	28,814	29,303	29,303
Electricity installation		6097	6097	6097	6097
Total (€)		55,055	55,055	35,400	35,400
Operating costs (€)					
Stocking and marketing		10,553	10,553	10,553	10,553
Feed		2186	2186	2186	2186
Fuel and electricity		1855	1855	1855	1855
Labour		2894	0	0	0
Maintenance		551	551	354	354
Total (€)		18,039	15,145	14,948	14,948
Income (€)					
	Fish	10,853	10,853	10,853	13,024
	Oysters	8199	8199	8199	9839
	Clams	574	574	574	689
Profit excluding depreciation (€)		1588	4481	4678	8603
Rate of return on capital costs (%)		2.9	8.1	13.2	24.3
Rate of return on operating costs (%)		9	30	31	58
Payback period (y)		34.7	12.3	7.6	4.1
NPV at:	5%	−41,685	−22,098	−2049	24,523
	10%	−41,737	−26,589	−7691	12,860
	15%	−41,286	−29,280	−11,373	4913
	20%	−40,546	−30,826	−13,787	−601
IRR (%) over:	10 years	–	–	3.6	19.4

Source: Bunting and Shpigel (2009).

appears possible to generate a reasonable internal rate of return (IRR) over 10 years (19.4 per cent).

Preliminary bioeconomic modelling assessment of integrated sea urchin, shrimp and *Salicornia* sp. production in Israel indicated that such a strategy could generate an IRR of 18 per cent over 10 years (Table 4.10). Subsequent scenarios tested using the model helped to identify critical aspects that could substantially increase financial returns, notably reducing urchin mortality and replicating *Salicornia* sp. yields achieved during trials under commercial conditions. Bioeconomic modelling represents a promising approach to structure financial assessments and to test potential future scenarios, but outcomes should be jointly assessed with prospective producers to account for technological and market constraints, financial and animal health risks and social and ethical concerns.

Household budgeting

Generally, it is assumed that it is not appropriate to ask people directly about the finances of their household as respondents may be unwilling to disclose such information. Understandably, people are concerned about making financial information public and what impact this might have on their status and perception by others or liability for taxes and other less transparent demands for payment. Consequently, it is supposed that financial data derived from even well-intentioned participatory rural appraisal activities would be incomplete, unreliable and misleading. Such presumptions, however, threaten to deny a great swathe of humanity the access to knowledge and insights regarding their own affairs and financial situation. Therefore, sensitive, appropriate and low-cost approaches to facilitate financial assessment of household activities and small-scale production are urgently needed to inform joint assessment and decision-making over allocation of livelihoods assets and livelihoods diversification opportunities, including aquaculture development.

Attempts to overcome barriers to financial assessments for households and small-scale producers should concentrate on building confidence in the process, avoid intrusive and extractive data collection and proceed in incremental stages. These would build trust and foster a mutually supportive environment. During preliminary meetings with households, a useful starting point would be to construct a profile of activities and assess their significance in the prevailing livelihood strategies. Subsequent meetings could be arranged to compile an inventory of capital costs necessary to undertake current livelihoods activities or implement proposed changes and operating costs required to sustain the various activities. Financial returns from productive activities then require assessment. Where goods and services are being sold, this could be achieved through recall, perhaps combined with logbooks or other appropriate means to record the magnitude and frequency of sales. Challenges arise when a significant proportion of cultured products and harvests from common property resources are consumed by households, given away or exchanged for other goods and services. Values can be assigned to items that are not sold based on local market prices where available or estimated by participants in the absence of such data, as is often required in the case of trees and livestock (Bosman et al., 1997; Moll, 2005).

Bioeconomic modelling was proposed as a potential means to facilitate the joint assessment of household budgets and future scenarios with people in highland areas of Asia that depend on aquatic resources in the EC HighARCS project. Livelihoods resource diagrams were developed initially to guide the assessment. It was essential that the approach adopted complement the assessment of livelihood assets with households and communities. Furthermore, the approach should contribute to evaluations of ecosystem services from highland aquatic resources and the assessment of potential impacts of action plans being formulated by stakeholders on livelihoods. Spreadsheet-based bioeconomic models were developed building on frameworks conceived previously for wastewater-fed aquaculture in the EKW (Bunting, 2007), integrated multi-trophic aquaculture in Europe (Bunting and Shpigel, 2009) and integrated shrimp–mangrove systems in Indonesia (Bunting et al., 2013).

Given the demands of assessing household budgets in highland areas of Asia, the main innovation was higher resolution in the modelling approach to deal with cash flows on a monthly rather than annual basis. A simple schematic representation of the model was produced to convey to stakeholders what was being assessed, how this was to be done and what application the modelling tool might have to assess future management and conservation scenarios. Based on information derived from household interviews and market assessment work, baseline scenarios were formulated for representative livelihood strategies. The sensitivity to changing input costs and market values of products was tested and future scenarios were explored to examine what implications these might have for household budgets. Scenarios centred on proposed elements of integrated action plans and addressed conservation, policy and livelihoods concerns, including enhanced catches from culture-based fisheries or stock enhancement programmes; increased revenue from catches and crop production certified as sustainable as a result of market premiums; reduced catches owing to closed seasons or gear restrictions; and additional revenue from appropriate livelihoods diversification.

Environmental assessments

Standardised tools such as life-cycle assessment (LCA) are emerging as potential means to assess a range of environmental impacts and permit potential trade-offs to be evaluated. This approach is critically reviewed in the following section and potential limitations are discussed. Environmental impact assessments are better suited perhaps to account for site-specific conditions with tailored assessments of local carrying capacity and stakeholder consultation. However, this approach is not necessarily appropriate for small-scale and poor producers. Consequently, participatory approaches to identify environmental impacts that might be associated with aquaculture development are considered and potential advantages in terms of identifying indirect social and economic impacts are assessed.

Life-Cycle Assessment

LCA can cover a range of potential impacts of industrial goods and services throughout the entire life-cycle of a product, and it has been applied to several aquaculture production systems. The life-cycle encompasses extraction of raw materials, manufacturing processes, consumption and final waste disposal by the consumer. Guidelines covering LCA principles, implementation framework and impact assessment have been developed by the International Organization for Standardization, namely: *International Standard 14040 Environmental Management – Life Cycle Assessment – Principles and Framework* (ISO, 1997) and *International Standard 14042 Environmental Management – Life Cycle Assessment – Life Cycle Impact Assessment* (ISO, 2000). Common LCA criteria for seafood production include abiotic resource appropriation, acidification potential, aquatic and terrestrial toxicology, biotic resource appropriation, energy use, eutrophication potential, global warming potential, human toxicity potential, ozone depletion and photochemical oxidant formation (Pelletier and Tyedmers, 2007). Not all sustainability criteria are necessarily covered by a particular LCA and to address specific concerns and needs others have been included, notably biodiversity measures and the Fish In: Fish Out (FIFO) ratio. Application of the LCA focuses on a functional unit of production. For food production this is often the weight of a product as presented to a customer; for example, 500 g of boned chicken breast or fillets of fish or a common measure of nutritional value (e.g. 100 g of protein or 100 joules of energy) to permit ready comparison across products and production systems.

Conducting a LCA of salmon aquaculture systems in Canada, Ayer and Tyedmers (2009) noted that the global warming potential of producing 1 tonne of live-weight in cages, floating bags, land-based flow-through and land-based recirculating systems was 2073, 1900, 2770 and 28,200 kg CO_2 equivalents, respectively. It was concluded that adopting recirculating systems would result in substantially higher material and energy inputs than conventional cage-based farming. Although the eutrophication potential associated with recirculation systems was less in this study, the perception of recirculating systems as more environmentally friendly may be misleading when broader impacts such as global warning are considered.

The notion that organic farming is benign has been challenged (Pelletier and Tyedmers, 2007). In the absence of fish meal and oil derived from reduction fisheries certified as sustainable, Pelletier and Tyedmers (2007) stated that conditions imposed by organic certification agencies mean that only by-products from fisheries for human consumption may be used and that this results in larger global warming impacts than aquaculture feeds derived from reduction fisheries. This is a consequence of the higher energy intensity of fisheries for species destined for human consumption and lower meal and oil yields. The authors concluded that 'current standards for organic salmon aquaculture ... fail to reduce the environmental impact of feed production for the suite of impact categories considered' (Pelletier and Tyedmers, 2007: 399).

Major challenges associated with LCA application are allocation of impacts across products and co-products and boundary setting. When co-products constitute the bulk of production, allocation decisions can result in most impacts being associated with these as opposed to the primary product. Allocation decisions might be made based on weight, energy content or economic value. The decision to allocate impacts to co-products often fails to acknowledge that non-use may result in the need for waste disposal with attendant impacts. Setting the boundary around the system or process to be assessed and choices over the relative allocation of impacts can seriously affect LCA results.

The nature and location of the manufacture or food producing activity greatly influence LCA outcomes. Data on potential LCA impacts derived from generic and proprietary databases may be highly context specific and have changed over time, in which case additional research would be warranted to establish case-specific and contemporary input data. The magnitude and distribution of LCA impacts with respect to spatial and temporal dimensions are not accounted for. Consequently, apparently more responsible or environmentally sound production strategies may exceed the capacity of the receiving environment to assimilate waste products. In contrast, seemingly more damaging practices may operate well within the environmental carrying capacity of the surrounding ecosystem and consequently be more sustainable.

Few individual aquaculture facilities, whether land based or in open water, have been subject to LCA assessment. Application of this approach across sectors and globally has, however, contributed to more comprehensive evaluation of environmental impacts of aquaculture development and has permitted comparisons between production systems and strategies and other food sectors (Hallström et al., 2019; Gephart et al., 2021; Crona et al., 2023). Outcomes from these assessments have highlighted research and development priorities to guide future sustainable aquaculture development. However, there are limitations of the LCA approach. These include the absence of criteria to account for biodiversity changes and destruction of sensitive habitats, impacts on ecological functioning and processes in the supporting or receiving environment, water appropriation and consumption, health and safety at work and freedom of association and collective bargaining, animal welfare, human food and nutrition security and well-being, equality and human rights, ethical considerations and socio-economic and demographic impacts.

Participatory impact assessment

Challenges facing aquaculture development and the assessment of possible environmental impacts include establishing cause-and-effect relationships, discerning direct and indirect impacts from normal variation and emerging trends and pre-empting unexpected consequences. An established approach to analysing the underlying causes of ecological degradation and livelihoods vulnerability is to formulate a problem tree with key informants or appropriately knowledgeable stakeholder groups. Focusing initially on the problem at hand, drivers of change and pressures giving rise to the phenomena in question are explored in depth, thus establishing root causes. A problem tree developed to better understand causes and consequences of mangrove loss in the Mahakam Delta, East Kalimantan, Indonesia, demonstrated that mangrove loss continued as a direct and indirect result of human activity in the delta and that consequences included loss of ecosystem services supporting societal systems and negative feedback for aquaculture producers (Figure 7.8).

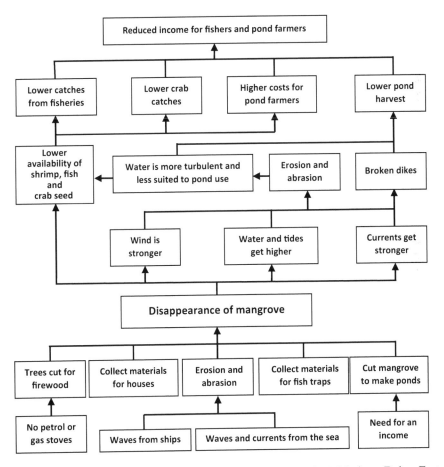

Figure 7.8 Problem tree formulated for mangrove loss in the Mahakam Delta, East Kalimantan, Indonesia.

Source: Bosma et al. (2007).

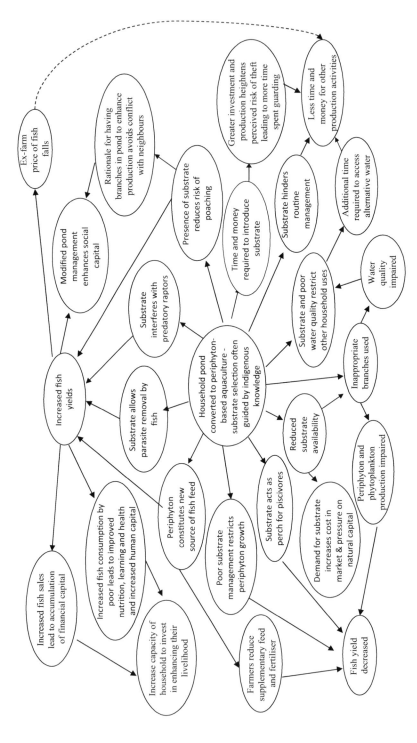

Figure 7.9 Impact diagram for adoption of periphyton-based pond aquaculture in Asia.

Source: Bunting et al. (2005a).

An alternative approach to identifying the consequences of proposed interventions in natural resources management and farming systems is to develop an impact diagram. Periphyton-based culture was proposed as a potential means to help small-scale farmers associated with the CARE Bangladesh LIFE (Locally Intensified Farming Enterprises) project (Bunting et al., 2005a). Development of an impact diagram (Figure 7.9) based on the synthesis of discussion with key informants and semi-structured interviews with prospective adopters yielded important insights concerning potential constraints and opportunities. Several scenarios were envisaged in trying to enhance resource use efficiency and production in which water quality and productivity might be negatively impacted. Positive impacts were foreseen in other circumstances for producers and society generally in terms of increased fish supplies and reduced prices. Secondary benefits, notably in terms of providing a deterrent to poaching, without making explicit concerns over theft to the community, were important too.

Summary

Promising approaches to stakeholder, institutional, market, value chain and food systems assessment in support of equitable and appropriate aquaculture development and aquatic resources management have been developed in various contexts. Participatory action planning has been deployed to facilitate joint assessment of options for aquatic resources management, but this may be questioned where it does not conform to constitutional norms. Application of the approach to the peri-urban EKW in West Bengal, India, demonstrated that stakeholder groups could work together to assess and prioritise actions needed to rehabilitate the system and consequently secure government investment to initiate the work. The feasibility of proposed actions was jointly assessed by stakeholders within the STEPS framework, which included evaluation of conditions required to ensure that actions are sustainable. Whilst planning and decision-making based on consensus may be desirable, it is not straightforward or always easy to discern. Consequently, the stakeholder Delphi approach was introduced as a means to promote consensus and test the strength of agreement amongst stakeholders. With joint assessment approaches there may be barriers associated with age, gender, poverty, social status or wealth that could prevent some groups from participating, and consequently safeguards are required to ensure that all stakeholder groups are represented and their views are included in the process. Acknowledging that systems must be financially viable and contribute to economic and social development, assessment approaches suited to aquaculture have been discussed, but action is needed to ensure that prevailing social–economic arrangements and policies favour sustainable production and penalise unsustainable producers and practices.

Notes

1 Peri-urban areas are often regarded as those physical spaces between urban and rural locations, but in practice the transition from urban through peri-urban to rural areas is defined by changing environmental, demographic, economic and social–psychological elements.
2 The eleven regions delineated for practical purposes had no legal or constitutional basis. A similar pragmatic approach is proposed for future planning activities as the wetland environment is not homogenous and the demands and expectations of stakeholders throughout the wetlands are not always the same.
3 If data are normally distributed or appropriate transformations are possible, more powerful parametric statistical tests may be appropriate.
4 Bioeconomic modelling has been invoked elsewhere to facilitate the combined assessment of fish stocks, their exploitation and fisheries management options.

References

Ayer, N.W. and Tyedmers, P.H. (2009) 'Assessing alternative aquaculture technologies: life cycle assessment of salmonid culture systems in Canada', Journal of Cleaner Production, vol 17, pp. 362–373.

Bosma, R.H., Sidik, A.S., Sugiharto, E., Fitriyana Budiarsa, A.A., Sumoharjo, et al. (2007) Situation of the Mangrove Ecosystem and the Related Community Livelihoods in Muara Badak, Mahakam Delta, East Kalimantan, Indonesia, Mulawarman University, Samarinda, East Kalimantan, Indonesia.

Bosma, R.H., Tendencia, E.A. and Bunting, S.W. (2012) 'Financial feasibility of green-water shrimp farming associated with mangrove compared to extensive shrimp culture in the Mahakam Delta, Indonesia', Asian Fisheries Science, vol 25, no 3, pp. 258–269.

Bosman, H.G., Moll, H.A.J. and Udo, H.M.J. (1997) 'Measuring and interpreting the benefits of goat keeping in tropical farm systems', Agricultural Systems, vol 53, pp. 349–372.

Bostock, J., Albalat, A., Bunting, S., Turner, W.A., Mensah, A.D. and Little, D.C. (2022) 'Mixed-sex Nile tilapia (*Oreochromis niloticus*) can perform competitively with mono-sex stocks in cage production', Aquaculture, vol 557, 738315.

Brugere, C. and Ridler, N. (2004) Global Aquaculture Outlook in the Next Decades: An Analysis of National and Aquaculture Production Forecasts to 2030, FAO, FAO Fisheries Circular No. 1001, Rome.

Bruneau, N.N., Thorburn, M.A. and Stevenson, R.M.W. (1999) 'Use of the Delphi panel method to assess expert perception of the accuracy of screening test systems for infectious pancreatic necrosis virus and infectious hematopoietic necrosis virus', Journal of Aquatic Animal Health, vol 11, pp. 139–147.

Bunting, S.W. (2001) A Design and Management Approach for Horizontally Integrated Aquaculture Systems, Institute of Aquaculture, University of Stirling, UK.

Bunting, S.W. (2002) 'Wastewater-reuse and poor livelihoods in peri-urban Kolkata', Paper presented at the 28th WEDC Conference on Sustainable Environmental Sanitation and Water Services, 18–22 November 2002, Kolkata, India.

Bunting, S.W. (2005) Second Market Analysis, European Commission Project GENESIS [EC INNOVATION IPS-2000-102], Institute of Aquaculture, University of Stirling, UK.

Bunting, S.W. (2007) 'Confronting the realities of wastewater aquaculture in peri-urban Kolkata with bioeconomic modelling', Water Research, vol 41, no 2, pp. 499–505.

Bunting, S.W. (2008) 'Horizontally integrated aquaculture development: exploring consensus on constraints and opportunities with a stakeholder Delphi', Aquaculture International, vol 16, pp. 153–169.

Bunting, S.W. (2010) 'Assessing the stakeholder Delphi for facilitating interactive participation and consensus building for sustainable aquaculture development', Society and Natural Resources, vol 23, pp. 758–775.

Bunting, S.W. (2014a) Stakeholder Delphi Assessment of UK Marine Aquaculture Sector Carbon Footprint Management, Final Report for The Crown Estate, University of Essex, Colchester, UK.

Bunting, S.W. (2014b) UK Marine Aquaculture Sector Carbon Budgeting, Final Report for The Crown Estate, University of Essex, Colchester, UK.

Bunting, S.W., Bosma, R.H., van Zwieten, P.A.M. and Sidik, A.S. (2013) 'Bioeconomic modelling of shrimp aquaculture strategies for the Mahakam Delta, Indonesia', Aquaculture Economics and Management, vol 17, pp. 51–70.

Bunting, S.W., Bostock, J., Leschen, W. and Little, D.C. (2023) 'Evaluating the potential of innovations across aquaculture product value chains for poverty alleviation in Bangladesh and India', Frontiers in Aquaculture, vol 2, no 3, pp. 1–23.

Bunting, S.W., Karim, M. and Wahab, M.A. (2005a) 'Periphyton-based aquaculture in Asia: livelihoods and sustainability' in M.E. Azim, M.C.J. Verdegem, A.A. van Dam and M.C.M. Beveridge (eds) Periphyton: Ecology, Exploitation and Management, CAB International, Wallingford, UK.

Bunting, S.W., Kundu, N. and Ahmed, N. (2017) 'Evaluating the contribution of diversified shrimp–rice agroecosystems in Bangladesh and West Bengal, India to social–ecological resilience', Ocean and Coastal Management, vol 148, pp. 63–74.

Bunting, S.W., Kundu, N., Punch, S. and Little, D.C. (2001) East Kolkata Wetlands and Livelihoods: Workshop Proceedings, Institute of Aquaculture, University of Stirling, Working Paper 2, DFID NRSP Project R7872, UK.

Bunting, S.W., Kundu, N., Saha, S., Lewins, R. and Pal, M. (2005b) EKW Water Management Action Plan and Preliminary Development Activities, Report to the East Kolkata Wetlands Management Committee, Institute of Aquaculture, Stirling, UK, and Institute of Environmental Science and Wetland Management, Kolkata, India.

Bunting, S.W. and Lewins, R. (2006) Urban and Peri-Urban Aquaculture Development on Bangladesh and West Bengal, India, University of Essex, interdisciplinary Centre for Environment and Society Occasional Paper 2006–2, Colchester, UK.

Bunting, S.W., Luo, S., Cai, K., Kundu, N., Lund, S., Mishra, R., et al. (2016) 'Integrated action planning for biodiversity conservation and sustainable use of highland aquatic resources: evaluating outcomes for the Beijiang River, China', Journal of Environmental Planning and Management, vol 59, pp. 1580–1609.

Bunting, S.W., Mishra, R., Smith, K.G. and Ray, D. (2015) 'Evaluating sustainable intensification and diversification options for agriculture-based livelihoods within an aquatic biodiversity conservation context in Buxa, West Bengal, India', International Journal of Agricultural Sustainability, vol 13, pp. 275–294.

Bunting, S.W. and Shpigel, M. (2009) 'Evaluating the economic potential of horizontally integrated land-based marine aquaculture', Aquaculture, vol 294, pp. 43–51.

Caffey, R.H. (1998) Quantifying Sustainability in Aquaculture Production, PhD thesis, School of Forestry, Wildlife and Fisheries, Louisiana State University, Baton Rouge.

Checkland, P. and Poulter, J. (2006) Learning for Action: A Short Definitive Account of Soft Systems Methodology and Its Use for Practitioners, Teachers and Students, John Wiley & Sons Ltd, Chichester, England.

Cohen, P.J., Simmance, F., Thilsted, S.H., Atkins, M., Barman, B.K., Bunting, S., et al. (2021) Advancing Research and Development Outcomes with Fish in Regional Food Systems, WorldFish, Penang, Malaysia.

Crona, B.I., Wassénius, E., Jonell, M., Koehn, J.Z., Short, R., Tigchelaar, M., et al. (2023) 'Four ways blue foods can help achieve food system ambitions across nations', Nature, vol 616, pp. 104–112.

de Bruyn, J., Wesana, J., Bunting, S.W., Thilsted, S.H. and Cohen, P.J. (2021) 'Fish acquisition and consumption in the African Great Lakes Region through a food environment lens: a scoping review', Nutrients, vol 13, 2408.

de la Caba, K., Guerrero, P., Trang, T., Cruz-Romero, M.C., Kerry, J., Fluhr, J., et al. (2019) 'From seafood waste to active seafood packaging: an emerging opportunity of the circular economy', Journal of Cleaner Production, vol 208, pp. 86–98.

DFID (2001) Sustainable Livelihoods Guidance Sheets, United Kingdom Government's Department for International Development, London.

Dulyapurk, V., Taparhudee, W., Yoonpundh, R. and Jumnongsong, S. (2007) Multidisciplinary Situation Appraisal of Mangrove Ecosystems in Thailand, Faculty of Fisheries, Kasetsart University, Bangkok.

Edwards, P. (2002) Institutional Assessment: Reviewing Policies, Processes and Stakeholder Positions at the Kolkata Peri-Urban Interface, Asian Institute of Technology, Working Paper 3, DFID NRSP Project R7872, Bangkok.

FAO (2008) Report of the Expert Consultation on Improving Planning and Policy Development in Aquaculture, FAO Fisheries Report No. 858, Food and Agriculture Organization of the United Nations, Rome.

FAO (2010) Report of the Regional Workshop on Methods for Aquaculture Policy Analysis, Development and Implementation in Selected Southeast Asian Countries, Bangkok, 9–11 December 2009, FAO, FAO Fisheries and Aquaculture Report No. 928, Rome.

Ferguson, P., Stone, T. and Young, J.A. (2005) Consumer Perceptions of Aquatic Products to Be Produced from GENESIS Integrated Systems: The UK Perspective, Department of Marketing and Stirling Aquaculture, University of Stirling, UK.

Gephart, J.A., Henriksson, P.J.G., Parker, R.W.R., Shepon, A., Gorospe, K.D., Bergman, K., et al. (2021) 'Environmental performance of blue foods', Nature, vol 597, no 7876, pp. 360–365.

Golden, C.D., Koehn, Z.J., Shepon, A., Passarelli, S., Free, C.M., Viana, D.F., et al. (2021) 'Aquatic foods to nourish nations', Nature, vol 598, pp. 315–320.

Hallström, E., Bergman, K., Mifflin, K., Parker, R., Tyedmers, P., Troell, M. and Ziegler, F. (2019) 'Combined climate and nutritional performance of seafoods', Journal of Cleaner Production, vol 230, pp. 402–411.

Haylor, G., Kumar, A., Mukherjee, R. and Savage, W. (2003) Indicators of Progress, Consensus-Building Process and Policy Recommendations, UK Government's Department for International Development, Network of Aquaculture Centres in Asia-Pacific, NRSP Project R8100, Bangkok.

HLPE (2020) Food Security and Nutrition: Building a Global Narrative towards 2030, High Level Panel of Experts, Rome.

ISO (1997) International Standard 14040 Environmental Management: Life Cycle Assessment – Principles and Framework, International Organization for Standardization, Geneva, Switzerland.

ISO (2000) International Standard 14042 Environmental Management: Life Cycle Assessment – Life Cycle Impact Assessment, International Organization for Standardization, Geneva, Switzerland.

Lewins, R., Coupe, S. and Murray, F. (2007) Voices from the Margins: Consensus Building and Planning with the Poor in Bangladesh, Practical Action Publishing, Rugby, UK.

Liu, Y., Bunting, S.W., Luo, S., Cai, K. and Yang, Q. (2019) 'Evaluating impacts of fish stock enhancement and biodiversity conservation actions on the livelihoods of small-scale fishers on the Beijiang River, China', Natural Resource Modeling, vol 32, no 1, e12195.

Lund, S., Banta, G.T. and Bunting, S.W. (2014) 'Applying stakeholder Delphi techniques for planning sustainable use of aquatic resources: experiences from upland China, India and Vietnam', Sustainability of Water Quality and Ecology, vol 3–4, pp. 14–24.

MEA (2005) Ecosystems and Human Well-Being: Wetlands and Water Synthesis, World Resources Institute, Washington, DC.

Mehar, M., Mekkawy, W., McDougall, C. and Benzie, J.A.H. (2022) 'Preferences for rohu fish (*L. rohita*) traits of women and men from farming households in Bangladesh and India', Aquaculture, vol 547, 737480.

Moll, H.A.J. (2005) 'Costs and benefits of livestock systems and the role of market and nonmarket relationships', Agricultural Economics, vol 32, pp. 181–193.

Muir, J.F. (1996) 'A systems approach to aquaculture and environmental management' in D.J. Baird, M.C.M. Beveridge, L.A. Kelly and J.F. Muir (eds) Aquaculture and Water Resource Management, Blackwell Science, Oxford, UK.

National Consortium on SRI (2012) Enhancing Employment and Sustaining Production: Framework for Integration of System of Rice Intensification (SRI) with Mahatma Gandhi National Rural Employment Guarantee Scheme (MGNREGS), National Consortium on SRI, Research and Resource Centre, Professional Assistance for Development Action, New Delhi.

Nguyen, N.H., Sonesson, A.K., Houston, R.D. and Moghadam, H. (2022) 'Editorial: applications of modern genetics and genomic technologies to enhance aquaculture breeding', Frontiers in Genetics, vol 13, 898857.

Pelletier, N. and Tyedmers, P. (2007) 'Feeding farmed salmon: is organic batter?', Aquaculture, vol 272, pp. 399–416.

Petrell, R.J., Mazhari Tabrizi, K., Harrison, P.J. and Druehl, L.D. (1993) 'Mathematical model of Laminaria production near a British Columbian salmon sea cage farm', Journal of Applied Phycology, vol 5, pp. 1–14.

Pretty, J.N. (1995) 'Participatory learning for sustainable agriculture', World Development, vol 23, pp. 1247–1263.

Punch, S., Bunting, S.W. and Kundu, N. (2002) Poor Livelihoods in Peri-Urban Kolkata: Focus Groups and Household Interviews, Department of Applied Social Science and Institute of Aquaculture, University of Stirling, Working Paper 5, DFID NRSP Project R7872, UK.

Sen, A. (2000) Development as Freedom, Oxford University Press, Oxford, UK.

Simmance, F.A., Kanyumba, L., Cohen, P., Njaya, F., Nankwenya, B., Manyungwa, C., et al. (2021) Sustaining and Improving the Contribution Small-Scale Fisheries Make to Healthy and Sustainable Food Systems in Malawi, WorldFish, Penang, Malaysia.

Spedding, C.R.W. (1988) An Introduction to Agricultural Systems, second edition, Elsevier Applied Science, London.

Thiao, D. and Bunting, S.W. (2022a) Impacts Socioéconomiques et Biologiques de l'industrie des Aliments Pour Animaux à Base de Poisson en Afrique Subsaharienne, FAO Circulaire sur les pêches et l'aquaculture 1236, FAO, Rome.

Thiao, D. and Bunting, S.W. (2022b) Socio-Economic and Biological Impacts of the Fish-Based Feed Industry for Sub-Saharan Africa, FAO Fisheries and Aquaculture Circular 1236, FAO, Rome.

Thilsted, S.H., Thorne-Lyman, A., Webb, P., Bogard, J.R., Subasinghe, R., Phillips, M.J., et al. (2016) 'Sustaining healthy diets: the role of capture fisheries and aquaculture for improving nutrition in the post-2015 era', Food Policy, vol 61, pp. 126–131.

Uphoff, N. (2012) 'Supporting food security in the 21st century through resource-conserving increases in agricultural production', Agriculture and Food Security, vol 1, no 18, pp. 1–12.

Willett, W., Rockström, J., Loken, B., Springmann, M., Lang, T., Vermeulen, S., et al. (2019) 'Food in the Anthropocene: the EAT–Lancet Commission on healthy diets from sustainable food systems', Lancet, vol 393, no 10170, pp. 447–492.

Yan, N. and Chen, X. (2015) 'Don't waste seafood waste', Nature, vol 524, pp. 155–157.

8   Promoting an enabling environment

Key points

The aims of this chapter on promoting an enabling environment are to:

- Review progress with implementation the FAO Code of Conduct for Responsible Fisheries, focusing on Article 9 'Aquaculture Development' globally.
- Review ongoing development of guiding principles for aquaculture sector planning.
- Critically review prospects for promoting sustainable aquaculture practices, examining the case of better management practices (BMPs).
- Consider the role of management practice guidelines and the evolution of different paradigms for invoking BMPs.
- Review the nature of third- and second-party certification schemes and how they are administered and note progress being made by global partnerships (i.e. Sustainable Shrimp Partnership in Ecuador) to promote BMPs and enhance traceability.
- Assess factors influencing buyer and consumer demand, including prevailing food systems and food environments, and consider associated implications for aquaculture development.
- Consider prospects for enhanced knowledge transfer and uptake promotion of sustainable aquaculture practices, trade and consumption, including the function of global networks and evidence programmes.
- Review education and training provision for sustainable aquaculture development.
- Present a framework for communication planning based on existing communication stakeholder knowledge, attitudes and practices.

A framework for sustainable aquaculture development globally

The FAO Code of Conduct for Responsible Fisheries (CCRF) established a framework to guide sustainable development globally (FAO, 1995). Recent assessment sponsored by the UK Government's Blue Planet fund and published under the auspices of the Blue Food Partnership, Friends of

DOI: 10.4324/9781003342823-8

Ocean Action and World Economic Forum (Bunting et al., 2022) reviewed progress in implementing the FAO CCRF with a focus on 'Article 9 – Aquaculture Development' (FAO, 1995: 23). Notable progress is apparent across sectors and value chain nodes, but greater attention is warranted on developing effective administrative and legal frameworks, evaluating effects on ecosystems integrity and genetic diversity, focusing on species and strategies that could contribute to food and nutrition security, evaluating contributions to sustainable livelihoods, regulating cumulative impacts associated with the proliferation of successful operations, protecting transboundary ecosystems and populations of aquatic animals and plants, cooperating internationally and regionally, enhancing biosecurity and conserving and optimally utilising genetic resources and ensuring interactive stakeholder participation in developing and promoting responsible aquaculture practices.

Guiding principles for aquaculture sector planning

'Article 9 – Aquaculture Development' of the CCRF (FAO, 1995: 23) specifies that 'states should establish, maintain and develop an appropriate legal and administrative framework which facilitates the development of responsible aquaculture' and 'produce and regularly update aquaculture development strategies and plans, as required, to ensure that aquaculture development is ecologically sustainable and to allow the rational use of resources shared by aquaculture and other activities'. Against a backdrop of limited progress in implementing such provisions in many countries and cognisant of 'uncoordinated growth' in some places and that 'with more suitable planning and policies, the sector could perform better, benefit more people with a reduced risk of detrimental effects on the environment and overall would contribute to poverty alleviation', the Director General of the United Nations' Food and Agriculture Organization convened an Expert Consultation on Improved Planning and Policy Development in Aquaculture (FAO, 2008: 1).

The resulting outline of the *Report of the Expert Consultation on Improving Planning and Policy Development in Aquaculture* (FAO, 2008) presents guidelines covering processes of policy formulation, policy implementation and supporting policy implementation, including, where appropriate, incentives to 'encourage good practices throughout the sector' (17). Although these guidelines were formulated through a process of expert consultation, it was recommended that further knowledge concerning 'case study information, "best practices" in policy formulation and implementation' and 'further analysis and assessments on the use of stakeholder participation methods in policy formulation' be undertaken (FAO, 2008: 4). Best practices and their potential for guiding responsible practices are commonly promoted across several domains, including aquaculture planning and development at national and regional scales and management and husbandry at local and farm levels. A critical review of this apparently promising mechanism to promote and attain responsible and sustainable aquaculture sector development follows.

Better management practices

The concept of best or better management practices (BMPs) is being supported, promoted, implemented and adopted within the realms of aquaculture development. One of its first proponents was the Network of Aquaculture Centres in Asia-Pacific (NACA). Under the STREAM (Support to Regional Aquatic Resources Management) initiative sponsored by NACA, DFID, FAO and VSO, the development of Better-Practice Guidelines (BPGs) was promoted to 'share lessons learned from local practice or research' within the Asia-Pacific region (STREAM, 2005a). Guidelines developed under this initiative covered self-help groups, a consensus-building process and information access assessments (STREAM, 2005b). The rationale for referring to 'better rather than best' management practices was that 'as we try to achieve best practices we gain new knowledge, develop new skills and sometimes discover new attitudes. By describing and sharing beneficial lessons we hope to provide guidance toward better practices' (STREAM, 2005a: 1). Elsewhere, the term BMPs is used in reference to supposed best management practices. Alluding throughout to best management practices, the review presented in *Changing the Face of the Waters* (World Bank, 2007: xv) made no distinction between best and better practice, and in the glossary BMPs were described as 'codes of practice and best management practices'. The distinction should perhaps not be overlooked, and possible implications of differences in conceptualisation and rationale are central to the debate presented here.

Best management practices and their identification, adoption and enhancement are central to environmental management systems (EMS) and their implementation. EMS thinking and practice evolved from total quality management in the chemical industry. EMS are now widely applied across industrial, commercial, service and public sectors globally and are governed and regulated according to international standards (ISO, 2023). The EMS framework was conceived as a cyclical process, initiated by an organisation making a public policy commitment to legal compliance, pollution prevention and continued improvement, followed by planning, implementation, monitoring and reviewing (see chapter 3). With regards to BMPs for aquaculture, the process whereby such an approach could become iterative and adaptive and sustained by the intended beneficiaries or target groups is often poorly defined. Proposed strategies and potential mechanisms are explored further in the following sections.

According to the World Bank (2007: 5), 'good governance [of aquaculture] will draw on codes of practice and best management practices (BMPs) to inform the implementation of policies and plans'. Best and better management practice guidelines have been prepared or are being developed by researchers, NGOs, producer associations, government departments and inter-governmental organisations. Given the complex and sometimes confusing nature of guidance on aquaculture development, the first part of this review outlines a hierarchy of practice, guidelines and standards, taking into account the voluntary, conditional and statutory nature of the arrangement and scale, in terms of both farm size and geographical scope at which they are deemed appropriate. Special attention is paid to guidelines that have been developed or are proposed but are not associated with certification or regulatory standards; motivations for adopting such guidelines are discussed, and potential advantages and disadvantages compared with regulatory or certification standards are considered.

Constraints and opportunities associated with BMP development are outlined in the second part of this review. Concerns regarding the inappropriate promotion and transfer of guidelines are discussed. Proposals to develop BMPs can become marred in intellectual debates regarding the primacy of scientific versus traditional ecological knowledge and the most efficient and effective means to achieve ecologically sound and socially responsible outcomes. Merits of continuing to differentiate between best and better practices are discussed, and potential areas of opportunity for BMPs developed in participatory ways and tailored to local environmental, social, economic, market and political settings are considered.

Aquaculture management practice guidelines

The Sub-Committee on Aquaculture established by the Food and Agriculture Organization of the United Nations has acknowledged the value of BMPs and certification for engendering consumer confidence but has queried the net benefit of some schemes for small producers and called for efforts to harmonise farming standards, notably for shrimp, and review certification procedures for acceptance and transparency globally (FAO, 2009). A summary of prominent best, good and better management practices formulated to guide aquaculture development in freshwater and marine, temperate and tropical, and developed and developing country contexts is presented in Table 8.1.

Reviewing prospects for environmental management of aquaculture development in coastal areas, Barg (1992) noted that options included formulating management frameworks to protect coastal environments, adopting integrated coastal area management, developing legislation specific to the governance of coastal aquaculture, better planning and management of coastal aquaculture and improving farm- or project-level environmental management. Guidelines for planning sustainable freshwater aquaculture in Vietnam were prepared under the auspices of the joint Government of Vietnam and Danish International Development Agency initiative on Fisheries Sector Programme Support and include ten principles. The first is to ensure peoples' participation in the planning process. The other principles are to facilitate cross-sector stakeholder involvement; use external facilitators; establish a steering committee;

Table 8.1 Management practice guidelines for aquaculture development

Initiative	Coverage	Rationale	Proponents
Global and regional initiatives			
Best environmental practice for monitoring and regulating marine aquaculture in Europe	Best Environmental Practices (BEPs) covering application of hydrodynamic and benthic modelling, controls on chemical use, genetic interactions with wild stocks, environmental monitoring; socio-economic policy issues were developed	The aim was to evaluate the scientific and social–economic basis for environmental practices relevant to marine aquaculture in Europe and propose recommendations for best environmental practices for monitoring and regulating marine aquaculture in the European Union	MARAQUA (Monitoring and Regulation of Marine Aquaculture) concerted action funded by the European Commission, Brussels
Best Aquaculture Practices	Best Aquaculture Practices (BAPs) facility certification programme developed for shrimp, tilapia and catfish production with standards for aspects of responsible aquaculture development, including community, environment and food safety issues	Objective to develop science-based aquaculture standards to support responsible aquaculture development, with standards covering farms, hatcheries and processing plants	Global Aquaculture Alliance, Missouri, USA
Aquaculture standards	Aquaculture standards are being defined for twelve farmed species through a series of multi-stakeholder 'Aquaculture Dialogues'. Standards for three species groups, tilapia, pangasius catfish and bivalves, have been published	An Aquaculture Stewardship Council (ASC) will be convened to oversee implementation, raise awareness and launch a labelling scheme for certified products. It is anticipated that certified products will command a market premium, thus benefitting producers adopting ecologically sound and socially responsible production	World Wildlife Fund, Washington, USA
Good aquaculture practices	Good aquaculture practices covering marine and freshwater species implemented in a regional scheme coordinated by the government of Hong Kong.	Facing pressure from cheaper imported aquatic products, the initiative was conceived to improve product quality and introduce a quality assurance scheme to reassure consumers.	Accredited Fish Farm Scheme, Agriculture, Fisheries and Conservation Department, Hong Kong

(Continued)

Table 8.1 (Continued)

Initiative	Coverage	Rationale	Proponents
Species and site specific			
Better management practices	Promote the development, validation and implementation of better management practices for striped catfish (*Pangasianodon hypophthalmus*) and assess the impact and prospects for scaling up existing guidelines for shrimp farming	Better management practices constitute the most practical, acceptable and economically viable approach for clusters of small farmers to meet demands for responsibly produced, quality and certified aquaculture products	Department of Primary Industries, Victoria, Australia; NACA, Bangkok; RIA No.2 and Can Tho University, Vietnam; with funding from the European Commission
Better management practices for marine finfish production in the Asia-Pacific	Development of better management practices for marine finfish production in the Asia-Pacific region, with a focus on ensuring that small-scale producers farming the majority of finfish in the region are enabled to participate in market access and certification schemes	Conceived to improve prevailing production practices to ensure environmental integrity and sustainability. Providing a positive and practical mechanism to joining formal certified schemes and meeting growing consumer demand for environmentally and socially responsible production	Directorate General of Aquaculture, Indonesia; NACA, Bangkok; with funding from ACIAR, Australia, and NACA
Better management practices for *tambak* farming in Aceh	Better management practices (BMPs) for shrimp farming developed to guide rehabilitation of *tambaks* for shrimp farming in Aceh following the 2005 tsunami	BMP formulation was guided by the *International Principles for Responsible Shrimp Farming* developed by the Consortium Program on Shrimp Farming and the Environment (FAO et al., 2006)	ADB, ACIAR, AwF, BRR, DKP, FAO, GTZ, IFC, MMAF, NACA, WWF consortium
Best management practices for the shellfish culture industry in southeastern Massachusetts	Best management practices for shellfish culture in southeastern Massachusetts for site selection and access; materials, operations and maintenance; improving survival and productivity; disease prevention and management; environmental quality	Shellfish producer organisation initiated best management practices	Southeastern Massachusetts Aquaculture Centre; Roger Williams University, Bristol, Rhode Island, USA

Community and development orientated

Better-Practice Guidelines	The STREAM initiative facilitated development of a series of Better-Practice Guides (BPGs) covering self-help groups, information access survey and consensus-building process	Short and colourful printed media developed to share knowledge of better practices for aquatic resources management with groups working closely with farmers	STREAM (Support to Regional Aquatic Resources Management), Bangkok, Thailand (STREAM, 2005b)
Good practices for community-based planning and management of shrimp aquaculture in Sumatra, Indonesia	Project in a coastal village (Pematang Pasir) in Lampung Province, Sumatra, to identify good practices for environmentally responsible and sustainable shrimp farming	Assumed that community-managed and voluntary approaches are needed owing to limited government resources and capability to develop, implement and enforce shrimp aquaculture regulations	Indonesian Coastal Resources Management Project (Tobey et al., 2002)
Best Practices Plan	Formulated in consultation with stakeholders in the East Kolkata Wetlands covering regulatory compliance; environmental, ecological, social, economic and educational objectives; principles of operation management; research; monitoring and surveillance; stakeholder consultation; post-harvest sector; periodic review triggers; institutional, legislative and regulatory assessment	Objective to promote continued wise use of the wetlands, increasing production and profitability and enhance the livelihoods of community groups living and working in the wetlands	East Kolkata Wetlands Management Authority, Kolkata, India; Asian Development Bank

integrate economic, social, environmental and technical factors; identify objectives for decision-makers; develop an action plan for implementation; continually monitor and adapt the plan; use scenarios to elaborate the intended direction of development; and ensure that planning guidelines and approaches are appropriate for local implementation, with limited external financial and technical inputs (Mathiesen et al., 2008).

Facing pressures associated with a burgeoning abalone aquaculture sector in Western Australia, Fisheries Western Australia developed policy guidelines for proponents and the authorities to ensure that abalone farming developed in an 'environmentally acceptable manner' (Fisheries Western Australia, 1999: 1). Issues covered included potential environmental impacts, broodstock collection, translocations, hatchery production and land-based and marine culture systems. The MARAQUA (Monitoring and Regulation of Marine Aquaculture) concerted action project funded by the European Commission undertook a scientific and socio-economic review of marine aquaculture practices in Europe and proposed a series of Best Environmental Practices (BEPs) covering application of hydrodynamic and benthic modelling, controls on chemical use, genetic interactions with wild stocks, environmental monitoring and socio-economic policy issues (Read et al., 2001). Best Aquaculture Practices (BAPs) were developed by the Global Aquaculture Alliance (GAA), St. Louis, Missouri, primarily to promote responsible aquaculture, encompassing issues of environmental protection, social responsibly, animal welfare, food safety and traceability (Anonymous, 2009b).

A code of good management practices for shrimp farming in Sri Lanka was prepared based on available technical guidelines and taking account of the FAO Code of Conduct for Responsible Fisheries (Siriwardena and Williams, 2003). Better management practices for shrimp farming were formulated to guide rehabilitation of shrimp-producing *tambaks* in Aceh following the 2004 tsunami (ADB et al., 2007). Formulation of the BMPs was guided by the *International Principles for Responsible Shrimp Farming* developed under the Consortium Program on Shrimp Farming and the Environment (FAO et al., 2006). Progress in achieving ecologically sound and socially responsible shrimp farming through development of the international principles was acknowledged with the World Bank Green Award 2006 (NACA, 2007). Given apparent benefits to producers and the environment, BMPs for *tambak* farming developed for Aceh were assessed in the context of the Mahakam Delta using a bioeconomic modelling approach (Bunting et al., 2013).

Projects funded by the European Commission and Australian Government have been commissioned to assess the development of better management practices for shrimp and striped catfish (*Pangasianodon hypophthalmus*) farming in Asia (Fisheries Victoria et al., 2009; NACA, 2009). GAPs for marine and freshwater culture are being implemented in a regional scheme conceived by the Government of Hong Kong (AFCD, 2023). Several producer associations have adopted codes of good practice; for example, the British Trout Association and Salmon Scotland (Salmon Scotland, 2023).

Best practices governing aquaculture development are an established concept in the USA where a number of sectors and state-level producer groups have formulated and adopted standards. Details for BAPs for shrimp farms developed by the Global Aquaculture Alliance (GAA, 2009), and best management practices prepared for the shellfish culture industry in south-eastern Massachusetts (Leavitt, 2009) are presented in Table 8.1.

Organic and fair trade schemes are well established for farming terrestrial plants and animals. Organic farming was estimated to be worth US$38.6 billion globally in 2006 (Willer et al., 2008). Whilst fair trade is perhaps less significant in economic terms, it is vitally important for certain small producer groups in developing countries. The welfare of farmed animals is also the focus of compassionate, free-range or freedom-based initiatives. Ecologically sound and socially just production standards for responsible and sustainable aquaculture products have begun to emerge more recently, but this process has often been controversial. Objections have been raised to restricting Atlantic salmon with an innate migratory instinct to cages, and concerns have been expressed over infection of wild fish with parasites, escapees and the discharge of untreated water (BBC, 2006). Regarding organic shrimp culture in Indonesia, commitments to conserve and replant mangroves, monitor discharge water quality and better train staff reportedly remained unfulfilled (Ronnback, 2003).

The World Wildlife Fund (WWF) implemented a series of eight 'aquaculture dialogues' starting in 2004 with the Salmon Aquaculture Dialogue, the objective being to define effective and credible standards to address the most significant environmental and social concerns and formulate good management practices to achieve compliance and alleviate negative impacts (Anonymous, 2009a). Collaborating with IDH, a Dutch initiative for sustainable trade, WWF initiated a process of constituting the Aquaculture Stewardship Council (ASC) that will oversee the implementation of the aquaculture standards, based on the premise that labelling products that conform to the standards will result in added-value products that 'translate social and environmental benefits into commercial benefits for individual farmers' (Anonymous, 2009a: 11). Previous initiatives coordinated by WWF include formation of the Forestry Stewardship Council and Marine Stewardship Council to oversee certification and labelling of forestry products and wild-caught seafood.

Provision of general advice on and guidance for environmental protection and risk reduction is common practice across many food-producing sectors. Within the agricultural sector in Europe, codes of conduct for pesticide handling, storage and application and guidelines for fertiliser and manure application rates and periods, waste disposal, crop storage and transport and environmental stewardship are routine (Crop Protection Association, 2007; HGCA, 2007; ACCS, 2008; CFE, 2009). Adoption of chemical handling and exposure control measures and post-harvest storage advice can help safe-guard farmer and public health; implementation of wildlife conservation and countryside stewardship measures have contributed to protecting endangered species and enhancing biodiversity. Together such measures help to ensure the image of the sector, instil consumer confidence and safeguard the market for produce.

Adherence to statutory guidelines and standards is demonstrated to buyers via scheme registration and is communicated to consumers via product labels such as the Red Tractor or LEAF insignia in UK agricultural production. Furthermore, eligibility for agricultural subsidies in the UK under the Single Farm Payment scheme funded by the European Community was conditional upon implementing environmental protection and food safety measures; greater payments are possible where more demanding entry- and higher-level standards are adopted. Affiliation with a recognised quality assurance scheme is often a prerequisite for selling to premium markets or for becoming a member of cooperatives undertaking collective storage and marketing. Although supposedly voluntary, it is increasingly apparent that without belonging to a quality assurance scheme, it is difficult to store, transport and sell agricultural and horticultural products from UK farms.

Guidance to enhance production, minimise costs, optimise resource use efficiency and ensure compliance with statutory requirements is often formulated and promoted by producer associations, trades unions, professional bodies and commercial organisations. Guidelines for aquaculture focusing on feeding rates, times and pellet type given the size of fish being cultured are often useful to optimise feed conversion rates and reduce waste, whilst advice on pest and disease control and treatment can significantly improve production, help to avoid mortalities and minimise risks to workers and consumers. Adoption of voluntary guidelines and standards as formulated by producer organisations, organic certification bodies, animal welfare charities and fair trade advocates generally go beyond quality assurance to address specific areas of environmental concern or broader social and ethical imperatives such as the right to associate freely and join trades unions or freedom from suffering for animals. The product is differentiated in the marketplace in recognition of this through scheme-specific labelling, with the expectation that it will command a price premium, thereby justifying higher production costs and rewarding considerate producers.

Development of generic guidelines is advocated as a cost-effective and efficient strategy as it potentially avoids the duplication of effort in farm trials, HACCP assessments and visits by extension agents and consultants; it may also restrict the need for farm inspections, product testing and remedial action. Advice could potentially be transferred usefully to other regional producer groups to improve performance and be drawn upon to address other production systems, species and issues through synthesis of existing knowledge. Indeed, advice from other sectors could be transferred to benefit aquaculture producers. Many smaller aquaculture producers could benefit significantly from general business and marketing advice by drawing on well-established principles and practices.

There are apparent conceptual, practical, economic and social–political limitations to the BMP approach that warrant further assessment. Evidence is emerging of competition between schemes that could contribute to more efficient administration and lower transaction costs, offer more choice to new and existing consumers or, worryingly, confuse and alienate consumers (Aarset et al., 2004). Reference to better as opposed to best practices in the context of aquaculture development was justified within the STREAM initiative by postulating that endeavouring to achieve best practices might result in equally valuable or unexpected new knowledge, skills and attitudes and that documenting this process and sharing lessons might yield better practices (STREAM, 2005a). The focus of the initiative was not specifying a gold standard but learning rather than repeating mistakes, promoting incremental improvements and ultimately supporting sustainable aquatic resources management and enhancing poor livelihoods. For those working directly with the poor (i.e. extension workers, field officers and community organisers), the guidelines were seen as a means to improve ways of working and encompassed knowledge, skills, capacity and practices. Earlier calls had been made for a new professionalism in development practice (Pretty, 1995), and this was arguably one of the first initiatives along these lines in the aquaculture development sector.

Whilst BPGs are aimed at practitioners and practice, BMPs are by definition focused on farm-level intervention to improve yields, optimise resource use efficiency and improve pest and disease management. Group formation, which is known to contribute significantly to positive development outcomes, notably strengthening social capital (Pretty and Ward, 2001; Pretty, 2003) and is often prescribed alongside aquaculture BMPs to facilitate enhanced disease control through coordinated regional approaches, promote the equitable management of natural resources, enable collective bargaining and allow for the establishment of investment and compensation funds (STREAM, 2005b; NACA, 2009). Within the discourse surrounding aquaculture, BMPs are now being recommended as the basis for local legislation or steppingstones to accreditation or certification under quality assurance or eco-labelling schemes (Fisheries Victoria et al., 2009). However, the route from adopting BMPs to joining a certification programme is not necessarily clear. Conceptual barriers, problems with harmonising standards and monitoring compliance, administrative and transaction costs and social, economic and ethical considerations pose serious challenges to such a proposition.

Discussion concerning the potential coverage and suitability of BMPs is reminiscent of the debate surrounding recommendation domains. Developed initially by the International Maize and Wheat Improvement Center in the early 1990s, recommendation domains were conceived as a means to target research and development investment and to select agricultural technology and farming systems suited to a particular geographic area or agroecosystem (Conroy and Sutherland, 2004). However, owing to variation in the agro-ecological environment, asset base of individual farmers, access to markets, services and infrastructure as well as cultural differences in decision-making, the validity and usefulness of recommendation domains was challenged (Okali et al., 1994). Producers may not be able to act on technical recommendations because of a lack of inputs, equipment, skills and expertise or illiteracy and innumeracy; providing credit and capacity-building might be seen as ways to promote adoption, but added transaction costs and ensuing debt may further deter famers. Guidance may be incorrectly interpreted or extrapolated, critical aspects and nuances may be missed and supposedly improved varieties may not be valued or accepted in the marketplace.

It is not always possible to account for the variability inherent in farming systems dependent on ecosystem services derived from ecologically based processes and situated in complex social–economic–political settings. Predetermined mixes of technological inputs have failed to generate uniform benefits and have even resulted in some farmers and farming systems becoming impoverished. Green revolution packages have produced short- to medium-term gains for many producers in Asia, but concern remains over the sustainability of these strategies. Farmers are locked into high-input monocultures that are associated with greater financial risk, undermine natural capital and have eroded social capital associated with collective and reciprocal action, for example, that characterised many traditional farming strategies. Adoption of management practices based on a technological package can result in unhealthy patron–client relationships that make it difficult for producers to revert to previous practices or selectively adopt package components. Farmers can find themselves contracted to buying inputs from the package provider and obliged to sell products at sub-optimal times and below market prices.

As noted in the case of the dialogue on catfish farming in the USA aimed at developing environmental and social standards for the industry 'producers from Mississippi, Arkansas and Louisiana are developing their own standards' (WWF, 2008: 3). This situation highlights questions concerning how buyers and consumers perceive different quality assurance and certification schemes, whether the underlying standards or BMPs matter or whether people out shopping and intent on buying responsibly or sustainably farmed produced are reassured by generic packaging statements or a labelling scheme apparently endorsed by the retailer. With several certification schemes for aquaculture products already in operation and new schemes on the horizon, calls for harmonisation of standards and labelling are probably set to intensify. Labelling for specific traits, origin of production or husbandry practices will probably continue to suffice for selected products and markets where provenance is particularly valued as a guarantee of quality and safety.

Scheme organisers may be keen to differentiate their approach from their rivals. In their 'Statement on WWF–GLOBALGAP Linkage', the GAA affirmed that their BAP-based approach is the 'global leader in the development of science-based aquaculture standards' and 'goes beyond standard-setting' providing 'educational and training programs to help government agencies and small producers in developing countries improve their practices' (Anonymous, 2009b: 24). Further work is perhaps required to assess the relative merits of such an approach compared with standards-based approaches developed though multi-stakeholder processes. The WWF (2008: 2) asserted that their standards 'will be the only measurable, science-based standards for aquaculture products' and that the scheme would 'identify and support the adoption and adaptation of better management practices that significantly reduce or eliminate' negative impacts on the environment and society (WWF, 2008: 3). Standards for bivalves, pangasius and tilapia state that a supplementary BMP manual will be developed 'explaining specific steps that can be taken by producers to achieve the standards' (WWF, 2009, 2010a, 2010b).

It is conceivable that producers may forego incremental improvements realised through local innovation and action research in pursuit of BMPs, incurring additional costs and facing greater uncertainty over financial returns, environmental and health risks and social acceptability. Adoption of generic practices may stifle innovation locally or supplant less risky and more appropriate traditional management practices. Compliance with certification standards may be burdensome, result in marginal gains and prevent producers from adopting risk-averse strategies or moving to diversify their livelihoods. Adopting different farming practices in certain cultures can result in resentment, undermine social capital and, in extreme cases, result in individuals and families being ostracised.

Proponents of BMPs do not usually evaluate possible costs and returns across the range of producers and production system scenarios. Without demonstrated benefits, producers may be reluctant to adopt such strategies. Wealthy and powerful producers adopting BMPs and able to invest resources to meet standards may stand to capture the biggest share of benefits and may actually undermine the price and market for smaller producers persevering with traditional farming approaches. Farmers operating marginal and poorly sited operations may be encouraged by the prospect of price premiums to continue farming when it may have been better if they had left the sector and the site were permitted to revert back to nature, bolstering stocks and flows of ecosystems services supporting society. Where smaller and less efficient producers move to adopt BMPs, potential market access and premiums may not be sufficient to compensate for transaction costs incurred, lower yields and the associated administrative burden; they may also become more vulnerable to environmental shocks and market trends.

Limitations to group formation for development purposes and constraints to the collective management and coordination of action concerning common property resources have been described extensively (Shackleton et al., 2002; Pretty, 2003; Zanetell and Knuth, 2004). Notable problems relevant to group formation in support of BMP adoption include deliberate exclusion of certain ethnic, gender or political groups and failure to include marginal and poor community members. Within group member households, additional burdens may not be shared equally, and women and children may end up doing more work. Community members excluded or missing from resource user groups could undermine collective action, compromising disease management plans; fail to support or interfere with regional habitat and species management plans; and propagate social disquiet and conflict. Some groups may be fundamentally opposed to the activity, making rational debate and constructive dialogue impossible. A focus on responsible, sustainable and ethical production may be at odds with prevailing norms, producer group priorities or policies aimed at immediate issues of poverty alleviation, regional food security and economic development. Extensification of production may indeed result in larger areas coming under aquaculture, compromising stocks and flows of ecosystem services supporting society.

Success resulting from the adoption of BMPs may lead to problems as rapid growth of a sector may result in a cumulative impact with negative environmental and social consequences. Prescribed standards and supporting guidance on management practices may weaken the case for establishing or continuing to fund research capacity able to develop locally appropriate management strategies and address emerging problems. It may be argued that adoption of standards and BMPs geared towards certification enshrines management practices that are unresponsive to underlying trends and external social and economic factors. As shrimp producers throughout Asia may return to traditional and extensive and semi-intensive low-input systems owing to the falling market value of shrimp, this may not be an option if they are locked into group management and decision-making and credit arrangements.

The review presented here raises questions over the primacy of knowledge and processes of how best to transfer information, scale-up promising outcomes from participatory research and promote more sustainable and responsible aquaculture practices. Two paradigms of BMPs are evident: those formulated to help farmers to improve their production efficiency and effectiveness and those designed to address broad social and environmental concerns associated with aquaculture development, in which the onus is on producers to 'mend their ways' so that they might benefit from continued market access and price premiums. For the former, in which the appropriateness of guidelines will be limited to particular aquatic ecosystems, production systems or regulatory regimes, the challenge is how to delineate and explain where such practices may be considered relevant. As it will not be possible to address all issues via generic guidelines, technical and extension support and continued research and development will be needed to address specific and emerging production constraints. There is a danger that those individuals and institutions with a vested interest in reinventing the wheel of aquaculture development may try to thwart the development and propagation of generic guidelines and practices. Where a need for improved social and environmental performance is identified outside the principal location for research and development or recommendation domain, BMPs and guidelines may be useful given appropriate testing, adaptation and uptake promotion.

Certification schemes serving the aquaculture sector are numerous and well established; however, new and existing schemes are competing for market share. Scheme managers may wish to differentiate themselves from their rivals in the marketplace, whilst concomitantly calls for harmonisation of standards and what constitutes an acceptable management practice are set to grow. Concern has been expressed about the reliability of certain schemes and failings in implementation, and standards checking by a small number of producers and inspectors could undermine global schemes and campaigns, bringing the credibility of guidelines into question. The role of authorities and natural resources managers in formulating policy and building capacity to produce an enabling institutional environment conducive to best practices in support of responsible and sustainable aquaculture development also demands further assessment.

Third- and second-party standards setting and certification

Schemes enabling producers and companies to demonstrate compliance with comprehensive standards (e.g. all legal and regulatory requirements, health and safety at work, staff competencies and best or better management practices) through a process of registration, record-keeping and verification by auditing have been established. When the audit is conducted by an independent organisation, this would be termed a 'third-party' scheme, but when there is a relationship between the parties, this would be a 'second-party' scheme. Analysis of aquaculture certification schemes shows a wide variation in which aspects are included (Bunting et al., 2022). Standards addressing animal welfare (see chapter 1) or organic production, for example, have a specific focus, whereas those having broader remits of responsible or sustainable production must cover all relevant aspects. This can include sourcing inputs (e.g. feeds) that are manufactured by firms certified under complimentary schemes (Tacon et al., 2021), operating within the carrying capacity of supporting ecosystem areas, ensuring inputs are utilised efficiently and that possible social impacts are considered and mitigated.

Third-party scheme compliance and accreditation permits products to be labelled with a distinctive logo as meeting the required standards. For aquatic foods this has been demonstrated to be an important consideration for retailers and consumers in Europe and North America (Bush et al., 2013). Therefore, producers operating outside these geographies may be compelled to join formal certification schemes. This constitutes a potential barrier to small-scale producers in low- and middle-income countries selling their produce to more lucrative markets. Clusters, cooperatives and groups of small-scale producers in a specific geography could potentially be assisted and help each other in adopting BMPs as an intermediate step to joining an established third- or second-party scheme (Haque et al., 2021). Collective action led by industry representatives, as witnessed with the Sustainable Shrimp Partnership in Ecuador, can help transform an entire sector through the adoption of more sustainable practices (Bunting et al., 2022). Part of this initiative assisted small- and medium-sized producers to work towards the standards and requirements of Aquaculture Stewardship Council (ASC) certification.

Most aquatic foods can reach consumers, predominantly through domestic markets, without the need for independent certification. Consequently, the scope for such schemes to transform aquatic food systems globally to more sustainable pathways is currently limited. Where aquatic food systems are in transition, from live fish sales and wet markets, to multiple retailers and ecommerce sites, it could be that third-party schemes will become more important or alternative transparent and trustworthy strategies may be appropriate. Emerging technologies – for example, AI (artificial intelligence), blockchain and QR codes (quick response codes) – could potentially help enable this to happen (The Fish Site, 2020; Bunting et al., 2022, 2023).

Consumer demand and production standards

BMP promotion has evolved in several instances to focus on prospects for enabling producers to join established labelling schemes or possibly establish independent initiatives to attract buyers and reassure consumers. Hence, efforts may switch from increasing production, minimising disease problems, maintaining water quality and avoiding environmental degradation and, consequently, self-pollution to meeting broader social and ethical demands (Figure 8.1). Buying behaviour has several dimensions ranging from

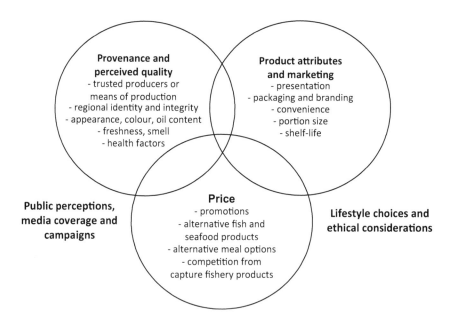

Figure 8.1 Factors influencing decision-making by aquaculture product buyers and consumers.

over-arching lifestyle choices, ethical considerations and prevailing public perceptions, which may be strongly influenced by media coverage and campaigns or celebrity chef endorsements, to more ephemeral concerns and product-specific attributes. Food system configurations and prevailing food cultures can dictate what aquatic foods are accessible to consumers and in what forms and at what times or seasons (HLPE, 2020). Food environments that people experience, or are subjected to, may govern, or strongly influence, preferences and buying behaviours (see chapter 1 and chapter 7; de Bruyn et al., 2021). Ultimately, diets may be a consequence of intra-household decision-making and the eating habits and behaviours of individuals.

Perceptions and buying behaviour can be transient and fickle, further compounding uncertainty over decision-making and investment concerning alternative production strategies. Ethical choices may be complicated by multiple imperatives, social justice, animal welfare and climate change mitigation, for instance, whilst poorer consumers may not have the purchasing power to exercise their preferred choice. The degree of alignment between labelling schemes and consumer preferences may not be clear, whilst standards and scheme guidelines are evolving based on new scientific understanding. Generic production standards and assurance schemes cover an increasing range of agricultural products and, in combination with statutory requirements, may increasingly meet buyer and consumer demand. Price is often the ultimate arbitrator, and consumers expressing a preference for organic produce may choose non-organic products when faced with a purchasing decision; price may also affect the frequency of purchasing ethical products and rigor of product selection across a shopping basket or with regards to product ingredients.

Supporting new development paradigms

Practical approaches to enhance knowledge transfer, improve training and education provision and create an enabling institutional environment for sustainable aquaculture are presented below, and notable examples of success from other sectors are reviewed.

Enhancing knowledge transfer and uptake promotion

Improvements to agricultural productivity were often supported by government extension services, whereby extension workers were based locally and visited farmers to advise them on adopting new varieties and implementing technology packages and often provided subsidised seed and inputs to encourage uptake. This mode of service delivery was subsequently adopted in many countries to achieve improvements in aquaculture production. Reviewing progress in this regard, the World Bank (2007: 51) noted that 'weak extension services have hampered a more effective diffusion of technology, particularly to small farmers' and that 'the traditional government-based research and extension system is not sufficiently responsive to the new challenges and opportunities, neither of aquaculture technologies and markets nor to the demand of farmers'.

Consequently, attention was given to enhancing the capacity of extension services to deliver a broader range of knowledge and options so that farmers could adopt appropriate strategies given their circumstances and resources. Alternative extension providers emerged and were sought to compensate for deficiencies in government services. Nationally constituted NGOs have been important agents for promoting and supporting aquaculture development, most notably perhaps BRAC and PROSHIKA in Bangladesh. Community-based organisations, including charities and religious groups, have worked locally to promote aquaculture development but face the same constraints as traditional extension services, notably limited resources and knowledge of recent advances.

International NGOs have adopted various roles in relation to aquaculture development, as advocates for sustainable aquaculture development, watchdogs and instigators of development programmes with aquaculture components to alleviate poverty and to enhance local food security. Working in Bangladesh during the last decade, CARE Bangladesh implemented programmes focused on enhancing natural resource-based livelihoods that included small-scale aquaculture components, cages in Cage Aquaculture for Greater Economic Security (CAGES), *beel* fisheries and *gher* farming in Greater Options for Local Development through Aquaculture (GOLDA) and enhanced household fishpond management in Locally Intensified Farming Enterprises (LIFE). During interviews with LIFE project participants, most demonstrated a good understanding of factors influencing periphyton-based aquaculture. Interviews demonstrated, however, that farmers working directly with CARE staff had a more detailed and comprehensive knowledge of enhanced management practices compared to those receiving training from local NGO partners.

Considering the limited time externally funded projects generally have to work directly with farmers, shortcomings in delivery by local NGO partners and limitations that might be foreseen in trying to disseminate information through actor networks (i.e. seed traders), it appears that stronger links and knowledge sharing with government extension services and the private sector should be promoted. Consequently, extension staff could be better equipped to deliver knowledge on promising approaches to farmers and to support them to adapt and refine such approaches during the early stages of adoption. Advocates propose that public–private initiatives (PPIs) offer a means to make more efficient use of public funds and achieve greater impacts and better performance levels. Such schemes have proved controversial, however, with opposition focused on the necessity of commercial operators to generate profits and seemingly poor value lease-back and supplementary service provision agreements. Partnerships between government and NGOs were highlighted as a possible PPI arrangement with potential to enhance extension service provision for aquaculture development (World Bank, 2007).

Capitalising on the potential of the private sector to deliver services to support small-scale producers, One-stop Aqua Shops (OASs) were established in eastern India as a single point for all aquaculture-related services in response to demand expressed through 'consultations and consensus-building with farmers, fishers and national fisheries policymakers, shapers and implementers' (NRSP, 2004: 5). Key attributes of OASs recounted by the Word Bank (2007) included serving as a contact point within communities for aquaculture suppliers, fisheries departments and rural banks; promoting aquaculture through self-help groups; helping establish links with villagers; providing information on input prices and suppliers, fisheries department schemes and information and application forms for micro-credit; arranging exposure visits and training for those contemplating aquaculture; and information on aquaculture in other countries and how this could be used to guide development locally. One-stop shops in support of aquaculture development were subsequently established in Pakistan and Vietnam, and Aqua Shops were being promoted in Kenya to support an emerging aquaculture sector (RIU, 2011).

Multinational companies, such as Cargill and Charoen Phokpand based in Thailand, have promoted technology packages to prospective producers in Asia and Central and South America. Farmers agreeing to adopt a package of measures are obliged to purchase inputs from the company and in return receive an assured farm-gate price for their production. A wide array of farmers have adopted such packages for tilapia farming in ponds and cages throughout Asia, including small-scale farmers in peri-urban areas of Ho Chi Minh City, Vietnam. Experiences appear mixed, however, with reports suggesting that production, in cages especially, is at risk from pollution, disease outbreaks and environmental perturbations and that some producers have not received expected levels of payment for their fish.

Increased dependence on external inputs (feed, seed, energy and information) and growing dependence on long-distance relationships with reduced interaction with local input suppliers and market actors and consumers appear to signify transition to a less resilient livelihood strategy (Figure 1.3). Vulnerability to input and energy cost fluctuations is heightened and, being committed to prices fixed and largely dictated by global production levels, there is little scope to undertake partial harvests to aid cash flows or mitigate for environmental perturbations or benefit from potential price premiums locally owing to product quality, freshness and provenance.

Scientific journals, libraries, scholarly societies and individual experts constitute conventional repositories for scientific knowledge, including that pertaining to aquaculture development. Public funds have been invested in consolidating scientific information on aquaculture development in accessible formats for professionals and practitioners, primarily in developing countries – for example, the CABI *Aquaculture Compendium* (CABI, 2006) – but criticism has been voiced concerning charging people to access such resources. Rapid developments with mobile network coverage in many developing countries offer new avenues to access scientific information, virtual practitioner and social networks and knowledge of BMPs and technological developments (Bunting et al., 2022, 2023). Information and communications technologies (ICTs) encompassing decision support systems (DSSs) in concert with appropriate mechanisms to facilitate joint assessment and decision-making with stakeholders have potential to address technical and disease problems and support sustainable aquaculture development. Drawing on knowledge concerning the development of programmes such as Livestock Guru (Heffernan and Nielsen, 2007), similar applications could be devised to guide improved water management and enhanced husbandry skills and disease management in aquaculture.

National and global networks and evidence programmes (e.g. producer organisations, portals, regulators and researchers, certification bodies, associations and sustainability reporting) have emerged as a diverse array of schemes that report on the sustainability of aquaculture (Bunting et al., 2022). A range of criteria (e.g. certification, licencing, animal health and welfare, water quality, escapees, feed, social, energy and GHGs, risks and recommendations) are covered by these initiatives. The scope and focus of each can vary significantly depending, for example, on the production system or value chain phase under consideration, stated purpose or target audience.

Training and capacity building

Mirroring concerns over traditional extension services, training delivered by government departments has been criticised variously as too generic or issue specific, top-down and not demand led, sporadic and based around short, one-off meetings and often arranged at inconvenient places or times. Focus groups with farmers in highland areas of north Bengal, India, who had received training on beekeeping, for example, demonstrated clear demand for continued interaction with experienced individuals to guide farmers and permit them to address issues as they arise. Costs obviously prohibit intensive mentoring and supervision, but better arrangements are required if programmes in support of livelihoods diversification are to deliver anticipated benefits to poor communities and alleviate pressure on natural resources.

Opportunities to experience demonstration activities and interact with knowledgeable and experienced people are often invaluable. Trips to view trials at field stations or demonstration systems, visits to see other commercial operations or farm walks and attending trade shows and fairs to meet and discuss with equipment and service providers are established means to promoting knowledge transfer and initiating adoption of new varieties and husbandry practices in conventional agricultural development. Alternative approaches to share knowledge and establish local support mechanisms within farming communities have been identified. The first involved the establishment of farmer field schools in which groups of twenty to twenty-five farmers are offered 'community-based, non-formal education' with 'discovery-based learning … related to agroecological principles in a participatory learning process throughout a crop cycle' (Braun et al., 2000: i). The second is the formation of Local Agricultural Research Committees (LARCs) that constitute a 'permanent agricultural research service staffed by a team of four or more volunteer farmers elected by the community' and 'create a link between local and formal research' (Braun et al., 2000: i).

Given deficiencies identified for sporadic and one-off training events, attention has come to focus on continuing professional development (CPD) supporting ongoing skills enhancement in response to emerging challenges and opportunities. Assessing the aquaculture industry in the UK, a survey of commercial producers conducted by the Sector Skills Council for the environment and land-based sector (Lantra, 2007: 15) identified several priorities for learning provision, notably, that 'bite-sized units of learning are needed and on-site training is preferable to off-site' and that 'considering the nature of aquaculture the industry needs courses delivered at convenient times, and in work-based and work-related environments, so new entrants and those already within the industry can participate'. Furthermore, providers should respond and innovate based on the needs of employers and associated organisations 'rather than deliver what has traditionally been delivered' (Lantra, 2007: 16). Similar sentiments are likely to be echoed when assessing the learning priorities of other groups of aquatic resource users, managers and professionals, including non-governmental organisations and government employees, whether in the UK or elsewhere.

Education

Elaborating the *Strategy for Aquaculture Development beyond 2000*, the first priority highlighted was 'Investing in People through Education and Training' (NACA and FAO, 2000: 7) in which it was noted that 'further investments in education and training are essential to build knowledge, skills and attitude of all people involved in the sector'. Specifically, a number of areas where enhancing human capacity could be made more responsive and cost-effective and associated institutional support could be strengthened were identified (Table 8.2).

Dedicated education schemes covering core aspects of aquaculture development have often developed in response to emerging commercial aquaculture sectors. However, costs associated with establishing such provision are significant, and transaction costs for students and those already in the industry wishing to enrol on such courses are increasingly prohibitive. Fees and travel and subsistence costs are considerable, and opportunity costs, notably lost wages and disruption to personal and professional relationships, compound the problem. Residential courses often come with limited scope for practical training and engaging with learning in a local context where future work and research is to be undertaken. Given this perspective, the focus placed on distance learning in the Bangkok Declaration seems appropriate. However, such provision presents new and interesting challenges.

Table 8.2 Priorities for education and training for aquaculture development

Domain	Key elements
Investing in people through education and training	• curriculum development employing participatory approaches • improved networking and cooperation between agencies and institutions • adopt multi-disciplinary and problem-based learning approaches • balance practical and theoretical approaches to train farmers and provide a more skilful and innovative workforce for the industry • use modern education, training and communication approaches, notably distance learning and the internet, to promote regional and international cooperation and networking in curriculum development and facilitate the exchange of experiences and development of supporting knowledge bases and resource materials
Strengthening institutional support	• provide training and education, research and extension services to support formulation of policy, regulatory frameworks and enforceable legislation, including incentives, notably economic, to enhance aquaculture management • target both government ministries and public sector agencies dealing with administration, education, research and development and representatives of private sector, NGO, consumer and stakeholder organisations and institutions • enhance the capacity of institutions to formulate and implement strategies targeted at poor people

Source: Developed from NACA and FAO (2000).

Distance learning provision relevant to aquaculture development and aquatic resources management (ARM) is critically reviewed below to assess opportunities and constraints associated with such provision; in this context, ARM refers to the conservation and wise use of wetlands. These subject areas present particular challenges concerning learning and teaching as it is often necessary to integrate expertise with practical knowledge and local experience across biological, ecological, social and economic domains with technical and institutional assessments. Potential strategies to improve access to, and the delivery of, distance learning in support of enhanced aquaculture development and ARM globally are then proposed and evaluated. Implications of adopting the QAA (1999) guidelines with respect to existing and possible future distance learning and teaching for enhanced aquaculture development and ARM are discussed.

Distance learning

Distance learning or flexible and distributed learning has been defined as 'educational provision leading to an award, or to specific credit toward an award, of an awarding institution delivered and/or supported and/or assessed through means which generally do not require the student to attend particular classes or events at particular times and particular locations' (QAA, 2004: 3). How distance learning is conceived and implemented varies immensely and boundaries between distance learning and other forms of higher education are becoming less defined as the character of institution-centred and collaborative provision changes and evolves (QAA, 1999). However, distance learning programmes share several common character-istics, principally that learners are physically separated, to some extent, from the institution responsible for learning, teaching and making the award. Consequently, this results in a different division of labour and responsibility compared with institution-centred programmes and raises concerns over how teaching and learning is managed and regulated to ensure quality of provision and assurance of academic standards. Particular areas for concern high-lighted in the guidelines were system design, notably the development of an integrated approach; establishment of academic standards and programme design, approval and review procedures; assurance of quality and standards in the management of programme delivery; student development, support, communication and representation; and student assessment (QAA, 1999).

Four 'dimensions of distance learning' were alluded to in the QAA (1999) guidelines. First was 'materials-based learning', referring to the learning resource materials provided to students studying at a distance. The nature of materials employed, modes of access available and scope of materials provided in different settings contribute to a rich range and diversity of potential learning resource materials. Second was 'programme components delivered by travelling teachers', referring to learning and teaching activities conducted by staff travelling to meet students at their place of study to provide orientation, deliver materials, conduct intensive teaching, provide tutorial support, facilitate student development and guidance, oversee assessment and gather feedback. Third was 'learning supported locally', acknowledging that support staff may need to be employed locally to support students. This may involve administrative responsibilities or specific teaching functions or a combination of the two. Fourth was 'learning supported from the providing institution remotely from the student' referring to the specific support and teaching components undertaken at a distance by the providing institution. The QAA (1999) described this in terms of modes of communica-tion and interaction, ranging from postal correspondence, telecommunica-tions and education technologies and conducted on a one-to-one basis or via telephone or online conferencing.

Each dimension demands separate consideration when evaluating prospects for enhanced aquaculture development and ARM through distance learning. Recent developments in online knowledge bases and repositories dedicated to aquatic resources management and aquaculture constitute valuable learning resources – for example, United Nations University (2008) OpenCourseWare Portal, *Aquaculture Compendium* (CABI, 2006b) and *oneFish community directory* (oneFish, 2008) – although quality control and continued access constitute immediate concerns. Local support for learning with respect to the emerging disciplines associated with participatory and integrated ARM is likely to constitute a potential limitation. A legacy of sectoral organisation in governance, administration and teaching in many developing countries means that appropriate support from practitioners with a multi- or interdisciplinary background is unrealistic. Furthermore, individuals and institutions with the appropriate skills and experience are likely to have other commitments, and securing the required support may be difficult and expensive. This might necessitate a preliminary programme of training and investment to build the capacity of local agents, tutors or institutions to provide students with the required support.

Considering the origins of distance learning, Henrichsen (2001) observed that it has a long history and that centuries earlier, instructional correspondence was exchanged between travelling royal families and their tutors. In the nineteenth century, postal tuition and correspondence courses became popular amongst the working classes in Europe and North America, whilst in the twentieth century telecommunications technologies ranging from radio to the internet have been applied variously in support of distance learning (Henrichsen, 2001). Enquiry concerning the early development of aquaculture in Europe and, significantly, the introduction across Europe of common carp (*Cyprinus carpio*) originating from the Danubian basin (see Balon, 1995) have led some to question the processes which gave rise to the pattern and timing of carp introductions and how knowledge was shared and gained. Hoffmann (1996: 669) surmised, 'How did long unwritten knowledge of pond design pass eastward and of carp pass westward?' Evidence shows that skilled laymen accustomed to working with water and fish and with expertise derived from knowledge accumulated over generations travelled long distances to guide landowners wishing to build fishponds and manage fisheries – an early example of distance learning for aquaculture development.

Reviewing the contemporary status of flexible and distributed learning (FDL), the QAA (2004: 7) described the variety of approaches employed as 'considerable' and noted that it 'embraces a continuum of pedagogical opportunities'. The characteristics and dimensions of this continuum ranged from on-site cohort groups engaged in predominantly face-to-face modes of learning to off-site, lone learners engaged in flexible and distributed learning and e-modes of learning.

Aquatic resources in the context of this assessment are regarded as both the physical and living aquatic resources, and their management is increasingly based on principles of participation and integration across sectors and disciplines. The Millennium Ecosystem Assessment (MEA, 2005) highlighted the importance of aquatic ecosystems services, including provisioning, regulating, supporting and cultural services in sustaining societal systems. The need to promote further and higher education in support of sustainable development, including that of aquatic resources, has been highlighted both nationally and internationally. The UK Government's sustainable development strategy emphasised an important role for education (UK Government, 2005) and 'The DfES Sustainable Development Action Plan 2005/06' (DfES, 2006) further highlighted the need to develop education for sustainable development.

Reviewing current provision of distance learning for aquaculture development and ARM shows that several important topics are addressed and an array of approaches are adopted. The institutions involved are varied, ranging from inter-governmental bodies; universities, whether operating alone or in collaboration; national and international non-governmental organisations; and professional associations to private sector institutions. Courses offered range from postgraduate degrees to short courses orientated towards professional development and dealing with specific topics.

The emergence of distance learning in the fields of aquaculture and ARM has in theory opened up opportunities for talented and committed students in developing countries to benefit from world-class learning and teaching. Often higher education provision in developing counties must compete for funding with other pressing priorities such as health care provision, ensuring food security and providing essential social services; consequently, public sector colleges and universities are often poorly equipped and unable to invest in enhancing teaching and learning provision. Local provision of quality higher education could help to counter the undesirable impacts of talented and motivated students leaving a country or region in pursuit of enhanced learning and teaching opportunities, whilst retaining talented individuals would help to build research capacity and permit research findings to feed back into learning activities to enhance quality.

Distance learning, and the more refined notion of flexible and distributed learning, whether developed by local institutions or by international institutions, alone or in partnership, may also help to overcome social and economic barriers to vulnerable and poor groups from entering into higher and further education; social norms and taboos in some cultures and countries can exclude females from formal education where it entails travelling away from home. Lower overheads for students associated with distance learning (reduced fees and avoidance of travel, accommodation and subsistence costs) should mean that students from poorer sectors of a society are better able to access further education opportunities. However, even access to international courses through distance learning will be beyond the reach of most in developing countries. International initiatives to develop distance learning provision are routinely linked to scholarships and grants to support learners. However, options to increase the sustainability of this remain unclear. There are also attendant concerns that developing such provision undermines prospects for enhancing learning and teaching provision locally. International development of flexible and distributed learning should include a strategy for capacity-building to permit the development and facilitation of effective flexible and distributed learning locally, to meet changing demand and address issues of sustainability, inequality and unequal access.

Principal amongst the concerns expressed about distance learning are worries that the quality of the learning experience and benefit to the learner are compromised in various ways. Discussing prospects for distance learning, Woodley (2001, cited in Biggs, 2003: 218) noted that 'classroom interaction is vital to learning and we will need to explore the ways in which the type of interaction that takes place in the classroom can be emulated online'. In this regard, the emergence of educational technology in support of flexible and distributed learning, and its appropriate application, is likely to be critical to help both practitioners and learners to reach their educational aims and objectives and to enhance learning and teaching. Managing learning at a distance can be assisted significantly by educational technology, allowing online access to information about departments, programmes, courses, regulations and staff and student records.

Educational technology, whether web-based or employed off-line, can help to engage people in appropriate and effective learning activities that match learning objectives and assessment tasks (Biggs, 2003). Activities can be synchronous, such as instant messaging or video-conferencing, or asynchronous, such as bulletin boards, email discussion groups or building online portfolios, blogs or wikis (McAndrew, 2006). Conceivably, a combination of individual and group-based activities conducted either synchronously or asynchronously would be desirable (Gulc, 2006). According to Stafani (2006), online assessment offers many advantages, including flexibility of access (time, place and question selection), equitability, promoting student-centred learning, providing immediate feedback to students and marks to staff, development of interactive assessment as learning experiences in themselves, enhanced student learning outcomes and cost and workload reductions. However, he also noted that it requires adequate resources and institutional procedures, staff and students with appropriate skills and experience, rigorous arrangements for administering online tests and measures to counter increased risks from plagiarism and cheating.

Adoption of QQA (2004) guidelines, precepts and exemplifying questions for flexible and distributed learning should help to avoid most of the constraints to distance learning identified above and permit the development of appropriate courses, whether for provision internationally to diverse student groups or to support new entrants and promote continuing professional development in industry (Jones and Thomas, 2005). A focus on better use of existing resources repositories and knowledge bases, in particular reputable open-access sources, would also help to avoid unnecessary costs, avoid effort duplication and provide students with access to contemporary and state-of-the-art learning materials.

Considering the varied nature of people's needs related to education and training in support of sustainable aquaculture development and ARM, a participatory process invoking the conceptual model of a continuum of approaches to flexible and distributed learning must be adopted. This seems critical to avoid a focus on the degree of separation, or distance, in the learning arrangement and to promote a more thorough assessment of the pedagogical opportunities to enhance learning and teaching in a specific situation, based on resources, existing capacity and needs. Such an approach also acknowledges that provision – that is, nature and scope of materials, modes of access, responsibilities and nature of support and provision locally – should evolve over time in response to changing needs and emerging issues (Jonker, 2005).

This review highlights possible constraints to distance learning but also identifies various mitigation and enhancement measures to help to obviate these. Sound principles have been elaborated elsewhere to guide the planning and implementation of flexible and distributed learning (QAA, 2004). The knowledge base concerning the most effective strategies and best practices to adopt given a particular setting is expanding; however, participatory approaches to curriculum development, considering the demands and expectations of all stakeholder groups, will be critical in elaborating approaches that are feasible, effective, equitable and ultimately self-sustaining. Further development of flexible and distributed learning is warranted in support of sustainable aquaculture development and ARM, in which robust and trustworthy multidisciplinary and participatory approaches are required to meet diverse needs and emerging challenges. In dealing with such issues, it is critical that learning and teaching reflect and respond to the local environmental, social and political setting. Furthermore, such provision will only be enhanced through greater opportunities to undertake learning and teaching in context.

Distance learning has the potential to address several Higher Education Academy professional values as well as to respond to the needs of students and the wider stakeholder community and to reduce transaction costs and constraints of social norms through respect to individual learners, commitment to development of learning communities and commitment to encouraging participation in higher education, acknowledging diversity and promoting equality of opportunity (HEA, 2009). However, considering the possible impact of international distance learning provision, Biggs (2003: 224) noted that 'implications for higher education can be immense, both educationally and politically. The competition provided by flexible delivery from prestigious overseas universities could even eliminate off-campus teaching at local universities, and seriously damage on-campus teaching.' Collaborative provision and a strategy for capacity-building locally will be critical to avoid a process of externally driven flexible and distributed learning coming to be regarded as counter-productive.

Communication planning

Communication planning is a critical and important, yet often overlooked, aspect of promoting and implementing sustainable aquaculture development programmes and initiatives. Guidelines developed by the Natural Resources Systems Programme (NRSP, 2003) of the UK Department for International Development specified the need to address:

- communication plan aims in relation to project or programme purpose;
- range and composition of communication stakeholders;
- current knowledge, attitudes and practices of communication stakeholders;
- research products and issues that communication should address;
- media and pathways most likely to prove effective.

Ten questions to facilitate effective communication planning were posed in the NRSP guidelines; these are summarised in Table 8.3.

Short- and long-term aims for the communication plan should be specified initially, with respect to the project purpose or the nature of activity planned. Current knowledge and relevant experiences demand consideration and must be reviewed and summarised. Responsibility should be assigned to implement the constituent parts of the communication plan and to monitor and evaluate progress. Communication stakeholders must be identified and characterised in terms of their roles and responsibilities and short- and long-term communication plan aims. Research with East Kolkata Wetland stakeholders demonstrated that several stakeholder groups could be identified with distinct communication requirements and short- and long-term project aims (Table 8.4). Particular

Table 8.3 Ten questions to guide effective communication planning

Question	Focus
Q1	What are the aims of the project's Communication Plan in relation to the project purpose?
Q2	Who within the project team will be responsible for the implementation of the Communication Plan?
Q3	Who are the communication stakeholders for the project?
Q4	What are the research products and other issues that the project team need to communicate about with the communication stakeholders?
Q5	What are the current knowledge, attitudes and practices (KAPs) of the communication stakeholders in relation to the products to be promoted?
Q6	What are the objectives of communicating about the products to the communication stakeholders (i.e. what might they want to be able to do once the project team have communicated with them)?
Q7	What media and channels might be used to communicate with the various communication stakeholders in relation to the research products (e.g. what is accessible to them, what are their preferences, what can be sustained after the project is over)?
Q8	How will the project team ensure that communication materials are useful (e.g. contain relevant information), usable (e.g. in a language they understand) and accessible (e.g. at a suitable time and place) for those with whom the project wishes to communicate during and after the project?
Q9	Are the proposed Communication Plan activities and materials included in the project budget?
Q10	How will the project team monitor and evaluate the implementation of the Communication Plan and its component parts?

Source: NRSP (2003).

Table 8.4 Communication stakeholders and why they were important

Communication aims related to stakeholders	Communication stakeholders	Why important to project
Short-term aims	Agricultural workers	Depend on paddy farming in the wetlands
	Horticultural workers	Depend on vegetable production in the wetlands
	Fish farm labourers	Depend on fish production in the wetlands
	Fish farm managers	Depend on fish production and have a role to play in managing water in the wetlands area
	Save Wetlands Committee	Members are primary stakeholders; organisation with broad support in wetlands area
	Fish Producers Association	Represents farmers who manage a significant area of water bodies in the wetlands
	CITU	Members are primary stakeholders; influential body in wetlands area
	Agricultural Development Officers	Responsible for supporting development in agricultural areas
	Kolkata Metropolitan Development Authority	Responsible for operation and maintenance of drainage infrastructure in KMC area
	Department of Irrigation and Waterways	Responsible for primary drainage system passing through wetlands and associated sluice gates
	Department of Environment	Responsible for the formulation of the wetlands management plan
	West Bengal Pollution Control Board	Responsible for monitoring pollution and the quality of water discharged from Kolkata
	Department of Fisheries	Provide guidance to cooperative fisheries, manage some fisheries in the wetlands and provide advice on technical aspects of fish production
	Kolkata Municipal Corporation	Powerful urban authority that administers part of the EKW with significant land ownership
	South and North-24-District Authorities	Powerful rural authority with administrative responsibility in part of the wetlands
	Salt Lake Authority	Powerful urban authority with administrative responsibility in part of the wetlands
	Asian Development Bank and DFID-India	Involved in infrastructure development focused on wastewater treatment that could impact EKW

(Continued)

Table 8.4 (Continued)

Communication aims related to stakeholders	Communication stakeholders	Why important to project
Long-term aims	*groups mentioned above Peri-urban communities around Kolkata (not involved in R8365) Peri-urban (PU) a wetland communities in West Bengal, India; and globally Stakeholders in India and globally that could benefit from new knowledge on PU action planning	Process could be employed in future negotiations Process could be adopted to address other issues; e.g. service provision, infrastructure development, livelihoods diversification Process could be used by communities for action planning to enhance livelihoods in other PU areas Knowledge of process could be used to implement participatory action planning in other PU settings and to raise awareness

Source: Bunting (2002).

aspects demanding consideration were inclusion of relevant information, use of easily understood language and making sure that materials and outputs were easily accessible to communication stakeholders.

A verification step, possibly facilitated through key informant meetings or stakeholder workshops, should be included to confirm and refine communication needs. Assessment of prevailing knowledge, attitudes and practices (KAP) with regards to activities and outputs foreseen constitutes a potentially useful approach. Outcomes of an assessment of KAP concerning participatory action planning in the EKW are presented in Table 8.5.

Appropriate communication media and pathways should be identified for the various communication stakeholders, taking into account accessibility, preferences and sustainability (Table 8.6). Pre-testing of draft materials and concepts with a sample of communication stakeholders is recommended and, where possible, support should be provided to community and stakeholder groups developing communication materials.

Following the formulation of the high-level communication plan, including potential communication activities and appropriate materials and pathways for communication stakeholders, it will be necessary to prioritise strategies given budget and time constraints.

Summary

Guidelines to direct responsible aquaculture sector planning have been established and an emerging paradigm of BMPs and codes of conduct, whether conceived and developed locally to aid clusters of small-scale producers to become more efficient and resilient or promoted more widely to assist producers comply with standards and certification scheme requirements, is evident. Barriers and constraints to adopting formal planning and BMPs have been identified, whilst it is also apparent that buyers and consumers are becoming increasingly influential in directing aquaculture development. There is an attendant risk, however, that transition to more intensive, market-orientated practices and increased demand for products from such culture systems may overwhelm the carrying capacity of supporting ecosystem areas and threaten environmental flows of ecosystem services. Opportunities to enhance knowledge transfer and uptake promotion of sustainable aquaculture practices are foreseen. There is a clear need for dedicated education and training provision and continuous professional development provision in support of sustainable aquaculture development. Better planned and executed communication strategies, with activities and outputs shaped by the knowledge demands and expectations of important stakeholder groups, are needed if producers, consumers, policymakers and responsible authorities are to fully comprehend the risks associated with prevailing aquaculture practices and effectively initiate the transition to sustainable aquaculture development pathways.

Table 8.5 Current knowledge, attitudes and practices (KAP) of communication stakeholders concerning participatory action planning (PAP)

Communication stakeholder	Knowledge of participatory action plan development	Attitude toward PAP*	Practice of participatory action plan development
Agricultural workers	Awareness among some farmers involved in PAP activities	+++	Some farmers involved but small proportion of total
Horticultural workers	Awareness among some labour union members	+++	Some union members participating in process
Fish farm labourers	Awareness among some labour union members (≈40%)	+++	Some union members participating in process
Fish farm managers	Awareness among some FPA members	+++	Farmworkers participating in project but not used previously
Save Wetlands Committee	Aware of ongoing PAP process	++++	Participating in current project but not used previously
Fish Producers Association	Aware of ongoing PAP process	++++	Participating in current project but not used previously
CITU	Aware of ongoing PAP process	++++	Participating in current project but not used previously
KMDA	Aware of ongoing PAP process	++	Action planning used in CEMSAP project (degree of participation unclear)
DoIW	Aware of ongoing PAP process	++	Participant in selected PAP activities
DoE	Aware of ongoing PAP process and worked on PD112	++++	Facilitating and participating in ongoing PAP process
WBPCB	Aware of ongoing action planning process	++	Participant in selected PAP activities
DoF	Aware of ongoing process and consulted in PD112	++	Participation limited
KMC	Aware of ongoing action planning process	++	Participation limited
North and South-24-District and Salt Lake Authority	—	++	Participation limited
Asian Development Bank [TA]	Aware of PAP approach being developed	+++	Participating in project activities and demand expressed for information/cooperation

DFID–India	Aware of ongoing PAP through meeting, project reports and ongoing dialogue	++	Demand expressed for further information/ cooperation
USAID	Aware of ongoing process	+	
Peri-urban interface/ wetland communities and NGOs in WB (not in R8365)	Good awareness among NGOs that have attended project workshops	++	Participation limited
PUI and wetland communities in India and globally	Very limited (a few communities participating in other projects aware of PAP)	+	
Stakeholders in India and globally that may benefit from knowledge on PAP	Very limited		

Source: Bunting (2002).

Notes
* Attitudes: + aware, ++ supportive, +++ positive and active in process, ++++ very positive.

Table 8.6 Media and channels regarded as appropriate for the communication stakeholders

Group	Composition	Media channels and formats appropriate for each group

Primary stakeholders	Fish, vegetable and rice farmers; employees and traders; community members	Field visits/farm walks
		Project meetings
		Presentations and representation at social/ political gatherings
		Participatory research activities (e.g. mapping, focus groups)
		Project workshops
		Print: Press conference for media coverage
		   Bengali booklet on EK System for EK schools
		   Opinion poll on EK System + press coverage
		  Bengali street drama (for East Kolkata communities)
		  Poster drama (for East Kolkata communities)
		  Procession and festivals: Activities for EK stakeholders
		TV: Documentary or serial
		   Interactive current affairs programmes
		   Current affairs news report of topical events
		Radio: Ask local celebrity to record 'sound bite'
		   Message that reinforces message in TV and print
		   Quiz programme in schools

Key stakeholders	GoWB agencies; Kolkata, Salt Lake S/N-24-Parganas authorities; stakeholder representatives; NGOs; DFID–India	Project meetings Telephone calls Project workshops Representation and discussion at apex meetings Press conference for media coverage Print: Project reports Draft and final action plans Presentations Internet: Project website, including project reports and updates with project activities
Regional	South and Southeast Asian policymakers; professionals; intergovernmental orgs; NGOs	Print: Press release and/or articles for international development and technical press Project reports Internet: Project website Articles for online magazines (e.g. *Urban Agriculture*)

(*Continued*)

Table 8.6 (Continued)

Group	Composition	Media channels and formats appropriate for each group
International	Donors (ADB, USAID, WB) and consultants; international organisations (IUCN, Ramsar, WWF, FAO, UN)	Contributions to email conferences Presentations at conferences in region Print: Press release and/or articles for international development and technical press Project reports Internet: Project website Articles for online magazines (e.g. *Urban Agriculture*) Contributions to email conferences Presentations at international conferences

Source: Bunting (2002).

References

Aarset, B., Beckmann, S., Bigne, E., Beveridge, M.C.M., Bjorndal, T., Bunting, J., et al. (2004) 'The European consumers' understanding and perceptions of the "organic" food regime. The case of aquaculture', British Food Journal, vol 106, pp. 93–105.

ACCS (2008) Cereals – Best Practice for Residue Minimisation, Assured Combinable Crops Information Sheet, Assured Combinable Crops, Surrey, UK.

ADB, ACIAR, AwF, BRR, DKP, FAO, GTZ, IFC, MMAF, NACA and WWF (2007) Practical Manual on Better Management Practices for Tambak Farming in Aceh, Asian Development Bank ETESP, Australian Centre for International Agriculture Research, Food and Agriculture Organization of the United Nations, International Finance Corporation of the World Bank Group, Banda Aceh, Indonesia.

AFCD (2023) Accredited Fish Farm Scheme, Agriculture, Fisheries and Conservation Department, Government of Honk Kong, https://www.afcd.gov.hk/english/fisheries/fish_aqu/fish_aqu_good/fish_aqu_good.html

Anonymous (2009a) 'ASC label to be launched in early 2011', Aquaculture Europe, vol 34, no 4, pp. 10–12.

Anonymous (2009b) 'GAA statement on WWF–GLOBALGAP linkage', Aquaculture Europe, vol 34, no 3, p. 24.

Balon, E.K. (1995) 'Origin and domestication of the wild carp, *Cyprinus carpio*: from Roman gourmets to the swimming flowers', Aquaculture, vol 129, pp. 3–48.

Barg, U.C. (1992) Guidelines for the Promotion of Environmental Management of Coastal Aquaculture Development, Food and Agriculture Organization of the United Nations, Rome.

BBC (2006) Concern over Organic Salmon Farms, British Broadcasting Corporation, http://news.bbc.co.uk/1/hi/sci/tech/5409434.stm

Biggs, J. (2003) Teaching for Quality Learning at University, The Society for Research into Higher Education and Open University Press, Maidenhead, UK.

Braun, A.R., Thiele, G. and Fernandez, M. (2000) Farmer Field Schools and Local Agricultural Research Committees: Complementary Platforms for Integrated

Decision-Making in Sustainable Agriculture, Agricultural Research & Extension Network, Overseas Development Institute, Network Paper No. 105, London.

Bunting, S.W. (2002) Renewable Natural Resource-Use in Livelihoods at the Calcutta Peri-Urban Interface, Final Technical Report, UK Government's Department for International Development, Institute of Aquaculture, NRSP Project R7872, Stirling, UK.

Bunting, S.W., Bosma, R.H., van Zwieten, P.A.M. and Sidik, A.S. (2013) 'Bioeconomic modeling of shrimp aquaculture strategies for the Mahakam Delta, Indonesia', Aquaculture Economics & Management, vol 17, no 1, pp. 51–70.

Bunting, S.W., Bostock, J., Leschen, W. and Little, D.C. (2023) 'Evaluating the potential of innovations across aquaculture product value chains for poverty alleviation in Bangladesh and India', Frontiers in Aquaculture, vol 2, no 3, pp. 1–23.

Bunting, S., Pounds, A., Immink, A., Zacarias, S., Bulcock, P., Murray, F. and Auchterlonie, N. (2022) The Road to Sustainable Aquaculture: On Current Knowledge and Priorities for Responsible Growth, World Economic Forum, Cologny, Switzerland.

Bush, S.R., Belton, B., Hall, D., Vandergeest, P., Murray, F.J., Ponte, S., et al. (2013) 'Certify sustainable aquaculture?', Science, vol 341, no 6150, pp. 1067–1068.

CABI (2006) Aquaculture Compendium, CAB International, Wallingford, UK.

CFE (2009) A Farmer's Guide to Voluntary Measures, Campaign for the Farmed Environment, Warwickshire, UK.

Conroy, C. and Sutherland, A. (2004) Participatory Technology Development with Resource-Poor Farmers: Maximising Impact through the Use of Recommendation Domains, Agricultural Research and Extension Network, AgREN Network Paper 133, London.

Crop Protection Association (2007) Best Practice Guides, Crop Protection Association, Peterborough, UK.

de Bruyn, J., Wesana, J., Bunting, S.W., Thilsted, S.H. and Cohen, P.J. (2021) 'Fish acquisition and consumption in the African Great Lakes Region through a food environment lens: a scoping review', Nutrients, vol 13, 2408.

DfES (2006) Learning for the Future, The DfES Sustainable Development Action Plan 2005/06, Department for Education and Skills, United Kingdom's Government, London.

FAO (1995) Code of Conduct for Responsible Fisheries, FAO, Rome.

FAO (2008) Report of the Expert Consultation on Improving Planning and Policy Development in Aquaculture, FAO Fisheries Report No. 858, FAO, Rome.

FAO (2009) The State of World Fisheries and Aquaculture 2008, Food and Agriculture Organization of the United Nations, Rome.

FAO, NACA, UNEP, WB and WWF (2006) International Principles for Responsible Shrimp Farming, Network of Aquaculture Centres in Asia-Pacific, Bangkok.

Fisheries Victoria, NACA, RIA No.2 and Can Tho University (2009) Better Management Practices (BMPs) for Striped Catfish (tra catfish): Farming Practices in the Mekong Delta, Vietnam, Fisheries Victoria, Victoria, Australia; Network of Aquaculture Centres in Asia-Pacific, Thailand; Research Institute for Aquaculture No. 2, Vietnam; Can Tho University, Vietnam.

Fisheries Western Australia (1999) Abalone Aquaculture in Western Australia: Policy Guideline, Fisheries Western Australia, Fisheries Management Paper No. 133, Perth.

The Fish Site (2020) A Practical Guide to Using AI in Aquaculture, https://thefishsite. com/articles/a-practical-guide-to-using-ai-in-aquaculture

GAA (2009) Aquaculture Facility Certification: Shrimp Farms Best Aquaculture Practices Certification Standards, Guidelines, Sample Application/Audit, Global Aquaculture Alliance, Portsmouth, NH.

Gulc, E. (2006) 'Blended learning', Centre for Biosciences Bulletin, vol 19, p. 5.

Haque, M.M., Alam, M.M., Hoque, M.S., Hasan, N.A., Nielsen, M., Hossain, M.I. and Frederiksen, M. (2021) 'Can Bangladeshi pangasius farmers comply with the requirements of aquaculture certification?', Aquaculture Reports, vol 21, 100811.

HEA (2009) Professional Standards Framework, Higher Education Academy, UK, www.heacademy.ac.uk/assets/York/documents/ourwork/professional/ ProfessionalStandardsFramework.pdf

Heffernan, C. and Nielsen, L. (2007) 'The livestock guru: the design and testing of a tool for knowledge transfer among the poor', Information Technologies and International Development, vol 4, pp. 113–121.

Henrichsen, L.E. (2001) 'Beyond adding telecommunications to a traditional course: insights into human and instructional factors affecting distance learning in TESOL' in L.E. Henrichen (ed) Distance-Learning Programs, Teachers of English to Speakers of Other Languages Inc., Alexandria, VA.

HGCA (2007) Predicting and Controlling Wheat Bulb Fly, Home Grown Cereals Authority, Topic Sheet 99, London.

HLPE (2020) Food Security and Nutrition: Building a Global Narrative towards 2030, High Level Panel of Experts, Rome.

Hoffmann, R.C. (1996) 'Economic development and aquatic ecosystems in medieval Europe', American Historical Review, vol 101, no 3, pp. 631–669.

ISO (2023) ISO 14000, International Organization for Standardization, Switzerland, https://www.imsm.com/gb/iso-14001

Jones, R. and Thomas, L. (2005) 'The 2003 UK Government Higher Education White Paper: a critical assessment of its implications for the access and widening participation agenda', Journal of Education Policy, vol 20, pp. 615–630.

Jonker, L. (2005) 'IWRM: what should we teach? A report on curriculum development at the University of the Western Cape, South Africa', Physics and Chemistry of the Earth, vol 30, pp. 881–885.

Lantra (2007) Aquaculture Industry: Sector Skills Agreement, The Sector Skills Council for the Environment and Land-Based Sector, Lantra, Stoneleigh Park, Warwickshire, UK.

Leavitt, D.F. (2009) Best Management Practices for the Shellfish Culture Industry in Southeastern Massachusetts, Southeastern Massachusetts Aquaculture Centre, Roger Williams University, Bristol.

Mathiesen, C., Griffiths, D., Dan, N.C., Dung, L.T.C., Fjalland, J., Huong, H.C., et al. (2008) 'Developing guidelines for sustainable freshwater aquaculture planning in Vietnam', Aquaculture Asia, vol 13, no 1, pp. 3–5.

McAndrew, T. (2006) 'Blogs, podcasts and wikis: publish and be damned?', Centre for Biosciences Bulletin, vol 18, pp. 6–7.

MEA (2005) Ecosystems and Human Well-Being: Wetlands and Water Synthesis, World Resources Institute, Washington, DC.

NACA (2007) Green Award Recognises NACA Work on Shrimp Farming, Network for Aquaculture Centres in the Asia-Pacific, Bangkok, https://enaca.org/?id=322& title=naca-newsletter-january-march-2007

NACA (2009) EU Supports Better Management Practices for Responsible Aquaculture, Network of Aquaculture Centres in Asia-Pacific, Bangkok, https://enaca.org/?id=333&title=naca-newsletter-october-december-2009

NACA and FAO (2000) Aquaculture Development beyond 2000: The Bangkok Declaration and Strategy, Conference on Aquaculture in the Third Millennium, Bangkok, 20–25 February 2000, Network of Aquaculture Centers Asia-Pacific, Bangkok, and FAO, Rome.

NRSP (2003) NRSP Project Communication Plan Guidelines, Natural Resources Systems Programme, HTS Development Ltd, Hemel Hempstead, UK.

NRSP (2004) 2003–2004 Research Highlights, Natural Resources Systems Programme, HTSPE Ltd, Hemel Hempstead, UK.

Okali, C., Sumberg, J. and Farrington, J. (1994) Farmer Participatory Research: Rhetoric and Reality, Intermediate Technology Publications, London.

oneFish (2008) oneFish Community Directory, www.onefish.org

Pretty, J.N. (1995) 'Participatory learning for sustainable agriculture', World Development, vol 23, pp. 1247–1263.

Pretty, J.N. (2003) 'Social capital and the collective management of resources', Science, vol 302, pp. 1912–1915.

Pretty, J.N. and Ward, H. (2001) 'Social capital and the environment', World Development, vol 29, pp. 209–227.

QAA (1999) Guidelines on the Quality Assurance of Distance Learning, The Quality Assurance Agency for Higher Education, Nottingham, UK.

QAA (2004) Section 2: Collaborative Provision and Flexible and Distributed Learning (Including e-Learning), Code of Practice for the Assurance of Academic Quality and Standards in Higher Education, The Quality Assurance Agency for Higher Education, Nottingham, UK.

Read, P.A., Fernandes, T.F. and Miller, K.L. (2001) 'The derivation of scientific guidelines for best environmental practice for the monitoring and regulation of marine aquaculture in Europe', Journal of Applied Ichthyology, vol 17, pp. 146–152.

RIU (2011) 'Kenya's first Aqua Shop opens for business', Research Into Use, www.researchintouse.com/news/110215aqua.html

Ronnback, P. (2003) Critical Analysis of Certified Organic Shrimp Aquaculture in Sidoarjo, Indonesia, Swedish Society for Nature Conservation, Stockholm, Sweden.

Salmon Scotland (2023) Code of Good Practice, Scottish Salmon Producers Association, https://www.salmonscotland.co.uk/code-of-good-practice

Shackleton, S., Campbell, B., Wollenberg, E. and Edmunds, D. (2002) Devolution and Community-Based Natural Resource Management: Creating Space for Local People to Participate and Benefit?, Overseas Development Institute, Natural Resources Perspectives 76, London.

Siriwardena, P.G.S.N. and Williams, R. (2003) Code of Better Management Practices for Shrimp Aquaculture in Sri Lanka, Report from the World Bank, NACA, WWF and FAO Consortium Program on Shrimp Farming and the Environment.

Stafani, L. (2006) Effective Use of IT: Guidance on Practice in the Biosciences, The Higher Education Academy Centre for Biosciences, Leeds, UK.

STREAM (2005a) Better-Practice Guideline, Support to Regional Aquatic Resources Management, Bangkok.

STREAM (2005b) 'Fourth NACA–STREAM regional conference and better-practice guidelines workshop', STREAM Update, vol 11, p. 2.

Tacon, A.G.J., Metian, M. and McNevin, A.A. (2021) 'Future feeds: suggested guidelines for sustainable development', Reviews in Fisheries Science and Aquaculture, vol 30, no 2, pp. 135–142.

Tobey, J., Poespitasari, H. and Wiryawan, B. (2002) Good Practices for Community-Based Planning and Management of Shrimp Aquaculture in Sumatra, Indonesia, Coastal Resources Center, University of Rhode Island, Kingston.

UK Government (2005) Securing the Future: The UK Government Sustainable Development Strategy, HMSO, Norwich, UK.

United Nations University (2008) OpenCourseWare Portal, https://archive.unu.edu/update/issue48_11.htm

Willer, H., Yussefi-Menzler, M. and Sorensen, N. (eds) (2008) The World of Organic Agriculture: Statistics and Emerging Trends 2008, International Federation of Organic Agriculture Movements, Bonn, Germany, and Research Institute of Organic Agriculture, Frick, Switzerland.

World Bank (2007) Changing the Face of the Waters: The Promise and Challenge of Sustainable Aquaculture, The World Bank, Washington, DC.

WWF (2008) The Aquaculture Dialogues, World Wildlife Fund, Washington, DC.

WWF (2009) International Standards for Responsible Tilapia Aquaculture, World Wildlife Fund, Washington, DC.

WWF (2010a) Bivalve Aquaculture Dialogue Standards, World Wildlife Fund, Washington, DC.

WWF (2010b) Pangasius Aquaculture Dialogue Standard, World Wildlife Fund, Washington, DC.

Zanetell, B.A. and Knuth, B.A. (2004) 'Participation rhetoric or community-based management reality? Influences on willingness to participate in a Venezuelan freshwater fishery', World Development, vol 32, pp. 793–807.

Index

Printed and bound by CPI Group (UK) Ltd, Croydon, CR0 4YY

17/10/2024

01775670-0012